AF705344

Fundamental Principles of Physics

Measurement and Mechanics

Fundamental Principles of Physics

Measurement and Mechanics

J C N Rajendra
BSc Hon(Jaffna), DPhil(Sussex), CPhys, MInstP, MIPSL
Senior Professor in Physics
The Open University of Sri Lanka

Kumaran Book House
Colombo - Chennai
2026

Fundamental Principles of Physics: Measurement and Mechanics
by J C N Rajendra ©

First Edition 2026

Published by
Kumaran Book House
39, 36th Lane, Colombo 6,
Tel.: 112364550, Email : books@kumarangroup.net
No.14, Anna 2nd Street, Therodum Veethi, Thiruverkadu, Chennai - 600077
Tel : +91 94 4480 894, E-mail: kumaranbookhouse@yahoo.com

Printed by
Kumaran Press Private Limited
39, 36th Lane, Colombo 6,
Tel.: 112364550, 113097608,
E-mail: books@kumarangroup.net

Publication No.: #1077

ISBN 978-624-6709-19-8

Dedicated to my wife
Late Dr Jeyabawani Nihal Rajendra

Preface

Physics lies at the heart of our understanding of the natural world. Among its various branches, mechanics is concerned with the study of motion, forces, and energy, which form the foundational framework upon which much of classical physics is built. This book, *Fundamental Principles of Physics: Measurement and Mechanics,* is designed as an accessible and structured introduction to this essential area, particularly for students preparing for the G.C.E. (Advanced Level) Physics examination and other introductory courses in mechanics.

The primary aim of this text is to make the fundamental principles of measurement and mechanics comprehensible to beginners by presenting them in a clear, concise, and pedagogically effective manner. The content spans from basic concepts such as scientific notation, units, and vectors to more advanced topics, including momentum, circular motion, and rotational dynamics. Each chapter is supported by carefully constructed illustrations, diagrams, and examples that clarify theoretical concepts and demonstrate their practical relevance.

Recognising the diverse learning needs of students, this book includes a variety of questions classified as conceptual, numerical, and practise exercises. These are tailored to strengthen understanding, enhance problem-solving skills, and build confidence in the application of physical principles. Answers are provided at the end of the book to facilitate self-assessment and independent study.

Although every effort has been made to ensure accuracy and clarity throughout the text, the author welcomes constructive feedback from readers. Suggestions for improvement, corrections of errors, and recommendations for future editions will be appreciated and acknowledged.

It is my hope that this book will serve as a valuable learning companion, inspire curiosity about the physical world, and foster a deeper appreciation of the beauty and logic inherent in the study of physics.

21 June 2025 J C N Rajendra

Acknowledgement

The development of this book has been a journey marked by the generous support, encouragement, and collaboration of many individuals for whom I owe my deepest gratitude.

The initial version of the manuscript originated as a series of handouts prepared for a seminar conducted for G.C.E. (A/L) Physics teachers. This formative experience was followed by numerous seminars and workshops organised for teachers and students in various districts in the northern and eastern provinces of Sri Lanka. The constructive feedback, enthusiasm, and continuous moral support received from these academic communities have been instrumental in shaping the content into its present form.

The author is especially grateful to the dedicated students and teachers who participated in these sessions. Their participation and encouragement served as a constant source of motivation throughout this effort.

The author expresses his sincere appreciation to Mr. N. Vageesamoorthy for his meticulous assistance in editing the language and equations of the manuscript. His thoughtful contributions have significantly improved the clarity and quality of the text.

Finally, I extend my heartfelt thanks to my daughter and son-in-law, whose unwavering love, support, and guidance have been with me in all my pursuits.

21 June 2025 J C N Rajendra

Contents

Chapter 1

Introduction to Physics

1.1 Introduction

The word physics comes from a Greek term meaning nature and historically became the term used to designate the study of natural phenomena. The study of physics gives the opportunity to learn how nature works. While learning physics, you will experience some richly rewarding and satisfying challenges. Our present understanding of the physical world has been built on foundations laid by scientific giants such as Galileo, Newton and Einstein. Their influence has expanded far beyond science to influence how we live and think. You often share some of the excitement of their discoveries when you learn to use physics to solve practical problems and to gain insight into everyday phenomena. Above all, you will see that physics has played a major role in the human intellect in its journey to understand the world we all live in. In this chapter, we shall learn some important insights relevant to the evolution of branches in physics and how physics is related to other sciences.

1.2 Scope and Excitement of Physics

Like all other sciences, physics is based on experimental observations and quantitative measurements. Physicists observe the phenomena of nature and try to find patterns that relate to these phenomena. These patterns are called **physical theories,** or when they are very well established and widely used, **physical laws.** A theory is an explanation of natural phenomena based on observation and accepted fundamental principles.

Scientists of all disciplines use the ideas of physics. Physics is the basis for most modern technology and the tools and instruments used in scientific, engineering, and medical research and development. Manufacturing industries are dominated by physics-based technology.

Physics is one of the oldest academic disciplines and perhaps the oldest through its inclusion of astronomy. Physics intersects with many interdisciplinary fields such as biophysics, medical physics, quantum chemistry, nanoscience, etc., and the boundaries of physics are not rigidly defined.

Advances in physics often enable advances in new technologies. For example, advances in understanding electromagnetism, quantum physics, solid-state physics, and nuclear physics led directly to the development of new products that have intensely transformed modern-day society, such as television, computers, domestic appliances, nuclear power, etc.

1.3 Physics and its Branches

Physics is a branch of science that deals with the properties', nature of matter and energy. It is one of the most important topics in all sciences. The branches of science have evolved to understand each and every underlying aspect of the physical world. Generally, people, having inquiring minds, have always had a great curiosity about how nature works. As a result, physics has evolved as more is learned about nature. In the beginning, the only sources of information were our senses, and therefore observed phenomena were classified according to the way they were sensed.

(i) The **light** was related to the act of vision, and optics was developed as a more or less independent science associated with this act.

(ii) The **sound** was related to the act of hearing, and acoustics developed as a correlative science.

(iii) The **heat** was related to another kind of physical sensation we designate as hot or cold. For many years the study of heat (called thermodynamics) was yet another autonomous branch of physics.

(iv) **Motion** is the most common of all directly observed phenomena and developed as mechanics, and this branch of science developed earlier than any other branch of physics. The laws of mechanics explained the motion of the planets caused by their gravitational interactions and the free fall of bodies.

(v) **Electromagnetism** is not being directly related to any sensory experience. Even though being responsible for most of them - it did not appear as an organized branch of physics until the nineteenth century.

Until the nineteenth century, physics appeared to be divided into a few branches of science called classical physics, which includes mechanics, thermodynamics, acoustics, optics, and electromagnetism. Although there is little or no connection between them, mechanics was, quite properly, the guiding principle for all.

Since the end of the nineteenth century, a profound conceptual revolution, assisted by refinement in experimental and observational methods, has taken place. This revolution, whose leaders were Max Planck and Albert Einstein, has modified our points of view and approaches or techniques used to address or solve physical problems as well as our understanding of natural phenomena, particularly the structure of matter, and constituted the theories of relativity and quantum mechanics. These new theories have provided a more unified view of natural phenomena and have evolved into what has been called 'modern' physics. Rather, it is a 'modern' or a new approach to the analysis of natural phenomena based on a deeper understanding of the structure of matter and the interactions among its components.

Most often, classical physics refers to pre-1900 physics, while modern physics refers to post-1900 physics, which incorporates elements of quantum mechanics and relativity. Modern physics includes the study of quantum mechanics, relativity, atoms, molecules, nuclei, elementary particles, and condensed matter. The following are some of the branches that describe the studies in modern physics.

(i) **Atomic and Nuclear Physics** is the study of the components, structure and behaviour of the nucleus of the atom.

(ii) **Quantum Physics** is the study of the discrete nature of phenomena at the atomic and subatomic levels. Its focus is on the discrete units of energy called quanta, as described by quantum theory.

(iii) **Relativistic Physics** is the study of phenomena that take place in the form of reference that is in motion with respect to an observer.

(iv) **Solid-State Physics** is the study of all properties of solid materials, including electrical conduction in crystals of semiconductors and metals. It is a study through methods such as quantum mechanics, crystallography, electromagnetism, and metallurgy.

(v) **Condensed Matter Physics** is the study of the properties of condensed materials (solids, liquids, those intermediate between them and dense gas) with the ultimate goal of developing new material with better properties. It is an extension of solid-state physics.

(vi) **Particle Physics** is the study of the fundamental subatomic particles, including both matter (and antimatter) and the carrier particles of the fundamental interactions.

(vii) **Low-Temperature Physics** is the study of the production and maintenance of temperatures down to almost absolute zero and the various phenomena that occur only at such temperatures.

1.4 How does physics related to other sciences?

Physics is the branch of science related to studying the basic laws of nature and their manifestations concerning different natural phenomena. As we have noticed earlier, it can also be referred to as the "fundamental science" because it constrains all the other significant branches of the sciences. It can be considered to be the study of the physical world and matter, along with its motion through space and time. It also pertains to the concepts of energy and force as well.

Physics is strongly related to mathematics. For physicists, mathematics is a tool used to answer questions raised in understanding nature. For mathematicians, physics can be a source of inspiration, with theoretical concepts such as general relativity and quantum theory providing a motivation for mathematicians to develop new tools. Physics and chemistry are closely related, and both study the physical world and the behaviour of matter. Many theories and concepts of physics are directly related to chemistry, such as X-ray diffraction, and radioactivity which has helped to distinguish between the various solids and modify the periodic table.

When scientists combine physics and biology, they learn more about biological systems at the molecular or atomic level. Biology is the study of living things in this universe. Without matter and energy, living beings would not exist. Therefore, for the study of living organisms, it becomes necessary to study the fundamental laws in physics. For example, physics helps us to explain how bats use sound waves to move in the dark. Many remarkable medical technologies, diagnostic tools, and treatment methods have emerged as a result of modern physics discoveries in the last century. This includes X-rays, radiation treatment, laser surgery, high-resolution ultrasound scans, computerized tomography (CT) scans, and magnetic resonance imaging.

In the field of astronomy, astronomers use principles of physics and mathematics to learn about the fundamental nature of the universe and its components, including the sun, moon, planets, stars, and galaxies. For example, optical telescopes have made it possible to study the motion of various planets and satellites in our solar system. Also, the Doppler effect predicted the expansion of the universe.

Atmospheric Physics applies physics principles to study atmospheric processes such as cloud formation, light scattering, and energy transfer. Atmospheric physics and meteorology use mathematical and physical models to understand the weather and climatic conditions.

Seismology is the study of earthquakes and seismic waves that move through and around the Earth. When an earthquake or explosion occurs, part of the energy is released as elastic waves transmitted through the Earth. Physics involves the study of energy, and earthquakes are produced by the energy that is released from the movement of the Earth's plates. Therefore, understanding physics is essential to realize how waves travel through the various layers of the Earth.

1.5 Physics, Technology and Society

The knowledge of physics has contributed the most to the birth of various technologies that make human life easier and perform specific tasks. It plays a vital role in many technologies, such as communication, medicine, space exploration, and many others.

Physics is an exciting intellectual exploration that inspires young people and expands the frontiers of our knowledge about nature. It has enabled us to understand the universe and the world around us and contributes to our everyday lives. Without physics, we would not have many of the things we use today, including transport, construction, and communications, or some of the most vital advances in healthcare, such as MRI scanners. Furthermore, communications and transportation such as aircraft have brought people around the world closer together.

Physics generates fundamental knowledge needed for future technological advances that will continue driving the world's economy. In addition, it improves our quality of life and contributes to the technological infrastructure and skills required to take advantage of scientific advances and discoveries.

Society's dependence on technology represents the importance of physics in daily life. Several aspects of modern society would not have been possible without the important scientific discoveries made in the past. These discoveries became the basis on which current technologies were developed. For example, innovations such as magnetism, electricity, conductors, and others made modern amenities like television, computers, phones, and other business and home technologies possible. Furthermore, knowledge of physics provides essential understanding for developing new instrumentation techniques for medical applications, such as computer tomography, magnetic resonance imaging, ultrasonic imaging, and laser surgery.

1.6 Applications of Physics in Everyday Life

The technology that we all use today is very much entrenched in physics. In fact, our way of life depends on it. The increased computational power and the need for a higher standard of living derived from physics-based technology allow us to pursue a deeper understanding of physics, which in turn again leads to improving the technology and so on. The applications of physics in everyday life are numerous. Following are some of the physics applications that we use in our day-to-day life.

Walking

Walking is a natural habit in our life, and it is interesting to see how physics is involved in this activity.

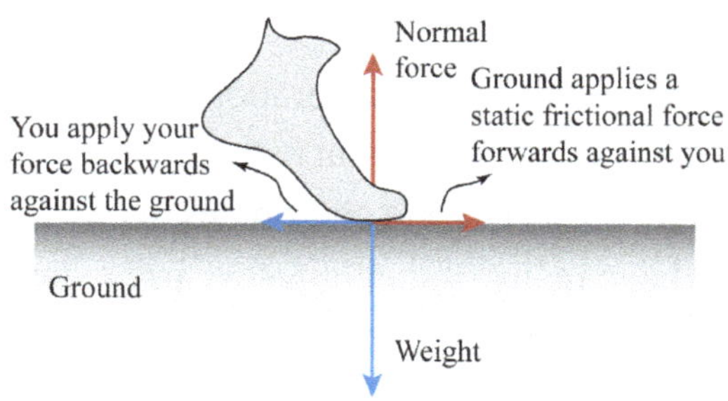

Figure 1.1: The friction and the reaction force from the ground while walking

When you walk in the park or on a tar road, you have a good grip without slipping due to the roughness between your foot or shoes and the surface of the ground/road. Furthermore, these situations are due to the concept called friction, which is studied in physics (figure 1.1).

Ballpoint Pen

Whether you are at your home, workplace or in your school, a ballpoint pen is one of the essential tools that you use almost every day. It uses the physics concepts while in use.

Figure 1.2: Ballpoint pen when it is in use

In a ballpoint pen, a ball is fixed on the tip of the pen (figure 1.2). As your pen moves across the paper, the ball on the tip of the pen turns due to friction with the paper and gravity forces the ink down onto the top of the ball where it is transferred onto the paper.

Transportation and Movement

Each day, you are faced with the need to travel or move from one place to another. If you consider common modes of transport, such as public transport, the car, train, or plane, physics played a huge role in the invention of all these modes. The transportation and movement that humans use in their daily lives use the laws of motion. The motion of all bodies is based on the presence of an external force acting on them, as stated in Newton's Laws of Motion, which describe the relationship between the motion of bodies and the external force acting on them.

It is interesting to see that engines in transportation applications, such as cars, represent the external driving force, allowing those applications to move and change positions. On the contrary, the brakes and public safety tools, such as the seat belt, represent an external force opposite to the engines to stop the movement of objects, thus resisting the occurrence of slips and accidents.

Mobile Phones

Mobile phones have paved their way in the modern world as extremely flexible, convenient and popular tools designed for communication. Due to their portability, reduced size and multiple integrated functions, they have become an almost indispensable accessory for the new generation.

An electronic telecommunication device is often referred to as a cellular phone or cell phone. The mobile phone connects to a wireless communication network through radio waves or satellite transmissions. Most mobile phones provide voice communication, short message service (SMS) and multimedia message service (MMS). The present smartphones have key features such as Internet connectivity, web browsing, email, embedded memory, wireless synchronization with other devices, Wi-Fi, touch screen, etc. Figure 1.3 shows the evolution of mobile technology during the last three decades.

Original: By Jojhnjoy - based on the work of Anders, Public Domain, https://commons.wikimedia.org/w/index.php?curid=36271517

Figure 1.3: Evolution of mobile technology during the last three decades

Your smartphone literally contains billions of transistors and other semiconductor elements. These can work as building blocks of digital electronic logic. Also, your smartphone contains a digital camera with a charge-coupled device (CCD) sensor. CCD is a light-sensitive integrated circuit that captures images by converting photons to electrons. This sensor is based on the photoelectric effect, which was explained by

Albert Einstein by introducing quanta of light, photons, which got him the physics Nobel prize. The batteries of your smartphone are getting better on the basis of a detailed understanding of electrochemistry and physics.

Digital Camera

Cameras are very common in our daily life. Initially, photographs were recorded images using the light-sensitive reaction of silver-based compounds such as silver chloride or silver bromide. Silver-based photographic paper was in common use until the arrival of digital photography in the 1980s, which is intimately connected to charge-coupled device (CCD) detectors. Colour is brought into play by putting red, blue, and green coloured filters over the pixels, resulting in coloured digital images. The underlying principle of operation of these cameras is the basic optics which we study in physics. The optics of a camera are pretty much the same as those of a single lens with an object distance that is significantly larger than the lens's focal distance.

Electronics is a big part of a digital camera. Figure 1.4 (a) shows a modern digital camera that has several lenses to produce a clear image. A CCD converts light signals into electronic signals, enabling electronic processing and storage of visual images. This is the basis for electronic imaging in all digital cameras [source: https://openstax.org/books/university-physics-volume-3/pages/2-6-the-camera].

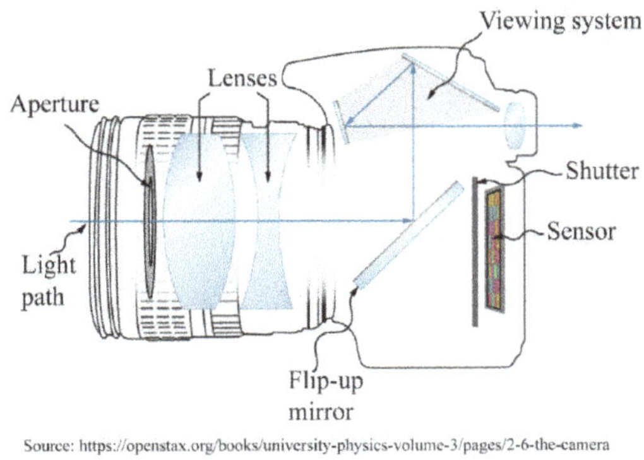

(a)

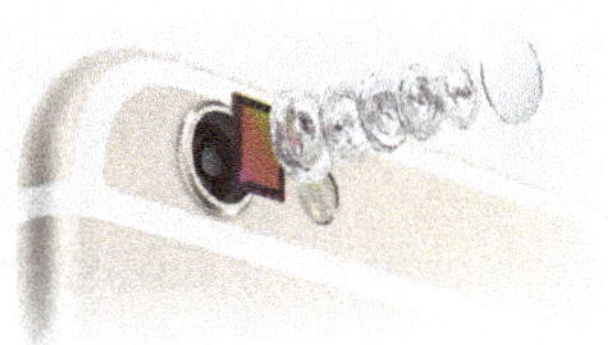

(b)

Figure 1.4: (a) A modern digital camera that has several lenses to produce a clear image.
(b) Typical internal arrangements of lenses that are assembled in a smartphone

Currently, the camera on a smartphone is often used by us. In particular, selfies are a common activity in our daily life. An average smartphone camera is equipped with a stationary wide-angle lens with a focal length equal to the thickness of the phone. Figure 1.4 (b) shows how the lenses are assembled in a smartphone. Unlike the normal camera system, you cannot change the lenses on your smartphone. The most common lenses used in smartphones are wide, ultra-wide, macro, and telephoto lenses. The image created by the lens is focused on the CCD detector mounted on the opposite side of the phone and stored as digital visual images.

Earphones/Headphones

In physics, sound is made up of air vibrations consisting of a series of high- and low-pressure zones. These are the cycles of a sound wave. Sound is the vibration of molecules. Even when we talk, the oscillations occur at different frequencies, and that is what our brains and ears hear. When sound enters your ear, your eardrum translates the air vibrations into mechanical vibrations of the tiny bones in the middle ear. These mechanical vibrations become fluid vibrations in your inner ear. Sensitive nerves then turn those vibrations into electrical signals that your brain interprets as sound.

(a) (b)

Figure 1.5: (a) Earphones/headphones that you normally use

(b) Typical internal structure of an earphone

Through physics, scientists have been able to understand how sound works and develop technology such as speakers and public address systems. Even the earphones that you use every day to enjoy your music can be credited to physics and its concepts. Figure 1.5 (a) shows a few earphones/headphones that you normally use.

Earphones/headphones that sit over your ears are just small speakers. They convert the electrical signals from your phone, record player, or computer into vibrations in the air that our ears interpret as sound. Figure 1.5 (b) shows the typical internal structure of an earphone. The concepts of magnetism, sound waves, and electromagnetism that you study in physics are used when making earphones/headphones.

1.7 Famous Physicists

Physicists are scientists who study the field of physics in great detail. They analyze the general happenings in nature and try to understand how the universe behaves. Physicists study the behaviour of the physical world at the most basic level and find practical ways to apply new knowledge gained from their research in areas of science and technology. On the basis of research through observation and experiments, they discover physical laws and propound theories that explain the functioning of natural forces like gravity, electromagnetism, and nuclear interactions.

Physicists are usually identified within three broad roles: (a) theoretical physicists, who develop theories or models of how particular aspects of the world work, (b) experimental physicists, who test these theories, determining their limits, and suggesting new approaches to them, and (c) applied physicists, who apply these findings in practical settings within the industry and the introduction of new technology. However, there are interactions between all three roles, and physicists generally have skills in each of these areas and have the necessary background in applications such as the development of medical equipment, electronic devices, nuclear reactors, etc. Physicists pioneered the pivotal discoveries of the 20th century that revolutionised the world, for example, the transistor, the laser, TV and radio, MRI and PET scans, quantum theory, and relativity.

During the earlier days, physics was considered a field of philosophy. It was only during the 19th century that physics was realized as a discipline separate from philosophy. Some of the major branches of physics are classical mechanics, quantum mechanics, the theory of relativity, and electromagnetism. Since physics covers a vast number of fields, most physicists specialize in one of the many subfields. However, the various subfields may overlap with each other and also with other sciences, resulting in interdisciplinary fields such as biophysics, geophysics, meteorology, etc.

The following are some of the world's most famous physicists and their great contributions to science and humanity.

Galileo Galilei (1564-1642)

Nationality: Italian

An Italian astronomer, engineer, and physicist, Galileo Galilei is widely regarded as the father of observational astronomy, the father of the scientific method, and the father of modern science. Created one of the first modern telescopes, Galileo revolutionised our understanding of the world supporting the work of Copernicus.

Isaac Newton (1642-1727)

Nationality : English

Sir Isaac Newton was an English mathematician, physicist, astronomer, alchemist, theologian and one of the most influential and popular scientists of all time. In his *Principia Mathematica,* published in 1687, he laid the foundations for classical mechanics, explaining the law of gravity and the laws of motion. These ideas have attracted the attention of scientists for more than three centuries.

Lord Kelvin (1824-1907)

Nationality: Irish

Lord Kelvin was a British mathematician, mathematical physicist and engineer born in Belfast. Absolute temperatures are stated in units of kelvin in his honour. He did important work in the mathematical analysis of electricity and formulation of the first and second laws of thermodynamics.

James Clerk Maxwell (1831-1879)

Nationality: Scottish

James Clerk Maxwell was a Scottish mathematician and physicist responsible for the classical theory of electromagnetic radiation. Maxwell's equations for electromagnetism have been called the "second great unification in physics", whereas the first one had been realised by Isaac Newton. He contributed to developing the Maxwell–Boltzmann distribution, a statistical means of describing aspects of the kinetic theory of gases.

Albert Einstein (1879-1955)

Nationality: German

The German-born Albert Einstein was one of the greatest minds to have ever lived. He is often thought to be the most important physicist of the 20th century. Einstein revolutionised modern physics with his general theory of relativity. Won Nobel Prize in Physics (1921) for his discovery of the Photoelectric effect, which formed the basis of Quantum Theory.

Ernest Rutherford (1871-1937)

Nationality: British

Ernest Rutherford was a New Zealand-born British physicist and chemist. He is remembered as the father of nuclear physics. His studies in radioactivity are what led others to explore nuclear physics. In 1908, Rutherford was awarded the Nobel Prize in chemistry for his work on radioactivity and the disintegration of elements.

Wilhelm Röntgen (1845 – 1923)

Nationality: German

Wilhelm Conrad Röntgen was a German physicist and mechanical engineer. He is best remembered for producing and detecting X-rays, for which he was honoured with the first Nobel Prize in Physics in 1901. His discovery of X-rays remains one of the greatest achievements in the field of medical science.

J. J. Thomson (1856-1940)

Nationality: British

J. J. Thomson was a British physicist credited with the discovery of the electron, the first subatomic particle to be discovered. He was awarded the Nobel Prize in Physics in 1906 for his work on the conduction of electricity in gases. Thomson is also credited with finding the first evidence for isotopes of a stable (non-radioactive) element in 1913.

Marie Curie (1867-1934)

Nationality: Polish

Marie Curie was a Polish physicist and chemist who conducted pioneering research on radioactivity. She was the first woman to win a Nobel Prize, the only person to win the Nobel Prize in two scientific fields and the first woman professor at the University of Paris. She is most famous for her findings on radioactivity.

Louis de Broglie (1892-1987)

Nationality: French

Louis de Broglie was a French physicist and aristocrat who made groundbreaking contributions to quantum theory. His de Broglie hypothesis, which suggests that all matters have wave properties, is one of the most important features in the theory of quantum mechanics. In 1929, de Broglie was honoured with the Nobel Prize for Physics for his work.

Max Planck (1858-1947)

Nationality: German

Max Planck was a German theoretical physicist whose discovery of energy quanta won him the Nobel Prize in Physics in 1918. He is remembered for originating the quantum theory of physics which revolutionized human understanding of atomic and subatomic processes.

Stephen Hawking (1942-2018)

Nationality: British

Stephen Hawking was an English theoretical physicist and cosmologist, who despite being afflicted with motor neurone disease that severely limited his physical abilities, was able to build a phenomenally successful career. He is widely known for his work involving black holes as well as gravitational singularities theorems in the framework of general relativity.

Chapter 2

Mathematics Review

2.1 Introduction

The study of physics requires basic knowledge of mathematics. Working knowledge of algebra, trigonometry and geometry is essential for the study of physics. The beauty of physics cannot be enjoyed if you do not have adequate proficiency in basic mathematical skills. The relevant topics in mathematics needed for algebra-based physics are summarised in this chapter. It is strongly recommended that you carefully study the content of this chapter, do the activities given, practise the end-of-chapter problems and become comfortable with them as quickly as possible. Thus, you could focus on the physics concepts and procedures that are being introduced without being distracted by unfamiliarity with the mathematics being used.

2.2 Exponentiation

Exponentiation is a mathematical operation written as x^n. It is generally stated as 'x raised to the power of n' or 'x to the power of n' or simply 'x to the n'. Here superscript n is called the **exponent** or the **power**, and x is called the **base.** If n is a positive integer and x is any real number, then it corresponds to repeated multiplication expressed as

$$x^n = \underbrace{x \times x \times x \times \ldots\ldots \times x}_{n \text{ times}}$$

When we write 5^4, the quantity 5^4 is equal to $5 \times 5 \times 5 \times 5 = 625$. Unique names are given for the operation when the exponent is 2 or 3. When the exponent is 2, we say that the quantity is **squared** (i.e. x^2 means x is squared). When the exponent is 3, the quantity is **cubed** (i.e. x^3 means x is cubed).

Any quantity raised to zero power is defined to be unity, this means $x^0 = 1$. If the exponent is 1, then $x^1 = x$, it means generally the exponent 1 is not written. Negative exponents are used for reciprocals, for example $x^{-5} = \frac{1}{x^5}$. An exponent can also be a fraction, for example $x^{1/5}$. The exponent $\frac{1}{2}$ is called a square root, and the exponent $\frac{1}{3}$ is called a cube root. Square root of x is denoted by $\sqrt{x}$ and can also be written as $x^{1/2}$. The cube root of x is denoted by $\sqrt[3]{x}$ and can also be written as is $x^{1/3}$.

The rules of exponentiation are summarized below.

(i) The product rule:

$$\left(x^m\right)\left(x^n\right) = x^{m+n}$$

Example: When $x = 2$, $m = 5$ and $n = 3$

$$\left(2^5\right)\left(2^3\right) = 2^{5+3} = 2^8 = 256$$

(ii) The quotient rule:

$$\frac{\left(x^m\right)}{\left(x^n\right)} = x^{m-n}, \text{ if } m = n, \frac{\left(x^m\right)}{\left(x^m\right)} = x^{m-m} = x^0 = 1$$

Example: When $x = 2$, $m = 5$ and $n = 3$

$$\frac{\left(2^5\right)}{\left(2^3\right)} = 2^{5-3} = 2^2 = 4$$

(iii) The first power rule:

$$\left(x^m\right)^n = x^{mn}$$

Example: When $x = 2$, $m = 2$ and $n = 3$

$$\left(2^2\right)^3 = 2^{2\times 3} = 2^6 = 64$$

(iv) Other power rules:

(a) $(xy)^n = (x)^n (y)^n$

(b) $\left(\frac{x}{y}\right)^n = \frac{x^n}{y^n}$

Example: When $x = 2$, $y = 5$ and $n = 3$

(i) $(2\times 5)^3 = (2^3)(5^3) = 8\times 125 = 1000$

(ii) $\left(\frac{2}{5}\right)^3 = \left(\frac{2^3}{5^3}\right) = \frac{8}{125}$

Suppose the base number is $(-x)$ and if we raised this base number to the power of n, we could write this as $(-x)^n$. This could be rewritten as $(-1)^n x^n$. Thus, $(-1)^n$ is $+1$ if n is even and $-$ if n is odd.

Activity 2.1

Simplify the expression $\frac{x^3 x}{x^{-4}\left(x^2\right)^3}$ and calculate its numerical value when $x = 2$.

2.3 Algebra

Algebra is an initial study of the most advanced forms of mathematics and provides a mathematical basis for many basic physics calculations. It demonstrates how to solve problems that include variables, allowing a user to find the value of one variable given conditional values for other variables in an equation. In physics, algebra is a fundamental mathematics skill both for its value when applying mathematics and its application in basic physics operations.

Solving Equations

Equations written in terms of symbols that represent quantities are frequently used in physics. An equation is a mathematical expression that contains an equal sign. There are two sides to an equation, with the left side equal to the right side. Equations often involve algebra and contain unknown, which we often represent with letters such as x or y. We could solve simple equations and more complicated equations to determine the value of these unknowns. Use the following rule to solve an equation:

An equation remains true if any valid operation performed on one side of the equation is also performed on the other side.

The operations could be

(i) adding or subtracting a number or symbol,

(ii) multiplying or dividing by a number or symbol, or

(iii) raising each side of the equation to the same power.

The following are examples of solving simple equations:

(i) Solve the equation $2x+5=x+8$ for x.

$$2x+5=x+8$$
$$2x+5-x=x+8-x$$
$$x+5=8$$
$$x+5-5=8-5$$
$$x=3$$

(ii) Solve the equation $2\sqrt{x}=10$ for x.

$$2\sqrt{x}=10$$
$$\frac{2\sqrt{x}}{2}=\frac{10}{2}$$
$$\sqrt{x}=5$$
$$x=5^2=25$$

Activity 2.2

(i) Solve the equation $3x^2+4=31$ for x.

(ii) Solve the equation $x=ut+\frac{1}{2}at^2$ for a.

Quadratic Equation

In algebra, a quadratic equation is any equation that can be rearranged in standard form as

$$ax^2+bx+c=0$$

where x represents an unknown, and a, b and c represent known numbers and $a \neq 0$. The numbers *a, b* and c are the coefficients of the equation. Such equations cannot be used in the previous simple methods to solve for x. Solutions to such equation are expressed by the **quadratic formula:**

$$x = \frac{-b \pm \sqrt{b^2 - 4ac}}{2a}$$

In general, a quadratic equation has two solutions, which may be real or complex numbers. If $b^2 - 4ac = 0$, then the two solutions are equal and real numbers. If $b^2 > 4ac$, then $b^2 - 4ac$ is positive, and the two solutions are unequal and real numbers. However, if $b^2 < 4ac$, then $b^2 - 4ac$ is negative and the solutions are unequal complex numbers and cannot represent physical quantities. In such a case, the quadratic equation has mathematical solutions, but no physical solutions.

Example: If x represents a physical quantity that takes only non-negative values, determine the value of x that satisfies the equation $x^2 + 4x = 5$.

The standard form of quadratic equation is $x^2 + 4x - 5 = 0$. Then, $a = 1$, $b = 4$ and $c = -5$.

The quadratic formula gives the solution as

$$x = \frac{-(4) \pm \sqrt{(4)^2 - 4(1)(-5)}}{2(1)}$$

$$= \frac{-4 \pm \sqrt{16 + 20}}{2} = \frac{-4 \pm 6}{2}$$

Therefore, $x = 1$ or $x = -5$. Since x represents a physical quantity that takes only non-negative values, then the negative solution is nonphysical and is discarded.

Activity 2.3

If x represents a physical quantity that takes only non-negative values, determine the value of x that satisfies the equation $x^2 + x = 1$. Take $\sqrt{5} = 2.24$.

Simultaneous Equations

Simultaneous equations are two or more algebraic equations that share common unknowns and are solved together. If a problem has two unknowns say x and y, then it takes two independent equations in x and y (i.e. two equations for x and y, where one equation is not simply a multiple of the other) to determine their values uniquely.

The most common method for solving simultaneous equations is the elimination method, which means that one of the unknowns will be removed from each equation and the remaining unknown (say x) can then be calculated. Thereafter, by adding this calculated value (say x) to one of the equations and the other remaining unknown (say y) could be calculated. This method of solving simultaneous equations is illustrated in the following example.

Example: Solve the following pair of equations for x and y.

$$x+2y=15$$

$$3x-5y=1$$

If the first equation is multiply by -3, then we get $-3x-6y=-45$
Adding this to the second equation successively gives,
$3x-5y+(-3x)+(-6y)=1+(-45)$; $-11y=-44$
this will give $y=4$;
If we substitue $y=4$ to one of the equations (say the first equation), then we get $x+2\times4=15$, $x+8=15$,
this will give $x = 7$.

Activity 2.4

Solve the following equations :

(i) $5x-4y=1$ and $5x-3y=2$

(ii) $3x+2y=12$ and $2x-y=1$

2.4 Direct, Inverse and Inverse-Square Relationship

The important part of physics is to describe and verify relationships among physical quantities. For example, the relationships could be two quantities that may be directly proportional to each other or inversely proportional to each other, or one quantity may be inversely proportional to the square of the other quantity.

Direct Relationship

Consider two quantities x and y. If an increase (or decrease) of the first quantity x causes an increase (or decrease) of the second quantity y by the same factor, the two quantities are said to be directly proportional to each other. Suppose, if $y$is directly proportional to x, the direct proportionality is written as $y \propto x$. The ratio $\frac{y}{x}$ is a constant, say k, and could be written as $\frac{y}{x} = k$, where k is the constant of proportionality.

For example, Ohm's law states that the current (I) through a conductor between two points is directly proportional to the voltage (V) across the two points. The usual mathematical equation that describes this relationship is $\frac{V}{I} = R$, where R is the constant of proportionality and is called the resistance of the conductor.

Activity 2.5

The two quantities x and y are directly proportional to each other. If $x = 2$ when $y = 8$, determine the value of x when $y = 12$.

Inverse Proportion

The two quantities x and y are said to be inversely proportional when one quantity (say x) increases, the other (y) decreases while their product stays the same. This means $y \propto \frac{1}{x}$, and thus $y = \frac{k}{x}$ or $yx = k$, where k is the constant of proportionality.

For example, according to Boyle's law (which you will study later in the physics course), if the temperature of an ideal gas is kept constant, its pressure, P, is inversely proportional to its volume, V. Thus their product PVremains constant.

Activity 2.6

A cylindrical container is fitted with an airtight piston and contains an ideal gas. Initially, the pressure of the inside gas is 12×10^4 Pa and the volume of the gas is 8×10^{-3} m^3. Assuming that the system is always at the temperature of 330 K, determine the volume of the gas when its pressure increases to 24×10^4 Pa.

Graphs of Direct and Inverse Proportionality Relationship

We have learnt how to represent the direct proportional and inversely proportional relationships of two quantities in the form of an equation. This can be expressed in another way by using graphs.

We have seen that if y is directly proportional to x, it could be written as $\frac{y}{x} = k$, thus $y = kx$, where k is the constant of proportionality. This shows that the graph of y *versus* x is a straight line passing through the origin as shown in figure 2.1.

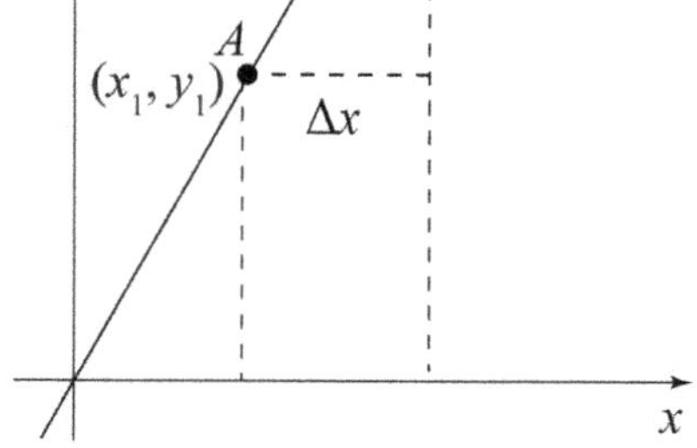

Figure 2.1: Straight line graph of y *versus* x passing through the origin

On the graph, if the point A moves along the straight line to the point B, the change of the quantity x is labelled as $\Delta x = x_2 - x_1$ and the corresponding change in y is labelled as $\Delta y = y_1 - y_2$, where (x_1, y_1) and (x_2, y_2) are coordinates of A and B respectively. The constant of proportionality between Δy and Δx is also k. Thus,

$$\Delta y = k\Delta x$$

The ratio $\frac{\Delta y}{\Delta x}$ is the steepness of the line and is called the slope or gradient of the line. Thus,

$$\text{slope} = \frac{\Delta y}{\Delta x} = k$$

As illustrated in figure 2.2, the slope of the straight line on the graph can be positive, negative, zero, or undefined. An undefined slope (or an infinite slope) means that x does not change, i.e. the line is parallel to the $y-$axis.

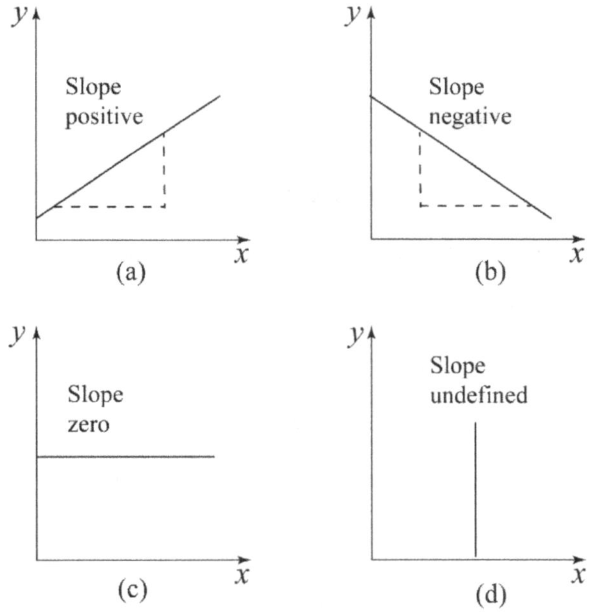

Figure 2.2: (a) slope positive; y increases as x increases (b) slope negative; y decreases as x increases (c) slope zero; y does not change, i.e. the line is parallel to the $x-$axis (d) slope undefined; x does not change, i.e. the line is parallel to the $y-$axis

We have also seen that if y is inversely proportional to x, it could be written as $xy = k$, where k is the constant of proportionality. The behaviour of y *versus* x

is illustrated in figure 2.3. That is, when x changes from x_1 to x_2, y changes from y_1 to y_2 so that $x_1y_1 = x_2y_2 = k$.

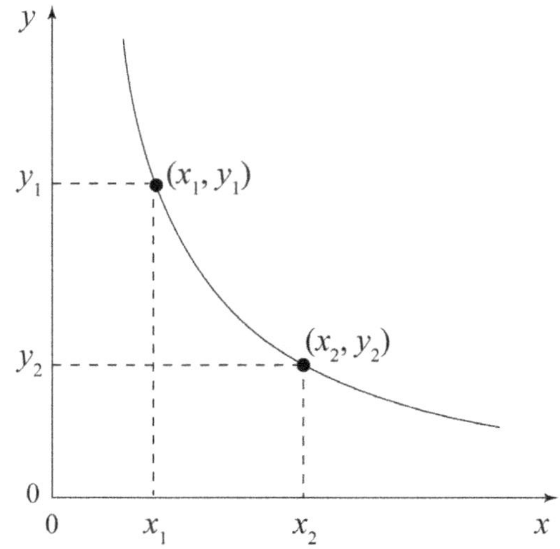

Figure 2.3: Illustrates the behaviour of y which is inversely proportional to x so that $xy = k$

Inverse Square Proportion

In the laws of nature, the inverse square dependence is common. The inverse-square law states that a specified physical quantity is inversely proportional to the square of the distance from the source of that physical quantity. For example, according to Newton's law of gravitation (which you will study later in the physics course), the force of gravity due to an object decreases as the inverse square of its distance from the object. If y varies inversely with the square of x, then we could write the inverse-square relationship as

$$y \propto \frac{1}{x^2} \quad \text{or} \quad y = k\left(\frac{1}{x^2}\right) \quad \text{or} \quad x^2 y = k$$

where k is the constant of proportionality.

The behaviour of y *versus* x is illustrated in figure 2.4. That is, when x changes from x_1 to x_2, y changes from y_1 to y_2 so that $x_1^2 y_1 = x_2^2 y_2 = k$, or $\frac{y_1}{y_2} = \frac{x_2^2}{x_1^2}$.

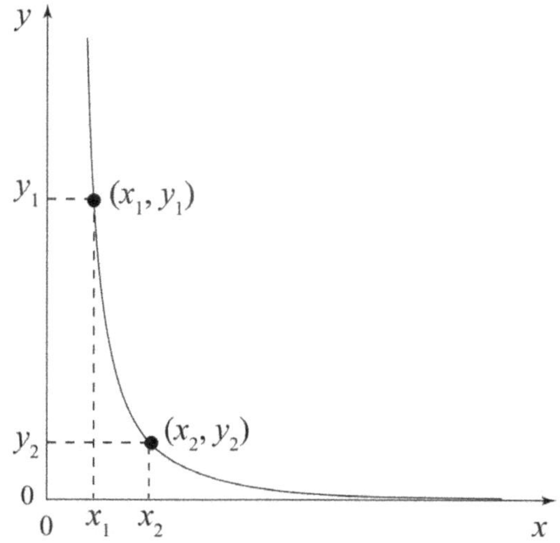

Figure 2.4: Illustrates the behaviour of y *versus* x, if y varies inversely with the square of x, so that $x^2 y = k$

Activity 2.7

The force of gravity on an object (the object's weight) varies inversely as the square of the distance from the centre of the earth. If an astronaut weighs 700 N when on the surface of the earth, determine the force of gravity on the astronaut when he is at a height of 6000 km from the surface of the earth. Assume that the astronaut is in orbit and he will float but gravity still acts on him. Take the radius of the earth as 6.38×10^6 m.

Logarithmic and Exponential Functions

The logarithm is the inverse function of exponentiation (see section 2.2). This means that the logarithm of a given number n is the exponent to which another fixed number, the base x, must be raised, to produce that number n. In the simplest case, the logarithm counts the number of occurrences of the same factor in repeated multiplication. For example, the number 1000 could be written as $1000 = 10\times10\times10 = 10^3$, the logarithm base 10 or **the base-10 logarithm** or the **common logarithm (log) of 1000 is 3.** Mathematically this could be written as $\log_{10}(1000) = 3$. In general, The logarithm of n to base x is denoted as $\log_x(n)$, or without parentheses, $\log_x n$, or even without the explicit base, $\log n$.

The antilog (which is also known as antilogarithms) of a number is the inverse technique of finding the logarithm of the same number. If we consider the case where y is the logarithm of a number with base x, i.e. $y = \log_x n$, then n is the antilog of y to the base x. Thus, if $y = \log_x n$, then $n = x^y$. For example, if $\log_{10} y = -3$, then $y = 10^{-3} = 0.001$.

It should be noted that $\log_{10} 1 = 0$, and the log of zero or of a negative number is not defined. The logarithm of a number is positive if the number is greater than 1. The log of a number is negative if the number is less than 1, but greater than zero.

Another logarithm that we frequently use in physics is the **natural logarithm,** where the base is the quantity e, which is approximately equal to 2.718. The natural logarithm of x is generally written as $\ln x$ or $\log_e x$. Parentheses are sometimes added for clarity, giving $\ln(x)$ or $\log_e(x)$. The natural logarithm of a number y is the power to which e must be raised to obtain y. That is, if $x = \ln y$, then $y = e^x$, which is called an **exponential function** and also sometimes written as $\exp(x)$. For example, if $\ln x = 4$, then $x = e^4$. The plot of the function e^x or $\exp(x)$ *versus* $x$is shown in figure 2.5. While, the plot of the function e^{-x} or $\exp(-x)$ *versus* x is a curve which is a reflection of figure 2.5 about the $y-$axis.

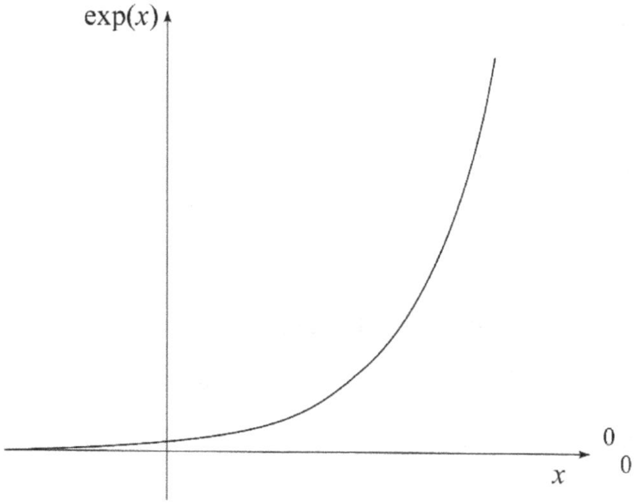

Figure 2.5: Plot of the function $\exp(x)$ *versus* x

It should be noted that the natural logarithm of e itself is 1 (since $e^1 = e$), i.e. $\ln e = 1$, and the natural logarithm of 1 is 0, i.e. $\ln 1 = 0$ (since $e^0 = 1$).

Logarithms with any choice of base, including base 10 or base e, obey the following simple rules:

(i) Product rule : $\log(M\,N) = \log M + \log N$

(ii) Quotient rule : $\log\left(\frac{M}{N}\right) = \log M - \log N$

(iii) Power rule : $\log\left(M^q\right) = q \log M$

In order to change the base from base a to base c, we can use the following logarithm change of base rule:

$$\log_a(M) = \frac{\log_c(M)}{\log_c(a)}$$

If the choice of bases are base 10 or base e, then $\ln(N) = \frac{\log_{10}(N)}{\log_{10}(e)}$.

Activity 2.8

Solve the equation $N = N_0 e^{-\alpha t}$ for t, α is a constant.

Simplify the expression $4\log x + 2\log y - 3\log(x-y)$

2.5 The Binomial Expansion

In physics, sometimes we end up with a quantity $(1+x)$ raised to the nth power, i.e. of the form $(1+x)^n$. This can be written as an infinite sum of terms, known as a **series expansion,** as shown below.

$$(1+x)^n = 1 + nx + \frac{n(n-1)}{2!}x^2 + ...$$

The above formula is useful in physics, mainly when x is very small compared to one, i.e. when $x \ll 1$. In this case, each successive term is much smaller than the preceeding term [for example, if $x = 0.01$ and $n = 3$, then the expansion of the first term is 1, the second term is $nx = 3 \times 0.1 = 0.03$, and the third term is $\frac{n(n-1)}{2!}x^2 = \frac{3\times 2}{1\times 2}(0.01)^2 = 0.0003$ and so on]. Thus, when the quantity x is small, very small compared to one, i.e. $x \ll 1$, we can ignore all but except the first two (or three) terms and can write as:

$$(1+x)^n \approx 1 + nx$$

This approximation often allows us to solve an equation easily that otherwise might be quite difficult. The following are a few examples that you may come across in your physics courses when the quantity x is very small compared to one, i.e. $x \ll 1$.

$$(1+x)^2 \approx 1 + 2x$$

$$\frac{1}{(1+x)} = (1+x)^{-1} \approx 1 - x$$

$$\sqrt{(1+x)} = (1+x)^{\frac{1}{2}} \approx 1 + \frac{1}{2}x$$

$$\frac{1}{\sqrt{(1+x)}} = (1+x)^{-\frac{1}{2}} \approx 1 - \frac{1}{2}x$$

Activity 2.9

Evaluate $\sqrt{1.02}$ using the binomial expansion, taking $x = 0.02$ which is much smaller than 1.

2.6 Areas and Volumes

Figure 2.6 illustrates the formulae for the areas (A), volumes (V), surface areas (S) and circumferences/perimeters (C) of geometric shapes frequently used in physics.

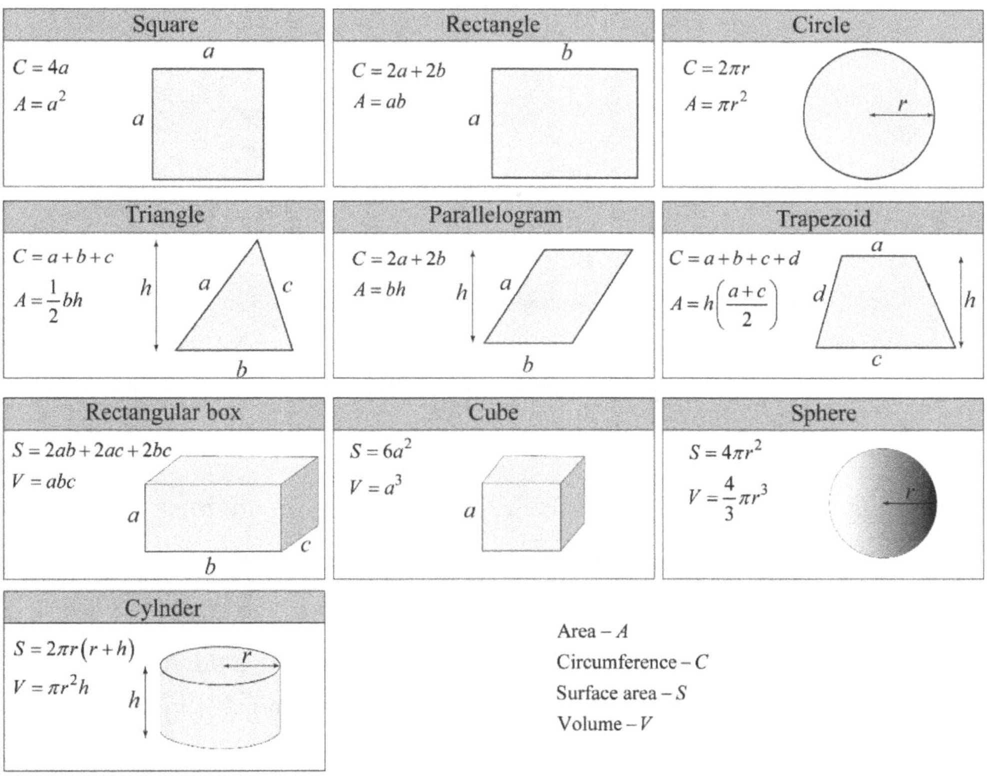

Figure 2.6: Areas, volumes, surface areas and circumferences, perimeters of geometric shapes

Activity 2.10

Calculate the total surface area and volume of a solid cylinder of radius 0.20 m and height 0.40 m.

2.7 Plane Geometry

We shall review some of the theorems that you would have studied in your early mathematics courses. Following are some useful results about angles:

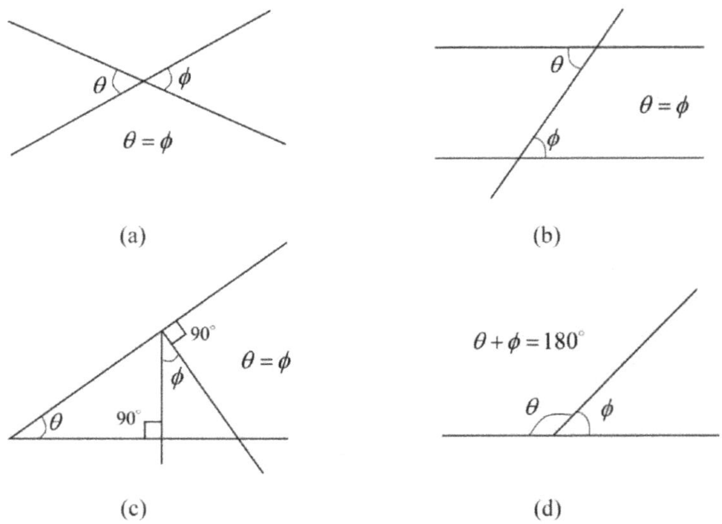

Figure 2.7: (a) Interior angles formed when two straight lines intersecting each other are equal, (b) when two parallel lines are intersected by a diagonal straight line (transversal), the alternate interior angles are equal (c) when the sides of one angle are each perpendicular to the corresponding sides of a second angle, then the two angles are equal (d) The sum of the angles on the side of a straight line is 180°

(i) Two angles are equal if any of the following conditions are true:

(a) When two straight lines intersect as shown in figure 2.7(a), the interior angles θ and ϕ formed are equal, i.e. $\theta = \phi$.

(b) When two parallel lines are intersected by a diagonal straight line as shown in figure 2.7(b), the alternate interior angles θ and ϕ formed are equal, i.e. $\theta \quad \phi$.

(c) When the sides of one angle (θ) are each perpendicular to the corresponding sides of a second angle (ϕ) as shown in figure 2.7(c), then the two angles θ and ϕ are equal, i.e. $\theta = \phi$.

(d) The sum of the angles θ and ϕ shown in in figure 2.7(c) on one side of a straight line is 180°, i.e. $\theta + \phi = 180^\circ$.

(ii) The sum of the angles in any plane triangle is 180 .

Similar Triangles

The triangles shown in figure 2.8 (a) and (b) are said to be similar if all three of their angles are equal, i.e. $\theta_1 = \phi_1$, $\theta_2 = \phi_2$ and $\theta_3 = \phi_3$. Similar triangles thus have the same shape but may be of different sizes and have different orientation.

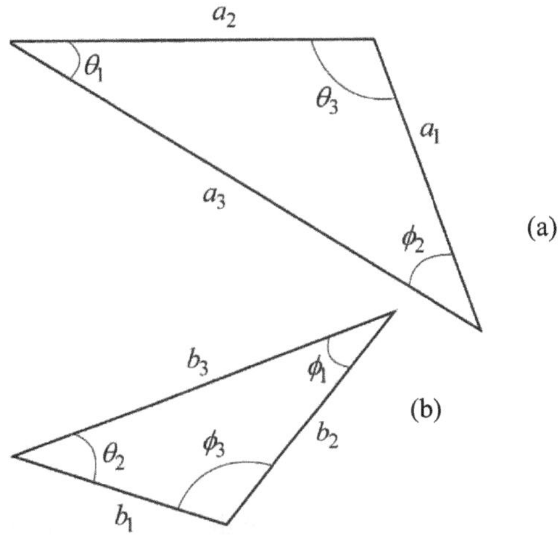

Figure 2.8: Two similar triangles: the same shape but may be of different sizes and have different orientations

The following are two useful results about similar triangles.

(i) Two triangles are similar if any two of their angles are equal. This is because their third angles must also be equal since the sum of the angles of a triangle is 180°.

(ii) The ratios of corresponding sides of two similar triangles shown in figure 2.8 (a) and (b) are equal. That is

$$\frac{a_1}{b_1} = \frac{a_2}{b_2} = \frac{a_3}{b_3}$$

If two similar triangles have the same size, they are said to be congruent. That is, if triangles are congruent, one can be placed precisely on top of each other.

Right Angled Triangles

An angle of a right triangle is 90° and is called a right angle. The two sides at the right angle are perpendicular to each other, as shown in figure 2.9. The other two acute angles (angles that are less than 90°) have a sum of 90°, i.e. $\theta + \phi = 90^\circ$. The side opposite the right angle is called the hypotenuse, as mentioned in figure 2.9.

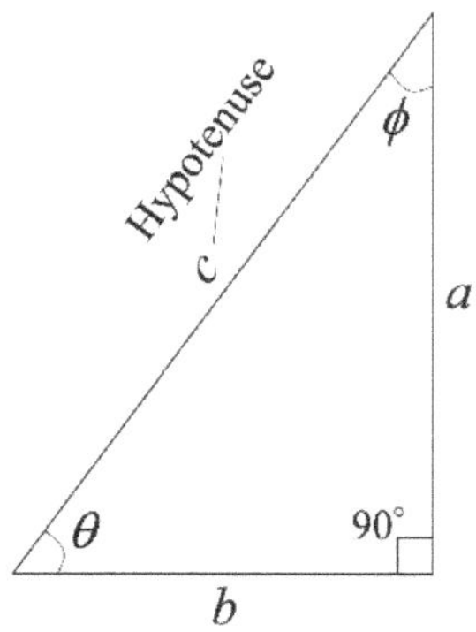

Figure 2.9: A right triangle having one angle 90°

In any right - angled triangle, the square of the length of the hypotenuse (i.e. c) is equal to the sum of the squares of the lengths of the other two sides (i.e. a and b). For the triangle shown in figure 2.9,

$$c^2 = a^2 + b^2$$

This formula is called the **Pythagorean Theorem.**

Trigonometric Functions

The trigonometric functions are real functions which relate an angle of a right-angled triangle to ratios of two side lengths.

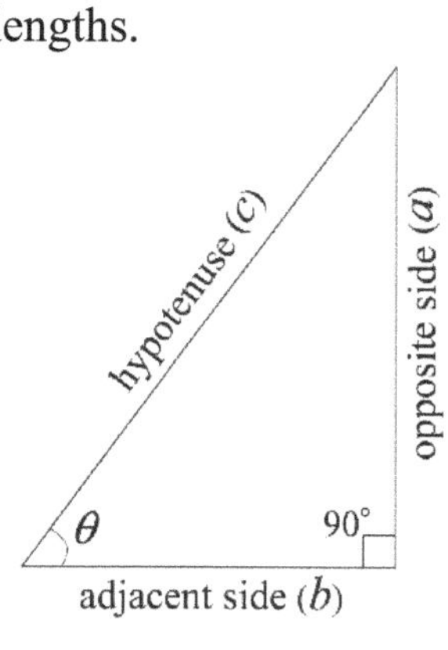

Figure 2.10: A right triangle showing hypotenuse, opposite side and adjacent side for angle θ

You may use these trigonometric functions in your physics courses very often. Most widely used trigonometric functions are the **sine,** the **cosine**, and the **tangent.** These are written as **sin, cos** and **tan** respectively. For the triangle shown in figure 2.10, these functions are defined as follows.

$$\sin\theta = \frac{\text{opposite side}}{\text{hypotenuse}} = \frac{a}{c}$$

$$\cos\theta = \frac{\text{adjacent side}}{\text{hypotenuse}} = \frac{b}{c}$$

$$\tan\theta = \frac{\text{opposite side}}{\text{adjacent side}} = \frac{a}{b}$$

From the above defined functions, we have, $\tan\theta = \dfrac{\sin\theta}{\cos\theta}$.

The reciprocals of the 'sine', the 'cosine', and the 'tangent' are the **secant**, the **cosecant,** and the **cotangent,** respectively and are written as **sec, csc** (or **cosec**) and **cot** respectively. For the triangle shown in figure 2.10, these functions are defined as follows.

$$\sec\theta = \frac{1}{\sin\theta} = \frac{\text{hypotenuse}}{\text{opposite side}} = \frac{c}{a}$$

$$\csc\theta = \frac{1}{\cos\theta} = \frac{\text{hypotenuse}}{\text{adjacent side}} = \frac{c}{b}$$

$$\cot\theta = \frac{1}{\tan\theta} = \frac{\text{adjacent side}}{\text{opposite side}} = \frac{b}{a}$$

Inverse trigonometric functions are the inverse functions of the trigonometric functions. That is, they are the inverses of functions 'sine', 'cosine', 'tangent', 'cotangent', 'secant', and 'cosecant'. The common notations for these inverse trigonometric functions are $\sin^{-1}\theta$ (or $\arcsin\theta$), $\cos^{-1}\theta$ (or $\arccos\theta$), $\tan^{-1}\theta$ (or $\arctan\theta$) etc. respectively. They are used to obtain an angle from any of the angle's trigonometric ratios and are widely used in physics, engineering and in navigations. Note that $\sin^{-1}\theta$ **does not** mean $\dfrac{1}{\sin\theta}$, it is an angle. In physics, angles are expressed in degrees or radians, where π radians = 180°.

Figure 2.11 shows the sign (+ or –) that sine, cosine and tangent take on for angles θ in four quadrants.

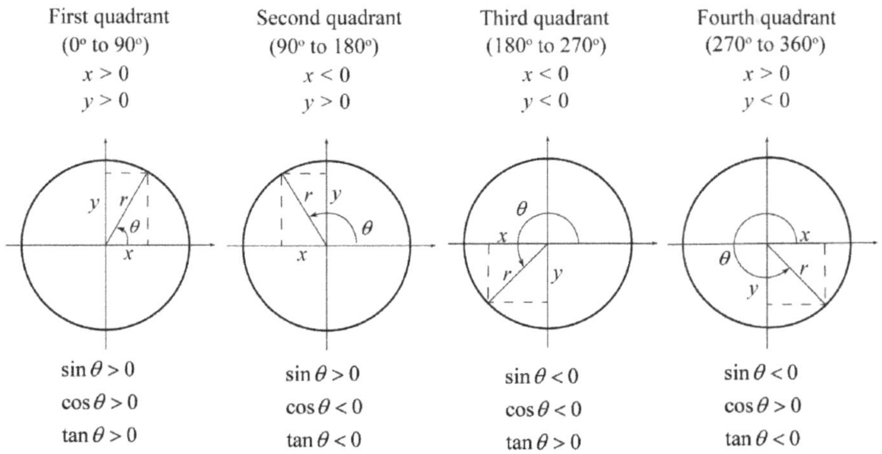

Figure 2.11: Signs of sine, cosine, and tangent take on for angles θ in four quadrants

The angles are measured counterclockwise from the x-axis as shown in figure 2.11, whereas the negative angles are measured clockwise, i.e from below the x-axis.

Figure 2.12 shows the variation of $\sin\theta$ and $\cos\theta$ against the angle θ in the range from $\;^\circ$ to 360°.

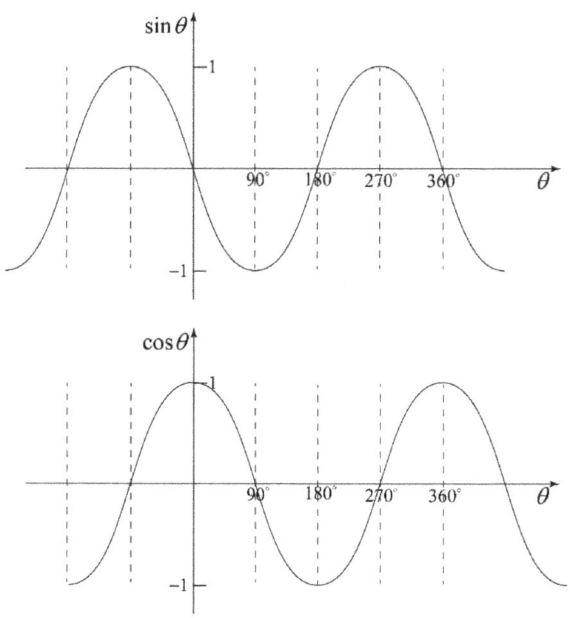

Figure 2.12: variation of $\sin\theta$ and $\cos\theta$ against the angle θ in the range from $0°$ to $360°$

As shown in figure 2.12, the graphs of $\sin\theta$ and $\cos\theta$ are of periodic function, with a period of 360°. Each function is positive and negative between angles 0° and 360° and the values vary between $+1$ and -1. Further, these two functions $\sin\theta$ and $\cos\theta$ are out of phase to each other. When one is zero, the other has its maximum or minimum value.

In trigonometry, the law of sines (or the sine law) is an equation relating the lengths of the sides of any triangle to the sines of its angles.

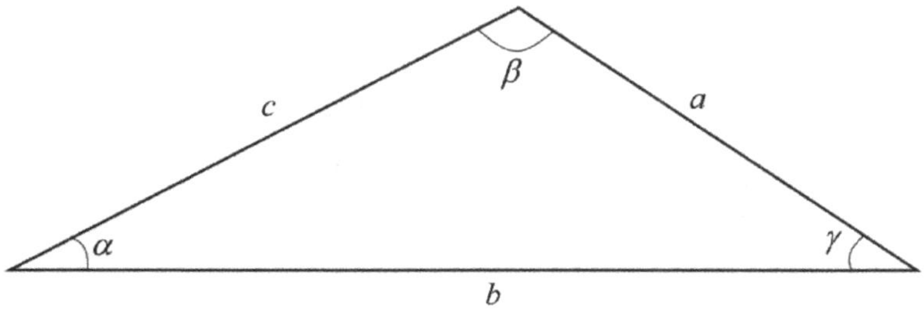

Figure 2.13: A triangle with the lengths of the sides *a, b, c,* with their opposite angles α, β, γ respectively

Accordingly, for any triangle (not necessarily a right triangle)

$$\frac{\sin\alpha}{a} = \frac{\sin\beta}{b} = \frac{\sin\gamma}{c}$$

where *a, b* and *c* are the lengths of the sides of a triangle, and α, β and γ are the opposite angles as shown in figure 2.13.

In trigonometry, the law of cosines (or the cosine rule) relates the lengths of the sides of a triangle to the cosine of one of its angles. If we consider figure 2.13, the law of cosines could be written as,

$$c^2 = a^2 + b^2 - 2ab\cos\gamma$$

where γ denotes the angle contained between the sides of lengths *a* and *b* and opposite the side of length *c*. For the same figure 2.13, the other two relations are:

$$a^2 = b^2 + c^2 - 2bc\cos\alpha \quad \text{and} \quad b^2 = a^2 + c^2 - 2ac\cos\beta$$

Useful trigonometric identities

The following are useful identities among the trigonometric functions.

$$\sin(-\theta) = -\sin\theta$$

$$\cos(-\theta) = \cos\theta$$

$$\sin(90^\circ - \theta) = \cos\theta$$

$$\cos(90^\circ - \theta) = \sin\theta$$

$$\sin(180^\circ - \theta) = \sin\theta$$

$$\cos(180^\circ - \theta) = -\cos\theta$$

$$\sin^2\theta + \cos^2\theta = 1$$

$$\sin 2\theta = 2\sin\theta\cos\theta$$

$$\cos 2\theta = \cos^2\theta - \sin^2\theta = 2\cos^2\theta - 1 = 1 - 2\sin^2\theta$$

$$\sin(\theta \pm \phi) = \sin\theta\cos\phi \pm \cos\theta\sin\phi$$

$$\cos(\theta \pm \phi) = \cos\theta\cos\phi \mp \sin\theta\sin\phi$$

$$\tan(\theta \pm \phi) = \frac{\tan\theta \pm \tan\phi}{1 \mp \tan\theta\tan\phi}$$

$$\sin\theta \pm \sin\phi = 2\sin\left(\frac{\theta \pm \phi}{2}\right)\cos\left(\frac{\theta \mp \phi}{2}\right)$$

When angle θ (in radians) is small,

$$\sin\theta \approx \theta \text{ and } \cos\theta \approx 1 - \frac{\theta^2}{2} \approx 1$$

Questions

NOTE: Although the solutions are given on the last page, it is in your own interest to attempt the questions on your own as an indication of how far you have understood the material of the section.

1. Use the exponent rules to simplify the following expressions and calculate its numerical value when $x = 2$ and $y = 1$:

   (i) $\left(-2x^2y^4\right)^2$

   (ii) $\left(\dfrac{6x^3}{3y^2}\right)^3$

   (iii) $\left(-\dfrac{x^2y^4}{xy^{-2}}\right)^2$

2. Simplify the expression $\dfrac{x^3y^{-3}xy^{\frac{4}{3}}}{x^{-4}y^{\frac{1}{3}}\left(x^2\right)^3}$ and calculate its numerical value when $x = 9$ and $y = 3$.

3. (i) If $x^4 = 81$, determine the value of x.

   (ii) Solve the equation $4.9t^2 + 2t - 20 = 0$

4. (i) Solve the equation $T = 2\pi\sqrt{\dfrac{l}{g}}$, for l

   (ii) Solve the equation $F = k\dfrac{m_1m_2}{r^2}$, for r

5. Solve the following equations.

   (i) $3x + y - 11 = 0$ and $2x + y - 8 = 0$

   (ii) $3x + 2y = 17$ and $4x - y = 30$

6. Use the equations $u = u_0 + at$ and $x = u_0 t + \frac{1}{2}at^2$ to obtain an expression for x that does not contain a.

7. According to the ideal gas equation, the volume of an ideal gas is directly proportional to its temperature in kelvin (K) if the pressure of the gas is constant. An ideal gas occupies a volume of 4 litres at 100 K. Determine its volume when it is heated to 200 K while held at a constant pressure.

8. The force needed to compress the coiled spring is directly proportional to the amount of compression. If a force of 80 N is needed to compress the spring by 0.02 m, determine the force required to compress the spring by 0.05 m.

9. Compute the total surface area and volume of a rectangular solid of length 0.20 m, width 0.15 m, and height 0.80 m.

10. In a stairway, each step has a length of 30 cm from the next lower step. If the stairway rises at an angle of $30°$ with the horizontal, what is the height of each step?

Chapter 3

Scientific Notation and Significant Figures

3.1 Introduction

Natural science, including physics, is concerned with describing, understanding, and predicting natural phenomena based on empirical evidence from observation and experimentation. The scientific method is a key feature of what makes the natural sciences so scientific. The **scientific method** is a process in which scientists try to investigate, verify, or construct an accurate and reliable version of any natural phenomenon.

Physics is an experimental science that always deals with a wide range of measurements, from very large to very small. Accuracy and precision are used in the context of measurement. There is a technical difference between accuracy and precision, and this will be discussed in a later chapter. In physics, both accuracy and precision are important, and both reflect how close a measurement is to an actual value. Furthermore, the numerical number that we obtain in a measurement or the result of a calculation must indicate the precision with which the number is known. The number of meaningful digits in a number that we acquire in a measurement or the result of a calculation is called the number of **significant figures** or **significant digits**. Generally, **scientific notation** is used to express very large or very small numbers that we obtain in a measurement or the result of a calculation. Scientific notation is often used in scientific calculations where the analysis must be precise.

3.2 Scientific Methodology

Generally, humans are curious to some extent. In our day-to-day life, we have the habit of looking around the environment in which we live. More often, we try to understand what we see. As we become serious about exploring nature, we become more organised and formal in collecting and analysing data. We attempt greater precision, perform controlled experiments (if possible), and write down ideas about how data may be organised. We then formulate models, theories, and laws based on the data we have collected and communicate those results with others. This is the foundation that describes the scientific method that scientists employ to decide scientific issues based on evidence from observation and experiment. Therefore, we could say that the scientific method is a process in which scientists try to investigate, verify, or construct an accurate and reliable version of any natural phenomenon. The scientific method can be applied broadly in science across many different fields, such as chemistry, physics, biology, etc. The scientific method uses a series of steps. The following are the basic steps that scientists employ to resolve scientific issues on the basis of evidence from observation and experiment.

Step 1: Observation

An investigation often begins with an observation. These observations could be a pattern or trend within the natural world. Observation may generate questions that we wish to answer. These careful observations may lead us to collect data from simple observations or experiments.

Step 2: Hypothesis

From the initial observation and data collected, a model is hypothesised. A hypothesis is a testable statement/prediction or a guess that explains something about the observation. In other words, it is an assumption that is made in order to draw out and test its logical or empirical consequences.

Step 3: Experiment

The hypothesis could be tested by designing and performing experiments. The aim of these experiments is to determine whether the hypothesis supports or contradicts the observations made in the real world. The confidence in the hypothesis increases or decreases based on the result of the experiments.

Step 4: Data analysis

During the experiment, data will be collected. These data are often analysed using statistical, mathematical, and/or graphical methods. The analysed data can then be interpreted and explained. The process of interpreting and explaining the analysed data involves looking for patterns, similarities, differences, trends, and other relationships and relating them to the hypothesis.

Step 5: Conclusion

On the basis of the analysis, conclusions could be drawn. The conclusion of the experiment might either support or contradict the hypothesis. If the data found in the analysis are consistent with the hypothesis, it is accepted. If the conclusion contradicts the hypothesis, the experimental procedure should be checked. Even after altering the procedure of the experiment, if the results still contradict the hypothesis, then the original hypothesis must be modified. Another experiment needs to be designed to test the modified hypothesis. This process needs to be continued until satisfactory and accepted results are obtained in support of the hypothesis.

A **scientific theory** is an explanation for patterns or trends in nature that have been repeatedly tested and validated in accordance with the scientific method by multiple researchers. In general, theories are tested under controlled conditions in an experiment. Suppose that controlled conditions are not amenable to experimental testing then theories are evaluated through principles of abductive reasoning (a form of logical inference). Established scientific theories would have undergone rigorous scrutiny and encompass scientific knowledge.

3.3 Scientific Notation

In physics, we frequently deal with numbers that are very large and very small. It can be cumbersome to write numbers in conventional decimal notation. When expressing a physical quantity, it is important to use a proper and meaningful number. These issues could be addressed by using scientific notation, which is a way of expressing numbers that are too large or too small to be conveniently written in decimal form. Using scientific notation, a quantity could be expressed as a decimal number with one digit between 1 and 10 to the left of the decimal point, multiplied by the appropriate power of 10. For example, the radius of the earth is approximately 6 380 000 m at the equator, and this could be written as 6.38×10^6 m.

Furthermore, scientific notation eliminates the need to write zeros to locate the decimal point correctly. For example, the Bohr radius of a hydrogen atom is 0.000 000 000 053 m, and this could be written as 5.3×10^{-11} m. Table 3.1 illustrates a few examples of decimal notation and its corresponding scientific notation.

Table 3.1: Decimal notation and its corresponding scientific notation

Decimal notation	Scientific notation
5	5×10^0
200	2×10^2
6 710.5678	6.7105678×10^3
−25 000	-2.5×10^4
0.07	7×10^{-2}
0.000 000 024 3	2.43×10^{-8}

3.4 Significant Figures

In science, a measurement or the result of a calculation must indicate the **precision** to which the number is known. The most basic way to indicate the precision of a quantity is to write it with the correct number of **significant figures** (also known as the significant digits). The significant figures are all the digits that are known accurately plus one estimated digit. Suppose that if we say that the distance between two neighbouring city centres, A and B, is 12 km, that does not mean that we know that the distance is exactly 12 km. Rather, the distance is 12 km to the nearest *kilometre*. Instead, if we say that the distance is 12.0 km, that would indicate that we know the distance to the nearest tenth of a *kilometre*. More significant figures mean greater precision.

The following are the rules that govern significant figures.

1. All non-zero digits are always significant.
   Example: The number 23.5 has three significant figures because all of the digits present are non-zero.

2. Zeros between two significant non-zero digits are significant.
   Example: 1025 has four significant figures. The zero is between 1 and 2.

3. Final or ending zeros written to the right of the decimal point are significant.
   Example: There are four significant figures in 82.00

4. Zeros written to the right of the decimal point for the purpose of spacing the decimal point are not significant.
   Example: The number 0.0032 has two significant figures.

5. One or more zeros written at the end of a number are ambiguous; they may or may not be significant. Rewriting the number in scientific notation is one way to remove ambiguity.

   Example: If we consider the number 9500, the numbers 9 and 5 are significant, but the zeros are ambiguous. The number 9500 can have two, three or four significant figures. If we assume the zeros are not significant, then the scientific notation of 9500 is 9.5×10^3 and has two significant figures. If both zeros were significant, then the scientific notation of 9500 is 9.500×10^3 and has four significant figures.

Activity 3.1

How many significant figures are in each of the following measurements?

(i) 5.64 g (ii) 0.120 g

(iii) 0.013 m (iv) 8.300×10^3 s

Significant Figures in Calculations

Although there is generally no difficulty in deciding how many significant figures a quantity should contain, problems can arise when calculations are made using these numbers. When calculators or computers are used to manipulate data, results often contain a great many digits that appear significant but are not. The results of a calculation cannot have a precision greater than that of the quantities used in it. Therefore, there are certain rules that must be followed for numerical calculations.

1. **Addition and Subtraction**

   When two or more quantities are added or subtracted, the result is as the *least precise* of the quantities used in the calculation. The least precise quantity has the fewest digits to the right of the decimal point.

   Example: Consider the numerical quantities 34.36005, 0.0568 and 1102.2. Among these quantities, the least precise quantity is 1102.2.

   The sum of these quantities is

$$34.36005 + 0.0568 + 1102.2 = 1136.61685$$

   The sum obtained in the calculation cannot be more precise than the least precise of the three quantities. Thus, all these digits obtained in the calculation cannot be written in the answer. The result of the sum would be reported as

$$34.36005 + 0.0568 + 1102.2 = \mathbf{1136.6}$$

Activity 3.2

Calculate the difference $358.32\ \text{m} - 5.824\ \text{m}$. How many significant figures are in the result?

2. **Multiplication and Division**

When quantities are multiplied or divided, the result has the same number of significant figures as the quantity with the *smallest number of significant figures.*

Example: Suppose the dimensions of a wooden block are recorded as 7.812 cm, 2.415 cm, and 1.01 cm. The volume of the block is the product of these three numbers. Thus the volume V of the block

$$V = 7.812 \times 2.415 \times 1.01\ \text{cm}^3 = 19.0546398\ \text{cm}^3 = 19.1\ \text{cm}^3$$

The smallest number of significant figures in the calculation is the dimension 1.01 cm, having only three significant figures. Thus, the result should have only three significant figures and the volume would be recorded as $19.1\ \text{cm}^3$.

Activity 3.3

If the mass of a block of volume $18.2\ \text{cm}^3$ is recorded as $25.318\ \text{g}$, calculate the density of the material of the block. How many significan figures are in the result?

Questions

> **NOTE: Although the solutions are given on the last page, it is in your own interest to attempt the questions on your own as an indication of how far you have understood the material of the section.**

1. What are some of the advantages of scientific notation?

2. Why is it important to write quantities with the correct number of significant figures?

3. Write these numbers in scientific notation:

   (i) the Sri Lanka population, 23 000 000.

   (ii) the diameter of a helium nucleus, 0.000 000 000 000 003 8 m.

4. How many significant figures are in each of the following measurements?

   (i) 6.28 m (ii) 0.720 kg (iii) 0.057 m

   (iv) 9.85×10^{2} g (v) 3.80×10^{5} m (vi) 6.400×10^{3} s

5. Write your answer to the following problems with the appropriate number of significant figures.

   (i) $2.723\times10^{6}\ \text{kg}+2.35\times10^{8}\ \text{kg}$

   (ii) $1.83\times10^{-5}\ \text{m}+8.5\times10^{-7}\ \text{m}$

   (iii) $9.65\times10^{7}\ \text{g}-3.849\times10^{5}\ \text{g}$

   (iv) $(3.2\ \text{m})\times(4.0\times10^{-3}\ \text{m})\times(1.3\times10^{-8}\ \text{m})$

   (v) $\dfrac{4.563\times10^{6}\ \text{m}}{2.8\times10^{-2}\ \text{s}}$

   (vi) $\dfrac{27.2\ \text{m}\times15.63\ \text{m}}{1.846\ \text{m}}$

6. Solve $\frac{7.33\text{ m}}{8.10\text{ ms}}$ and express the answer in metres per second $\left(\text{m s}^{-1}\right)$ with the appropriate number of significant figures.

7. Find the volume of a rectangular solid having dimensions 10.2 cm, $8.24\text{ cm}$and 1.8 cm. Write your answer with the appropriate number of significant figures.

8. If the radius, r, of a circle is 5.15 cm, then determine the diameter of the circle with the appropriate number of significant figures.

9. Determine the circumference of a circle of diameter 10.3 cm. Assume that π contains at least four digits.

10. Suppose a hollow sphere has an inner radius of 3.550 cm and an outer radius of 5.21 cm. Determine the volume of the sphere. Write your answer with the appropriate number of significant figures. Assume that π contains at least four digits.

Chapter 4

Measurements and Units

4.1 Introduction

As we have learned in previous chapters, the study of physics gives us the opportunity to learn how nature works. In other words, it is the study of natural phenomena. To describe natural phenomena, we take measurements of various aspects of nature. Each measurement is associated with a physical quantity, such as the length of an object. If we report the results of the measurement to someone who wishes to reproduce this measurement, a standard of a unit must be defined. Measurement standards or units used by different people in different places (throughout the universe) must yield the same result. Furthermore, the standards used for the measurements must not change with time.

In this chapter, we will learn important preliminaries that we need throughout our studies, such as standards and units. Importantly, the systems of units that are used to describe physical quantities will be discussed in detail.

4.2 Standards and Units

Like all other sciences, physics is based on experiments and requires measurements. Generally, the result of measurements is expressed in numbers. Any number that is used to describe a physical phenomenon quantitatively is called a **physical quantity**. In experiments, each measurement is associated with a physical quantity, such as length. In mechanics, length, mass, and time are called fundamental physical quantities. All other quantities in mechanics can be expressed in terms of these three.

When measuring any physical quantity, we always compare it with some reference standard, and this **standard** must be well-defined. Such a standard defines a **unit** with a unique name for the physical quantity in use and allows for the comparison of measurements made at different times and places. For example, when we say that a wooden block is 15 metres long, we mean that it is 15 times as long as a metre stick, which we define to be 1 metre long. In this case, we define a standard which defines exactly how long one metre is, and a metre becomes a unit of length. It is important that whatever standard is chosen must be readily accessible and possess some property that can be measured reliably. In addition, the standards used for measurements must not change with time.

Throughout history, ancient civilisations have used measurements for agricultural, construction, and trade purposes. They developed their own standard units that could only have been applied to a single community or small region. With the development of technology and the growing importance of trade, modernised, simplified, and uniform systems of weights and measures were developed. Under the terms of the Metre Convention, new international prototypes of the metre and kilogram were assembled and formally adopted by the first General Conference on Weights and Measures (CGPM) in 1889. [The CGPM is the supreme authority of the International Bureau of Weights and Measures (BIPM), the intergovernmental organisation established in 1875.] In 1960, the 11th meeting of the CGPM formally defined and established the **International System of Units,** universally abbreviated as **SI** (the abbreviation for its French name Système International) which is the modern form of the **metric system**. The SI consists of two classes of units, namely **base units** and **derived units.**

In 2018, a set of extensive changes were made for SI at the 26th CGPM meeting and adopted in May 2019. This is the most recent and perhaps the most significant revision of the SI since its establishment. The revised definition of SI units is established in terms of a set of seven defining constants. The complete system of units could be derived from the fixed values of these defining constants, expressed in units of SI. These particular constants were chosen after being identified as the best choice, taking into account the previous definition of the SI, which was based on seven base units, and progress in science. Table 4.1 illustrates the seven defining constants that specify the exact numerical value of each constant when its value is expressed in the corresponding SI unit. Fixing the exact numerical value, the unit becomes defined since the product of the numerical value and the unit has to equal the value of the constant, which is postulated to be invariant [source: BIPM, The International System of Units (SI), 9th edition, 2019].

Table 4.1: The seven defining constants of the SI and the seven corresponding units they define

Defining constants	Symbol	Numerical value	Unit
hyperfine transition frequency of Cs	ν_{Cs}	9 192 631 770	Hz
speed of light in vacuum	c	299 792 458	$m\ s^{-1}$
Planck constant	h	$6.626\ 070\ 15 \times 10^{-34}$	J s
elementary charge	e	$1.602\ 176\ 634 \times 10^{-19}$	C
Boltzmann constant	k	$1.380\ 649 \times 10^{-23}$	$J\ K^{-1}$
Avogadro constant	N_A	$6.022\ 140\ 76 \times 10^{23}$	mol^{-1}
luminous efficacy	K_{cd}	683	$lm\ W^{-1}$

[Source: BIPM, The International System of Units (SI), 9th edition, 2019]

4.3 Definitions of the SI units

As we learnt in the previous section, the SI was classified into two classes of units, namely base units and derived units, before the adoption of the current definitions in 2019. Fixing the numerical values of seven defining constants (table 4.1) to revise the SI has led, in principle, to recognise that the classification of base units and derived units is not required since all units (including base units and derived units) may be constructed directly from the defining constants. However, the concept of base units and derived units is still maintained because of their usefulness and historical value. Table 4.2 illustrates the seven base units with special names and symbols that have been defined by the international committee. [Source: BIPM, The International System of Units (SI), 9th edition, 2019].

Table 4.2: SI base units

Base quantity		SI base unit	
Name	**Typical Symbol**	**Name**	**Symbol**
time	t	second	s
length	l, x, r, etc.	metre	m
mass	m	kilogram	kg
electric current	I,i	ampere	A
thermodynamic temperature	T	kelvin	K
amount of substance	n	mole	mol
luminous intensity	I_v	candela	cd

Units of Time, Length and Mass

Most of the physical quantities that we come across in this book can be expressed in terms of the base units of time, length, and mass. Therefore, we shall first focus on defining these three base units. The definition of the remaining base units and derived units is given in table A.1 of Appendix A,.

Time

The SI base unit of time is the 'second', and its definition has been changed over time to make it more precise. Initially it was defined as 1/86 400 of a mean solar day. This was changed in 1967 based on an atomic clock which measures vibrations of cesium 133 $\left({}^{133}\text{Cs}\right)$ atom. Cesium atoms can be made to vibrate steadily and these vibrations can be readily observed and counted. Therefore, the second was defined as the time required for 9 192 631 770 of these vibrations to occur.

According to the new SI standard adopted in 2019, the second was redefined taking the fixed numerical value of the cesium frequency $\Delta\nu_{\text{Cs}}$, the unperturbed ground-state hyperfine transition frequency of the cesium 133 atom, to be $9\ 192\ 631\ 770$ when expressed in the unit Hz, which is equal to s^{-1}.

This definition implies the exact relation $\Delta\nu_{\text{Cs}} = 9\ 192\ 631\ 770\ \text{Hz}$.

Therefore,

$$1\ \text{Hz} = \frac{\Delta\nu_{\text{Cs}}}{9\ 192\ 631\ 770} \qquad \text{or} \qquad 1\ \text{s} = \frac{9\ 192\ 631\ 770}{\Delta\nu_{\text{Cs}}}$$

Length

The SI base unit of length is the 'metre', and its definition has been changed over time to make it more precise. Initially, in 1791 the metre was defined as 1/10 000 000 $1/10\ 000\ 000$ the distance from the equator to the north pole. This definition was improved in 1889 by redefining the metre to be the distance between two engraved lines on a platinum-iridium bar kept near Paris. In 1960, the metre was redefined even more accurately in terms of the wavelength of light. Thus, metre was redefined as $1\ 650\ 763.73$ wavelengths of orange light emitted by krypton atoms. In 1983, the metre was given the definition after knowing the speed of light to be exactly $299\ 792\ 458\ \text{m s}^{-1}$.

According to the new SI standard adopted in 2019, the metre was redefined taking the fixed numerical value of the speed of light in vacuum c to be $299\ 792\ 458$ when expressed in the unit m s^{-1}, where the second is defined in terms of the cesium frequency $\Delta\nu_{\text{Cs}}$.

This definition implies the exact relation $c = 299\ 792\ 458\ \text{m s}^{-1}$.

Therefore,

$$1\ \text{m} \aleph \left(\frac{c}{299\ 792\ 458}\right) \text{s} \quad \frac{9\ 192\ 631\ 770}{299\ 792\ 458} \frac{c}{\Delta\nu_{\text{Cs}}} \quad 30.663\ 319 \frac{c}{\Delta\nu_{\text{Cs}}}$$

Accordingly, metre is defined as the distance light travels in a vacuum in $1/299\ 792\ 456$ of a second (figure 4.1). This definition has greater accuracy than the earlier definitions.

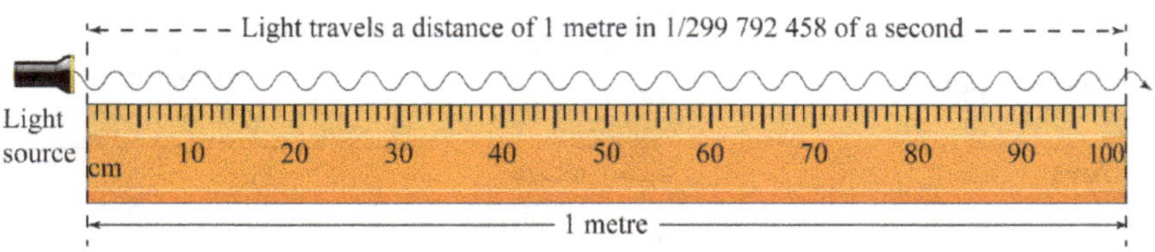

Figure 4.1: The metre is defined to be the distance light travels in 1/299 792 458 of a second in a vacuum

Mass

The SI base unit of mass is the 'kilogram', and its definition has been changed over time to make it more precise. The 1889 definition of the kilogram was simply the mass of the international prototype of the kilogram, an artefact made of platinum-iridium alloy still kept at the BIPM, near Paris.

According to the new SI standard adopted in 2019, the kilogram was redefined taking the fixed numerical value of the Planck constant, h, to be $6.626\ 070\ 15 \times 10^{-34}$ when expressed in the unit J s, which is equal to $\text{kg m}^2\ \text{s}^{-1}$, where the metre and the second are defined in terms of c and $\Delta\nu_{\text{Cs}}$.

This definition implies the exact relation $h = 6.626\ 070\ 15 \times 10^{-34}\ \text{kg m}^2\ \text{s}^{-1}$.

Therefore, $1\ \text{kg} = \dfrac{h}{6.626\ 070\ 15 \times 10^{-34}\ \text{m}^2\ \text{s}^{-1}}$

which is equal to

$$1\ \text{kg} = \frac{(299\ 792\ 458)^2}{(6.626\ 070\ 15\times10^{-34})(9\ 192\ 631\ 770)}\frac{h\Delta\nu_{\text{Cs}}}{c^2}$$

$$\approx 1.475\ 5214\times10^{40}\frac{h\Delta\nu_{\text{Cs}}}{c^2}$$

Derived Units

Derived units are expressed algebraically in terms of base units or other derived units. The symbols for derived units are obtained by means of the mathematical operations of multiplication and division. Table 4.3 illustrates a few examples of SI derived units expressed in terms of SI base units.

Table 4.3: Examples of SI derived units expressed in terms of SI base units

Derived quantity	SI derived unit	
	Name	**Derived unit symbol**
Area	square metre	m^2
Volume	cubic metre	m^3
Speed, velocity	metre per second	$m\ s^{-1}$
Acceleration	metre per second squared	$m\ s^{-2}$
Wave number	reciprocal metre	m^{-1}
Density, mass density	kilogram per cubic metre	$kg\ m^{-3}$
Specific volume	cubic metre per kilogram	$m^3\ kg^{-1}$
Current density	ampere per square metre	$A\ m^{-2}$
Magnetic field strength	ampere per metre	$A\ m^{-1}$
Luminance	candela per square metre	$cd\ m^{-2}$
Amount of substance concentration	mole per cubic metre	$mol\ m^{-3}$

Metric Prefixes

Metric prefixes or SI prefixes are a set of prefixes that have been adopted for use with the SI units. The SI prefixes refer to powers of 10. They are convenient for expressing the values of quantities that are much larger than or much smaller than the SI unit. Table 4.4 illustrates the names and symbols of these multiple and submultiple prefixes.

Table 4.4: Metric prefixes

Factor	Name	Symbol	Factor	Name	Symbol
10^1	deca	da	10^{-1}	deci	d
10^2	hecto	h	10^{-2}	centi	c
10^3	kilo	k	10^{-3}	milli	m
10^6	mega	M	10^{-6}	micro	µ
10^9	giga	G	10^{-9}	nano	n
10^{12}	tera	T	10^{-12}	pico	p
10^{15}	peta	P	10^{-15}	femto	f
10^{18}	exa	E	10^{-18}	atto	a
10^{21}	zetta	Z	10^{-21}	zepto	z
10^{24}	yotta	Y	10^{-24}	yocto	y

When the prefixes are used, the prefix name and the unit name are combined to form a single word. Similarly, the prefix symbol and the unit symbol are written without any space to form a single symbol, which may itself be raised to any power. For example, we may write: kilometre as km; microvolt as µV; femtosecond as fs. Table 4.5 shows some examples of the use of multiples of 10 and their prefixes with units of length, mass, and time.

Table 4.5: Few examples of the use of multiples of 10 and their prefixes

Length	Mass	Time
1 nanometre = 1 nm = 10^{-9} m	1 microgram = 1 µg = 10^{-9} kg	1 nanosecond = 1 ns = 10^{-9} s
1 micrometre = 1 µm = 10^{-6} m	1 milligram = 1 mg = 10^{-6} kg	1 microsecond = 1 µs = 10^{-6} s
1 millimetre = 1 mm = 10^{-3} m	1 gram = 1 g = 10^{-3} kg	1 millisecond = 1 ms = 10^{-3} s
1 centimetre = 1 cm = 10^{-2} m	1 kilogram = 1 kg = 10^3 kg	1 minute = 1 min = 60 s
1 kilometre = 1 km = 10^3 m		1 hour = 1 h = 3600 s
		1 day = 24 h = 86 400 s

Orders of Magnitude

The orders of magnitude is a system of classification determined by size, typically in powers of ten to express quantities with a wide range of values. Table 4. 6 shows approximate values of some measured lengths, mass, and time.

Table 4.6: Approximate values of some measured lengths, mass and time

Approximate values of some measured lengths

Length (or Distance)	**In metres**
Diameter of a proton	10^{-15} m
Diameter of an atomic nucleus	10^{-14} m
Size of virus	10^{-7} m
Size of cells of most living organisms	10^{-5} m
Size of smallest dust particles	10^{-4} m
Length of a house fly	5×10^{-3} m
Height of 4-year-old child	10^{0} m
Length of a football field	10^{2} m
Height of Mount Everest	8.9×10^{3} m
Mean radius of the Earth	6.37×10^{6} m
Mean distance from the Earth to the Moon	3.84×10^{8} m
Mean orbit radius of the Earth about the Sun	1.50×10^{11} m
One light-year	9.46×10^{15} m
Distance from the Earth to the nearest large galaxy (Andromeda)	2×10^{22} m
Distance from the Earth to the most remote galaxies	10^{26} m

Approximate masses of some objects

Mass of an object	**In kilograms**
Electron	9.11×10^{-31} kg
Hydrogen atom	1.67×10^{-27} kg
Bacterium	10^{-15} kg
Mosquito	10^{-5} kg
Paper clip	10^{-3} kg
Litre of water	10^{0} kg
Human	10^{2} kg
Moon	7.35×10^{22} kg
Earth	5.97×10^{24} kg
Sun	1.99×10^{30} kg
Observable Universe	10^{53} kg

Approximate values of some time intervals

Time interval	**In seconds**
Duration of a nuclear collision	10^{-22} s
Period of visible light waves	10^{-15} s
Period of vibration of an atom in a solid	10^{-13} s
Lifetime of muon	10^{-6} s
Time required for sound to travel 0.343 m in air at 20°C	10^{-3} s
Time interval between human heartbeats	8×10^{-1} s
One day	8.6×10^{4} s
One year	3.2×10^{7} s
Span of recorded history	10^{11} s
Age of the Earth	10^{17} s

4.4 Conversion of Units

In previous sections, we learned that any physical quantity consists of a number and a unit. In addition to SI units, the British system of units is also used in some regions of the world, including Sri Lanka. British units are occasionally used in mechanics and thermodynamics. However, throughout this book, SI units will be used in all examples and problems. Sometimes, you may come across situations where you need to convert units from one measurement system to another or convert within a system, for example, suppose that you measure the length of a table in British units for length 'inch' and want to express in SI unit for length 'metre'. In such situations, it is convenient to use conversion factors. The following are some conversion factors that you may use frequently. A complete list of conversion factors is given in Appendix A.

$1\text{ m} = 39.37\text{ in} = 3.281\text{ ft}$ $1\text{ ft} = 0.3048\text{ m} = 3.48\text{ cm}$

$1\text{ mile} = 1.609\text{ m} = 1.609\text{ km}$ $1\text{ in} = 0.0254\text{ m} = 2.54\text{ cm}$

Example: If the height of Mount Everest is $8\,848.86\text{ m}$ its height in feet, is

$$8\,848.86\text{ m} = \left(8\,848.86\ \cancel{\text{m}}\right)\left(3.281\ \frac{\text{ft}}{\cancel{\text{m}}}\right) = 29\,033.11\text{ ft}$$

Activity 4.1

On a southern highway, a car travels at a speed of 65.0 mi h^{-1}. Is the driver exceeding the speed limit of 100.0 km h^{-1}?

Activity 4.2

If a solid wooden block has a volume of 2 cubic inches, what is its volume in cubic centimetres?

4.5 Dimensional Analysis

The word dimension denotes the physical nature of a quantity. If we take the physical quantity, say the length of an object, it could be measured in metres, inches, miles, etc., and each can be converted into any other. All express the dimension of length. Table 4.7 illustrates the symbols used for the base quantities and the symbols used to denote their dimension.

Table 4.7: Base quantities and dimensions used in the SI

Base quantity	Typical symbol for quantity	Symbol for dimension
time	t	T
length	l, x, r, etc.	L
mass	m	M
electric current	I, i	I
thermodynamic temperature	T	Θ
amount of substance	n	N
luminous intensity	I_V	J

Each of the seven base quantities used in the SI has its own dimension. Usually, the dimension of any quantity Q is denoted either by $[Q]$ or $\dim Q$. In this book, square brackets [ ] will be used to denote the dimensions of a physical quantity. Accordingly [M], [L], and [T] are used to stand for mass, length, and time dimensions, respectively.

In physics, any equation relating to physical quantities must be dimensionally consistent, which means that every term in an expression of an equation must have the same dimensions. In particular, the expressions on each side of the equality in an equation must have the same dimensions. For this analysis, a procedure called dimensional analysis is commonly used. Dimensions can be treated as algebraic quantities, and quantities can be added or subtracted only if they have the same dimensions. Standard mathematical functions, such as trigonometric functions (such as sine and cosine), logarithms, or exponential functions that appear in the equation, must be dimensionless.

During the process of dimensional analysing, dimensions of the derived quantities are written as products of powers of the dimensions of the base quantities using the equations that relate the derived quantities to the base quantities. In general, the dimension of any quantity Q is written in the form of a dimensional product (equation 4.4).

$$[Q] = \text{T}^a\text{L}^b\text{M}^c\text{I}^d\Theta^e\text{N}^f\text{J}^g \tag{4.4}$$

where the exponents a, b, c, d, e, f, g are called dimensional exponents and are small integers, which can be positive, negative, or zero.

Example: Consider the equation $v = u + at$, where u is the initial velocity, a is the uniform acceleration for a time t, and v is the final velocity. Show the equation $v = u + at$ is dimensionally correct.

Dimension of v, $[v] = \mathrm{LT}^{-1}$

Dimension of u, $[u] = \mathrm{LT}^{-1}$

Dimension of at, $[at] = [a] \times [t] = \mathrm{LT}^{-2}\mathrm{T} = \mathrm{LT}^{-1}$

Therefore, dimensions $[v] = [u] = [at]$. Hence, we have the same dimensions on both sides of the equation $v = u + at$ and it is dimensionally correct.

Activity 4.3

Consider the equation $s = ut + \frac{1}{2}at^2$, where u is the initial velocity, s is the distance travelled durig the time t, and a is the uniform acceleration. Show the equation is dimensionally correct

Activity 4.4

Consider a situation where the acceleration a of a particle moving with uniform speed v in a circle of radius r is proportional to some power of r, say r^{α}, and some power of v, say v^{β}. Determine the values of α and β and write the simplest form of an equation for the acceleration.

Questions

NOTE: Although the solutions are given on the last page, it is in your own interest to attempt the questions on your own as an indication of how far you have understood the material of the section.

1. If the density of iron is $7.86\ \mathrm{g\ cm^{-3}}$, convert this to $\mathrm{kg\ m^{-3}}$.

2. An auditorium measures length, width, and height $50\ \mathrm{m}$, $25\ \mathrm{m}$, $12\ \mathrm{m}$ respectively. Calculate the volume of the auditorium in cubic feet.

3. According to the label on a bottle of coconut oil, the volume of the contents is 0.425 litre. Determine this volume in cubic inches. Take $1\ \text{litre} = 1000\ \mathrm{cm^3}$ and $1\ \mathrm{in} = 2.54\ \mathrm{cm}$.

4. The distance from your home to the hospital is $10\ \text{mi}$, and it usually takes $20\ \text{min}$ to travel by bus this distance. Calculate the average speed in metres per second. (Note: The average speed is the distance travelled divided by the time of travel).

5. A person wishing to know the height of a mountain measures the angle of elevation of the mountain top as 10°. After walking $1.5\ \text{km}$ closer to the mountain on level ground, the person finds the angle to be 12°.

   (i) Draw suitable diagrams to illustrate the problem, neglecting the height of the person's eyes above the ground.

   (ii) If the symbol h represents the mountain height and the symbol d represent the person's original distance from the mountain, label the diagrams.

   (iii) Using (i) and (ii), write two trigonometric equations relating the two selected variables h and d.

   (iv) Determine the height h.

6. Estimate how much water there is in a lake shown below. Assume that the lake is roughly circular, about $1\ \text{km}$ across, and has an avaerage depth of about 10m.

7. Some countries, often use miles per hour (mi h^{-1}) when considering speed, but the SI unit of speed is m s^{-1}. What is the conversion factor for changing m s^{-1} to mile h^{-1}? If you want to make a quick approximation of the speed in mi h^{-1} given the speed in m s^{-1}, what might be the easiest conversion factor to use?

8. Three astronauts are in a spherical space station. If each of the astronauts breathes about 450 cm^3 of air with each breath, approximately what volume of air (in cubic metres) do these astronauts breathe in a year? Assume that each astronaut can take 10 breaths in one minute.

9. A light-year is a distance light travels in one year at a speed 2.998×10^8 m s^{-1}.

   (i) How many metres are there in 1.00 light-year?

   (ii) An astronomical unit (AU) is the average distance from the Sun to Earth, 1.50×10^8 km. How many AU are there in 1.00 light-year?

   (iii) What is the speed of light in AU h^{-1}.

10. An equation for mechanical potential energy U is given by $U = mgh$. If U is in joules, with m in kg, h in m, and g in m s^{-2}, find the combination of SI base units that are equivalent to joules.

11. Consider the physical quantities s, t, v, u and a with dimensions $[s] = \text{L}$, $[t] = \text{T}$, $[v] = \text{LT}^{-1}$, $[u] = \text{LT}^{-1}$ and $[a] = \text{LT}^{-2}$. Determine whether each of the following equations is dimensionally consistent.

    (i) $v^2 = u^2 + 2as$ (ii) $s = ut + \frac{1}{2}at$ (iii) $v = \sin\left(\dfrac{at^2}{s}\right)$

12. The position x of a particle moving under uniform acceleration a is some function of time t and acceleration a. If we write this position as $x = ka^{\alpha}t^{\beta}$ where k is a dimensionless constant, using dimensional analysis, determine the value of α and β when the expression is dimensionally consistent. Can this analysis give the value of k?

13. Consider the equation $d = At^3 + Bt$ describes the motion of a particular object, with d having the dimension of length, t having the dimension of time and *A*, *B* being constant. If the equation is dimensionally consistent, determine the dimensions of the constants *A* and *B*.

14. An equation for the period *T* of a planet (the time to make one orbit about the Sun) is expressed by $T^2 = \dfrac{4\pi^2 r^3}{GM}$, where *T* is in s, *r* is in metre, *G* is in m^3 kg^{-1} s^{-2} and *M* is in kg. Show that the equation is dimensionally correct.

15. Use dimensional analysis to determine how the period T of a swinging pendulum (the elapsed time for a complete cycle of motion) shown below, depends on some, or all, of these properties: the length l of the pendulum, the mass m of the pendulum bob, and the gravitational field strength g measured in $\mathrm{m\,s^{-2}}$. Assume that the maximum angle that the string makes with the vertical has no effect on the period.

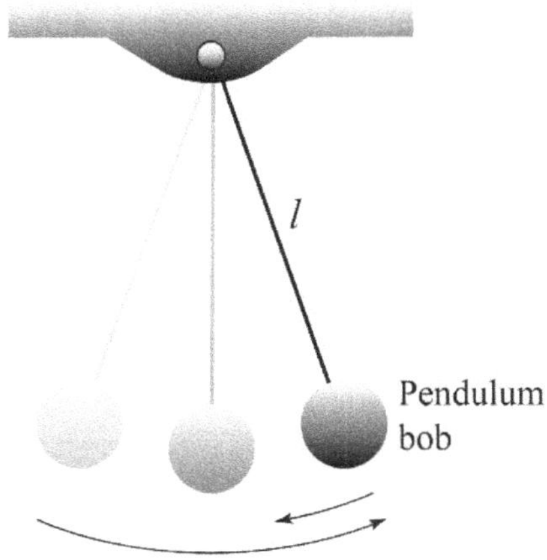

Chapter 5

Practical Skills

5.1 Introduction

All natural science subjects involve not only theory but also practical work. The very fundamental nature of science is that theory can be tested by experiments, so the ability to carry out experiments logically and scientifically is essential.

Although all branches of science require experimentation, controlled laboratory experiments are of central importance in physics. Generally, controlled experiments are performed in laboratories. The development of practical skills to carry out experiments in the laboratory is vital. The practical skills and abilities we learned in the laboratories will be very useful in our daily lives and encourage safe practice. In addition, these practical skills help develop attitudes such as concern for accuracy and precision, initiative, inventiveness, and a spirit of inquiry.

5.2 Errors and Uncertainties

Often, verification of a physical law or the determination of a physical quantity involves measurements. When making a measurement, we always try to find the true or actual value of a quantity. You could find this true value only if you are perfect enough to make the measurement. However, no measurement can ever be perfect, and there will always be some uncertainty. Let us consider a situation where a set of readings was taken from the scale on a voltmeter, a stop watch, or a metre stick. These measurements may be used in a sequence of analyses to find a quantity or a law under study. Any uncertainty in these readings would result in uncertainty in the final result. Therefore, any basic laboratory technique should consider the nature of the uncertainty in individual measurements and how the uncertainties in two or more measurements are propagated to determine the uncertainty in the quantity or law investigated. Such uncertainties are often called experimental errors. An error is the difference between a measured result and the true value; the true value is the value that would have been obtained in an ideal measurement. Uncertainty is the reported value that lies within the range of values in which the true value is asserted. Measured quantities are quantified by associating them with an uncertainty. Suppose that if the value of a measured quantity Q has uncertainity ΔQ, which is called the absolute uncertainty in the measured value. Q can be expressed as $Q \pm \Delta Q$. For example, if we have the best estimate of a length L is 2.59 cm, but due to uncertainty, the length might be as small as 2.57 cm or as large as 2.61 cm. L can be expressed in the units of the measured quantity with its uncertainty as, $L = 2.59 \pm 0.02$ cm.

Activity 5.1

Two independent experiments give two sets of data with the expressed results and uncertainties of 2.5 ± 0.1 cm and 2.5 ± 0.2 cm, respectively. Which result is less uncertain? Explain your answer.

5.3 Types of experimental errors

In order to determine the uncertainty for a measurement, the nature of the errors affecting the experiment must be examined. Many different types of error can occur in an experiment, but they will generally fall into one of two categories; they are (i) random errors and (ii) systematic errors.

Systematic errors

A systematic error is an error that will occur consistently in only one direction each time the experiment is performed, i.e., the value of the measurement will always be greater (or lesser) than the true value. These errors are due to identifiable causes and can, in principle, be eliminated. Errors of this type result in measured values that are consistently too high or too low. Systematic errors may be of four kinds:

Instrumental – When the instrument being used does not function properly, causing errors in the experiment. For example, a poorly calibrated instrument such as a thermometer reads 103°C when immersed in boiling water and 3°C when immersed in ice water at atmospheric pressure. Such a thermometer would result in measured values that are consistently too high.

Observational – This may occur mainly due to parallax, which refers to the change in the apparent position of an object when viewed from different points. For example, parallax in reading a metre scale.

Environmental – This may occur when the surrounding environment (such as a laboratory) causes interruptions in the experiment. For example, if you are trying to measure the mass of mango fruit on a scale and your laboratory is windy, the wind may cause the scale to read incorrectly.

Theoretical – This may occur due to simplifications of the model or approximations in the equations describing it. For example, if a frictional force is acting during the experiment but such a force is not included in the theory, then the theoretical and experimental results will consistently disagree.

In measuring instruments, you may get another type of systematic error called the zero error. This type of systematic error occurs when a measuring instrument gives a reading when the true reading is zero. For example, the zero on a metre ruler might not be at the very beginning of the ruler, which will introduce a fixed error in any reading unless it is allowed.

Random errors

A random error is one which varies and which is equally likely to be positive or negative. Random errors are always present in an experiment, and in the absence of systematic errors, they cause successive readings to spread about the true value of the quantity. The sources of random errors cannot always be identified. Possible sources of random errors are:

Observational – This may occur due to the judgement of an observer when reading the scale of a measuring device to the smallest division. For example, you measure the circumference of an object using a measuring tape. But your measuring tape is only accurate to the nearest half-centimetre, so you round-off each measurement up or down when you record data.

Environmental – This may occur due to unpredictable fluctuations in line voltage, temperature, or mechanical vibrations of the equipment. For example, you repeatedly open and close the door when the pressure is measured in a laboratory, which causes fluctuations in the reading.

In principle, an experimentalist could identify and eliminate systematic errors. Random errors can often be quantified by statistical analysis. Therefore, the effects of random errors on the quantity or physical law under investigation can usually be determined.

Figure 5.1: Measurement (a) with random errors only and (b) with systematic and random errors. Each mark indicates the result of a measurement

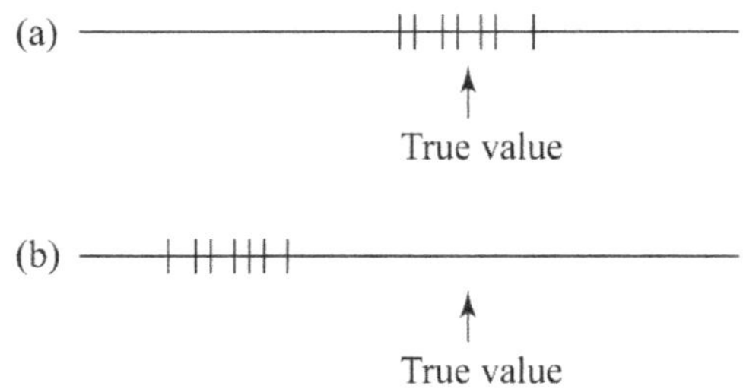

To explain the distinction between random errors and systematic errors, consider the following example. Suppose that you measure a physical quantity repeated several (say seven) times under the same conditions. If there are only random errors, then the measured values will spread about the true value (some will be too high and others too low), as shown in figure 5.1(a). If there is also a systematic error in addition to the random errors, then the measured values will spread, not about the true value but about some displaced value, as shown in figure 5.1(b).

In experiments, systematic errors can be minimised by taking sensible precautions, such as checking for zero errors and avoiding parallax errors, and by drawing a suitable graph. In particular, you should always check for zero errors before using any instrument and make allowance for them if any are present. This is true even for digital instruments, which we often assume will be accurate, and thus do not bother to check for any zero error. Random errors can be minimised by taking the average of a number of repeat measurements and by drawing a graph that, in effect, averages a range of values.

5.4 Accuracy and Precision

As mentioned in the previous section, when making a measurement, we always try to find the true value (or actual value) of a quantity. Accuracy and precision are two important factors to consider when making measurements. Both accuracy and precision reflect how close a measurement is to a true value. Accuracy and precision may have similar meanings in English, but there is an important distinction in experimental measurements, and the difference must be understood.

The accuracy of a measurement signifies how close a measurement is to the true value of the quantity being measured. Precision is a measure of how close a series of measurements are to one another. Precision is independent of accuracy. That means it is possible to be very precise but not very accurate, and it is also possible to be accurate without being precise. The best quality and reliable measurements are considered accurate, precise, and reproducible.

To demonstrate the difference between precision and accuracy, let us take an example with a dart board. Think of the centre (bull's eye) of a dart board as the true value. In this situation, precision means getting all the darts close together and accuracy means getting them where they should be (at the centre).

Figure 5.2: Dart boards showing different accuracy and precision scenarios

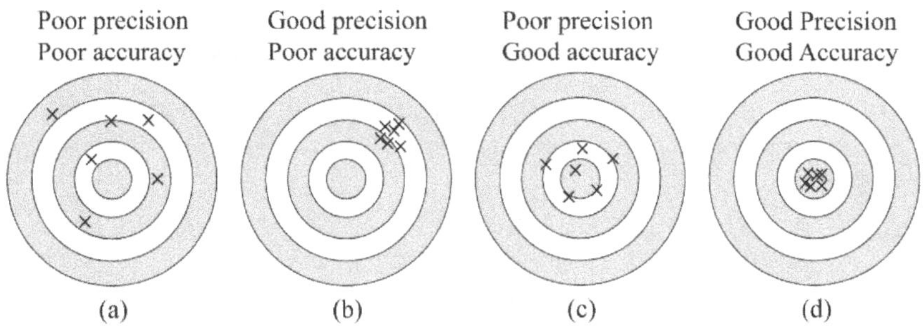

Figure 5.2 shows the different accuracy and precision scenarios we could get. Consequently,

- if the darts are neither close to the centre nor close to each other, there is neither accuracy nor precision [figure 5.2 (a)].
- if all darts land very close together but far from the centre, there is precision but not accuracy [figure 5.2 (b)].
- if the darts are all about an equal distance from and spaced equally around the centre, there is accuracy because the average of the darts is in the centre. This means that it is accurate, but not precise [figure 5.2 (c)].
- if the darts land close to the centre and close together, there is accuracy and precision [figure 5.2 (d)].

If your measurements are high precision but have poor accuracy, then this gives rise to a systematic error. If your measurements have poor precision and are scattered around the true value but have high accuracy, then this leads to random errors.

Activity 5.2

Choose the most appropriate option for the following question.

A voltmeter has a zero error. This will affect

(i) neither the precision nor the accuracy of the readings.
(ii) only the precision of the readings.
(iii) only the accuracy of the readings.
(iv) both the precision and the accuracy of the readings.

Activity 5.3

Choose the most appropriate option for the following question.
A student has taken reading of a constant potential difference in an experiment for four times. The readings are

2.186 V, 2.188 V, 2.187 V, 2.186 V

Although the voltmeter has a zero error, during the analysis, the student averages these readings without considering the zero error on the voltmeter. The average measurement of the potential difference is

(i) precise and accurate.
(ii) precise but not accurate.
(iii) accurate but not precise.
(iv) not accurate and not precise.

5.5 Least Count and Instrument Limit of Error (ILE)

The smallest division marked on a measuring instrument is the least count (LC). Therefore, a metre ruler will have the least count of 1.0 mm , and a digital stop watch might have the least count of 0.01 s. The instrument limit of error (ILE) is the precision with which a measuring device can be read and is always equal to or smaller than the least count. Measuring tools are calibrated against internationally accepted standards. The ILE is generally taken to be the least count or some fraction of it (in most cases $1/2$ of the least count). There are no hard and fast rules. Instead, it would be best if you were guided by common sense. If the space between the scale divisions is large, you may be comfortable enough to estimate to $1/2$ the least count. If the scale divisions are very close, you may only be able to estimate to the least count. In practice, you may notice that the ability to use high precision scales is limited by the spacing between the marks. Thus, in a metre ruler, it is easy to have a least count of 1 mm, more difficult to have a least count of 0.5 mm, and practically impossible to have a least count of 0.05mm.

Principle of Vernier and Vernier Scale

In order to increase precision in measurements, an auxiliary scale called the vernier scale is used. The vernier scale subdivides the least count from the main scale into 10, 50 or 100 subdivisions. Vernier scales are found on a wide range of measuring instruments (e.g. vernier calliper, micrometer screw gauge, travelling microscope, etc.). Any instrument that uses a vernier will have two scales: the main scale and the vernier scale. Measurements are made by combining the readings from these two scales. The vernier scale can be moved along the main sale in a measuring instrument with the vernier scale.

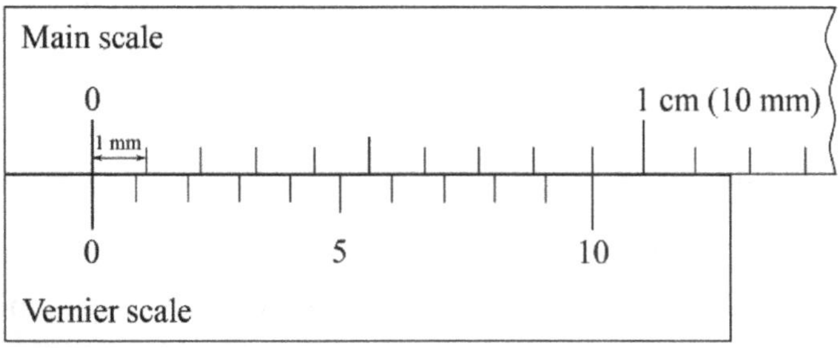

Figure 5.3: Principle of the vernier scale: the vernier scale and the main scale

Figure 5.3 illustrates the main scale and the vernier scale of a measuring instrument. If we closely look at figure 5.3, the vernier scale is constructed by dividing the 9 mm range from the main scale into 10 equal parts. When the instrument is adjusted, the front ends of both scales coincide. If there is no zero error, the 0th mark of the main scale will be aligned with the 0th mark of the vernier scale, and the 9th mark of the main scale will coincide with the 10th mark of the vernier scale. Here, we can observe that one main scale division (MSD) does not have the same length as one vernier scale division (VSD).

That is $$9\ \text{MSD} \quad 10\ \text{VSD}.$$

This means, $$1\ \text{VSD} = \frac{9}{10}\ \text{MSD}$$

Since one MSD is equal to $1\ \text{mm}$, we have

$$1\ \text{VSD} = \frac{9}{10} \times 1\ \text{mm} = 0.9\ \text{mm}$$

The least count (LC) is defined as the difference between the length of one MSD and the length of one VSD. i.e.,

$$1\ \text{LC} = 1\ \text{MSD} - 1\ \text{VSD}$$

Therefore, the scale shown in figure 5.3 has the least count of,

$$\text{LC} = 1\ \text{mm} - 0.9\ \text{mm} = 0.1\ \text{mm} = 0.01\ \text{cm}$$

Activity 5.4

In a measuring instrument, the smallest division of the main scale is 0.5 mm. The vernier scale of that measuring instrument is constructed by dividing the $49\ \text{mm}$ range from the main scale into 50 equal parts. Calculate the least count of the vernier.

Extended Vernier Scale

The extended vernier scale is similar to the vernier scale discussed earlier. It used to be designed to read further small measurements with an easily readable vernier scale with calibration gaps not so close to each other. Figure 5.4 illustrates such an extended vernier scale, where the vernier scale is constructed by dividing the $39\ \text{mm}$ range from the main scale into 20 equal parts.

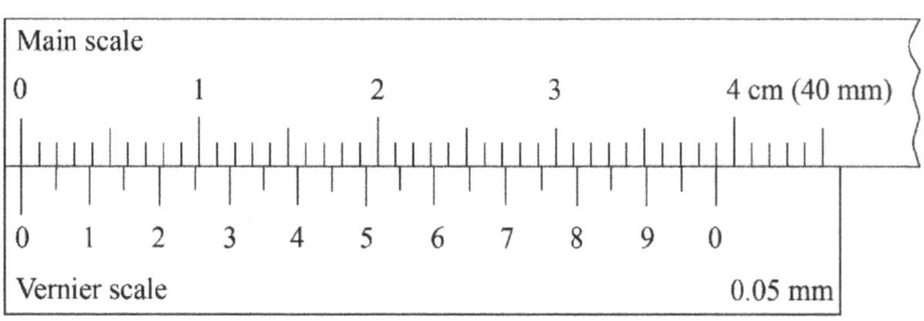

Figure 5.4: Extended vernier scale: the vernier scale and the main scale

The least count of the vernier shown in figure 5.4 is the difference between two divisions of the main scale and one division of the vernier scale.

According to extended vernier scale shown in figure 5.4, we have

$$39 \text{ MSD} = 20 \text{ VSD}$$

This means,

$$1 \text{ VSD} = \frac{39}{20} \text{ MSD}$$

Since one MSD is equal to 1 mm, we have

$$1 \text{ VSD} = \frac{39}{20} \times 1 \text{ mm}$$

The least count (LC) is the difference between the length of two MSD and the length of one VSD. i.e.,

$$1 \text{ LC} = 2 \text{ MSD} - 1 \text{ VSD}$$

Therefore, the scale shown in figure 5.4 has the least count of,

$$\text{LC} = 2 \text{ mm} - \frac{39}{20} \text{ mm} = \frac{1}{20} \text{ mm} = 0.05 \text{ mm}$$

Circular Vernier Scale

Circular verniers are commonly used in instruments such as spectrometer and theodolite. The spectrometer is provided with a circular scale for measuring angles of rotation. However, for accurate measurement of angles, there is a vernier scale sliding close to the main scale. Figure 5.5 shows the circular scale calibrated in half a degree divisions.

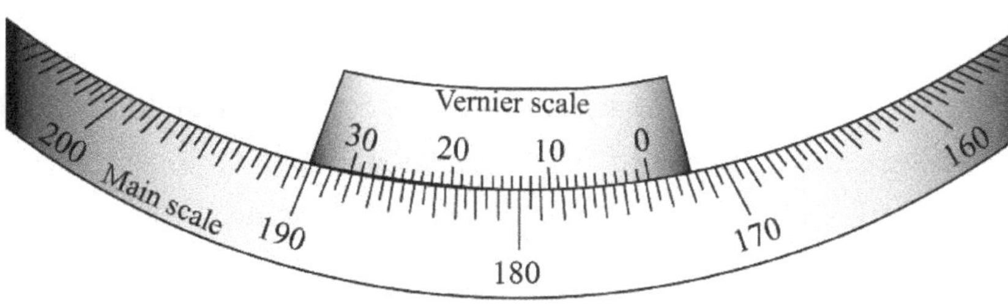

Figure 5.5: Circular scale calibrated in half a degree divisions

The entire circle is divided into 360 degrees. The vernier scale has 30 divisions which are included in 29 divisions or $14^{\circ}30'$ of the circular scale. Therefore,

$$1 \text{ MSD} = \text{ half a degree} = 30'$$

$$30 \text{ VSD} = 29 \text{ MSD}$$

This means, $$1 \text{ VSD} = \frac{29}{30} \text{ MSD}$$

Since one MSD is equal to $30'$, we have

$$1 \text{ VSD} = \left(\frac{29}{30}\right) \times 30' = 29'$$

The least count (LC) is the difference between the length of two MSD and the length of one VSD. i.e.,

$$1 \text{ LC} = 1 \text{ MSD} - 1 \text{ VSD}$$

Therefore, the scale shown in figure 5.5 has the least count of,

$$\text{LC} = 30' - 29' = 1' = \frac{1}{60} \text{ degree}$$

Hence, the reading shown in figure 5.5 will be determined as follows.

The 0th mark of the vernier scale lies between $173^{\circ}30'$ and 174°. The circular scale can read upto half a degree or $30'$. In figure 5.5, the 12th mark of the vernier coincides with a mark of the circular scale (main scale).

Thus the reading will be $173^{\circ}30' + 12' = 173^{\circ}42'$.

Positive Zero Error and Negative Zero Error

Generally, the 0th mark of the main scale should coincide with the 0th mark of the vernier scale, as shown in figure 5.6 (a). However, due to various reasons, it might happen that the 0th marks of the two scales do not coincide. Then we say that it has a zero error.

If the 0th mark of the vernier scale is slightly moved towards the right side of the 0th mark of the main scale, as shown in figure 5.6 (b), then we say that it has a positive zero error. On the other hand, if the 0th mark of the vernier scale is slightly moved towards the left side of the 0th mark of the main scale, as shown in figure 5.6 (c), then we say that it has a negative zero error.

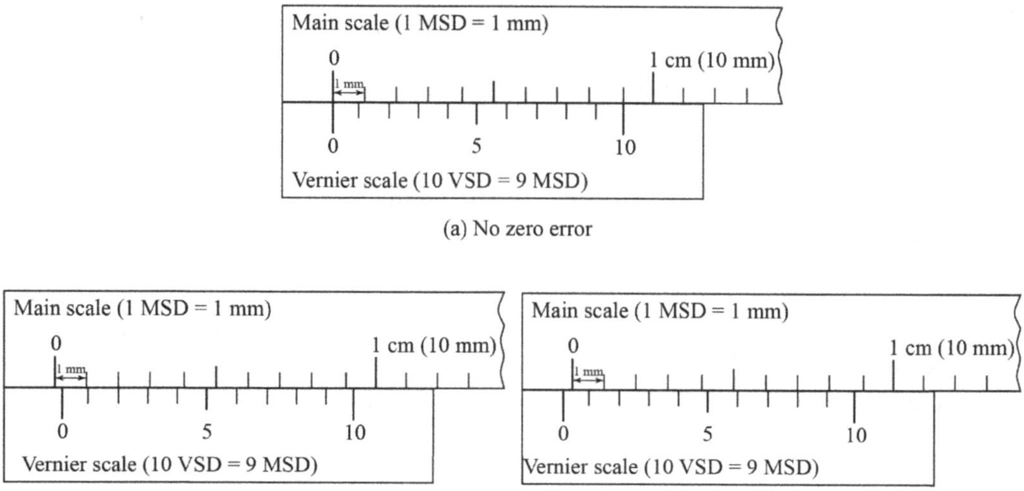

(a) No zero error

(b) Positive zero error

(c) Negative zero error

Figure 5.6: Zero error in a measuring device having a vernier scale

The main scale reading (MSR) is the first reading of the mark on the main scale immediately to the left of the 0th marking of the vernier scale. The division on the vernier scale that coincides with any mark on the main scale is the vernier scale reading (VSR). Then the measured value is given by

Measured value = MSR + VSR × LC

If we consider figure 5.6 (b),

the LC of the given measuring instrument is 0.01 cm.

$$\text{the MSR} = 0\text{ cm}, \text{ and the VSR} = 2.$$

Thus, the positive zero error is given by

$$\text{zero error} = \text{MSR} + \text{VSR} \times \text{LC} = 0 + 2 \times 0.01\text{ cm} = 0.02\text{ cm}.$$

If we consider figure 5.6 (c),

the LC of the given measuring instrument is 0.01 cm.

Usually, it is the first reading of the mark on the main scale immediately to the left of the 0th marking of the vernier scale. But in this case, there are no marks on the main scale to the left of the 0th mark of the vernier scale. If we carefully observe the estimation of the distance between the 0th mark of the vernier scale and the 0th mark of the main scale, in this case, it seems to be less than 1 mm, so we can take

$$\text{MSR} = -1\text{ mm} = -0.1\text{ cm}, \text{ and the } \text{VSR} = 8.$$

Thus, the negative zero error is given as

$$\text{zero error} = \text{MSR} + \text{VSR} \times \text{LC} = -0.1\text{ cm} + 8 \times 0.01\text{ cm} = -0.02\text{ cm}$$

In general, if we use the measuring device having a vernier scale with zero error° the required measurement will be;

$$\text{Measured value} = \text{MSR} + \text{VSR} \times \text{LC} - \text{zero error}$$

Figure 5.7 shows the situation where the measuring device with a vernier scale with a positive zero error of 0.02 cm has been used to measure the length of a wooden strip.

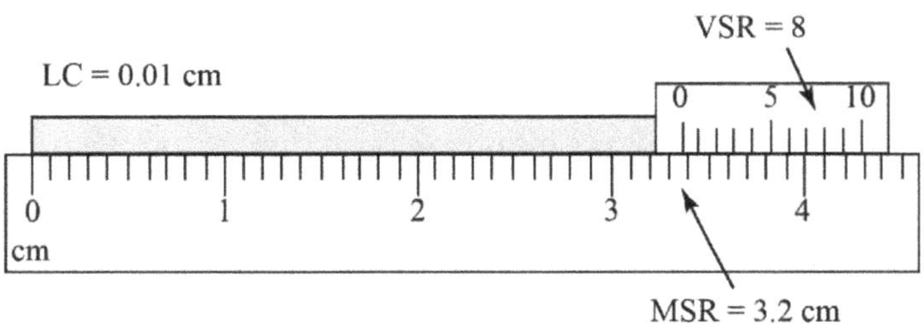

Figure 5.7: Measuring the length of a wooden strip

The main scale works just like a ruler. The 0th mark on the vernier is compared with the main scale, and the result is written down. Use the mark next to the zero, not the mark next to the edge of the vernier. Record the value of the main scale mark that is just to the left of the vernier 0th mark, as shown in figure 5.7. Hence the main scale reading is 3.3 cm. Now, look closely at the vernier scale in figure 5.7. Choose which vernier mark is closest to matching the main scale mark. In figure 5.7, it is clear that the vernier mark 8 coincides with the mark on the main scale. Thus, by combining the two readings along with the zero error of the instrument, the final measured value of the length of the wooden strip is

$$\text{Measured value} = (3.3 + 8 \times 0.01 - 0.02)\text{ cm} = 3.36\text{ cm}$$

You will learn more about the vernier measurements in the latter section of this chapter.

5.6 Error Analysis

A key part of the scientific method is recognising the errors that are present in the experimental technique being used, and working to reduce these as much as possible. In previous section you learned about the type of errors that you may come across during an experiment. You also should know how to calculate errors in quantities that are combined in different ways and how to estimate errors from graphs (see section).

The error is the difference between the measurement result and the true or accepted value. This is not a mistake in measurement. If the quantity measured has a standard, accepted or true value, then the accuracy of the experimental value is given by a ratio of the error to the true or accepted value and the percentage error (error %) is evaluated as

$$\text{Percentage error (error \%)} = \frac{|\text{experimental value} - \text{accepted value}|}{\text{accepted value}} \times 100\%$$

Percentage errors mean how accurate our results are when we measure something. Smaller percentage errors mean that we are close to the true/accepted value. The magnitude of the percent errors will depend heavily on the overall precision of the measuring instruments. This means that in some cases an error of 5% or 10% might be acceptable. In other cases such an error would indicate a very poorly run experiment. Thus, the success of an experiment, in terms of percent error, can only be judged when the method and instrumentation of the experiment are taken into consideration.

When we take a measurement, we often would like to know how precise the measurement is. We need some way to judge it, which is done by finding the percentage uncertainty or the percentage error of the measurement. We will refer to the percentage uncertainty of measurement as the ratio between the measurement's uncertainty and its measured value multiplied by 100%. Suppose that a measurement is reported as $Z \pm \Delta Z$, where Z is the measured quantity and ΔZ the measurement's uncertainty, the ratio $\frac{\Delta Z}{Z}$ is the fractional uncertainty in the measurement and $\frac{\Delta Z}{Z} \times 100\%$ is the percentage uncertainty in the measurement.

Activity 5.5

Consider a measurement of the length of the side of a cube, given as 50 ± 1 mm.

(i) What is the absolute uncertainty in the measured value?

(ii) Determine the fractional uncertainty in the measurement of length.

(iii) Determine the percentage uncertainty in the measurement of length.

When you repeat a measurement, you may often get different results. There is uncertainty in the measurement that you have taken. It is important to determine the uncertainty in the measurements so that its effect can be considered when drawing conclusions about experimental results. If the readings were repeated several times, the uncertainty might be half the range of the repeats. This means that if you have a set of repeated measurements of a quantity, say $\{x_1, x_2, x_3, ... x_n\}$, then the range is the difference between the maximum and minimum values of the measurement, i.e., $(x_{max} - x_{min})$ and the uncertainty Δx of the measurement is $\Delta x = \frac{x_{max} - x_{min}}{2}$. This way of determining the uncertainty in the measurements deals with random errors made in the readings, but does not account for systematic errors. However, if the repeated readings are all the same, do not assume that the uncertainty is zero. The uncertainity can never be less than the value you obtain for the smallest scale division. If the uncertainty is predictable, i.e. systematic, then the uncertainty should be subtracted from each reading, for example if there is a zero error on an instrument.

5.7 Recording data and Drawing graphs

Graphs are vital tools for analysing and displaying data. There are many quantities that can be measured in a physics experiment, and usually all but two of them are controlled so that they do not change during the course of the experiment. Of the two controlled quantities, we change one quantity (known as the independent variable) and see what happens to another (known as the dependent variable). From the data obtained, it will be interesting to investigate how one variable depends on another and to find the relationship between these measured variables. One good way to do this is to plot the graph of the data and then analyse the graph. An independent variable is usually plotted on the x-axis, and a dependent variable is usually plotted on the y-axis. Graphs are very useful for seeing the pattern in the relationship between two variables rather than seeing it in a table of numerical values. Graphs can also provide extra information, such as gradients (slopes), intercepts, and the areas under graphs often have important meaning. Figure 5.8 illustrates the terminology commonly associated with graphs.

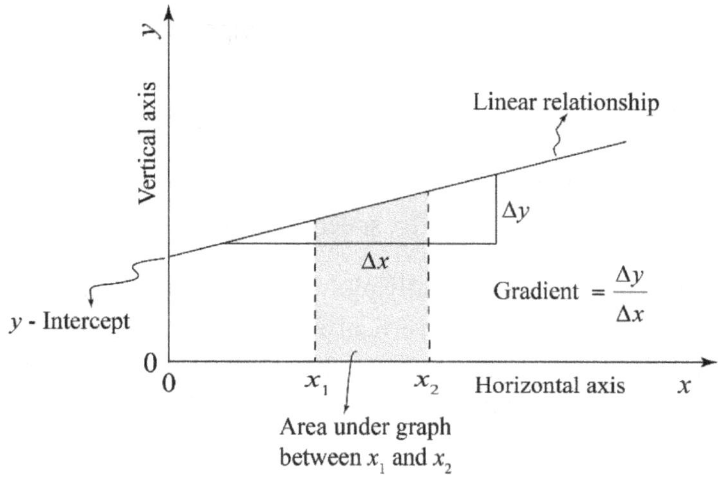

Figure 5.8: Terminology used in graphs

Recording Data

It is important to develop the skill to record data in a clear and concise way. Tables are a way to organise your data in a structured way while engaged in an experiment. Generally, the numerical results will be recorded in a table. Here are a few recommendations for making a table.

- Plan your table, especially how many columns and rows you will need. Try to anticipate what readings you will take. Remember to include columns for the results of any calculations. For example, you may sometimes need to calculate the average number of repeated readings.

- Each table should be drawn neatly and have a title that reflects your reason for recording the data. If you use more than one table in your report, then number them Table 1, Table 2, etc.

- Each column of a table must be labelled with the quantity being measured and the unit in which it is measured.

- Record your data carefully. Remember, for example, that writing 14.0 cm and 14 cm is not the same. In the first case, you are recording a reading to the nearest millimetre, and in the second case, you only measured to the nearest centimetre.

- If any computation is made, it has to be reported in a separate column in the same table with the correct significant figure.

Drawing Graphs

In physics, drawing good quality graphs is very important. It would be helpful to consider the following general guidelines when drawing graphs.

- If you are drawing the graph by hand, always use a pencil.
- Each axis should be labelled with the quantity being measured and the unit in which it is measured (e.g. force/N, speed/ $\mathrm{m\ s^{-1}}$).
- The independent variable is usually plotted on the horizontal axis (x -axis), and the dependent variable is on the vertical axis (y -axis). Sometimes, the choice of what to plot on each axis is made so that the gradient of the graph has a particular meaning. If time is one of the variables, it is often plotted on the horizontal axis.
- Choose appropriate scales for the x, y axes so that the graph will not be too small on the page but will cover a fair portion of the page in each direction. The main divisions of the x, y axes scales should be selected in such a way that they can be easily subdivided and all the values can be included. Subdivisions such as 1, 2, 5 and 10 are the best; 4 is sometimes used but never use 3,7, or 9, since these make it very difficult to read values from the graph. The divisions on the x and y axes scales need not be alike.
- x, y axes scales usually start at zero so that the origin, the point (0,0) is included in the graph. This is often important when interpreting a graph. However, this is not always sensible, especially if it would mean that all the readings were restricted to a small part of the graph.
- The data points on the graph should be neat and small. If points are used (instead of crosses), drawing a small circle around them can ensure that they are not overlooked, especially if the line goes through them.
- The more data points that can be plotted, the more precisely the line that represents the relationship can be drawn. Generally, at least six points are needed, although this may not always be possible.
- Use error bars to indicate errors in measurements.

- When all the points have been plotted, a pattern will usually be clear and a smooth and thin line of best-fit can be drawn (see figure 5.9). These lines are called trend lines. Use a ruler or straight edge to draw these best-fit trend lines. These trend lines may be a straight line (called linear) or a curve.

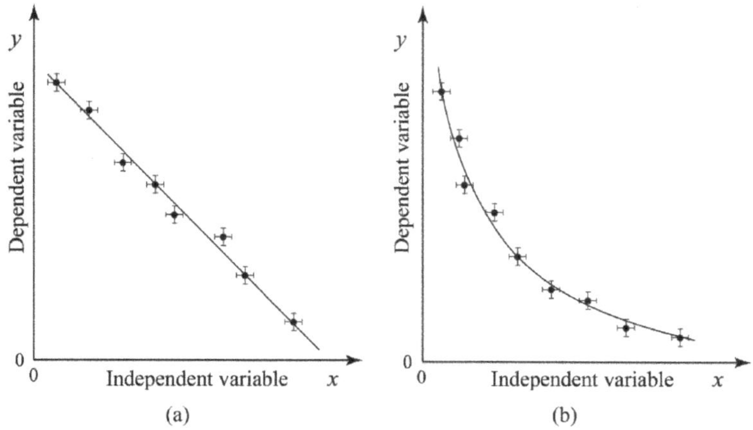

Figure 5.9: Drawing best-fit lines (a) straight line (b) curve

- Usually, a line of best-fit is drawn to cover a specific range of measurements recorded in an experiment (see figure 5.10). Within that range, other values could be predicted by interpolation, as shown in figure 5.8. Outside the range of measurements, if we want to predict what would happen, we need to extend the best-fit line, called extrapolation (figure 5.10).

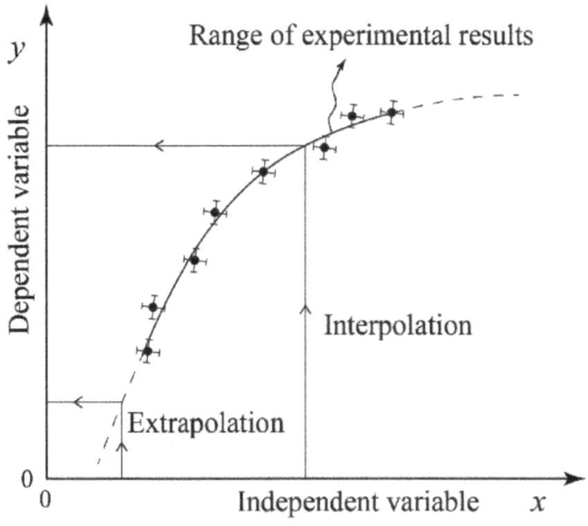

Figure 5.10: Interpolating and extrapolating to find the intercept

Predictions made by extrapolation should be treated with care, because it may be wrong to assume that the behaviour seen within the range of measurements also applies outside that range.

- Drawing graphs by hand is a skill that you should practise. However, knowing how to use a computer program to plot graphs is also a very valuable and time-saving skill. A graph generated by a computer must be evaluated by the same standards as a hand-drawn graph.

Straight-line Graphs

The simplest possible relationship between two variables (between independent and dependent variables) is that they are directly proportional to each other.

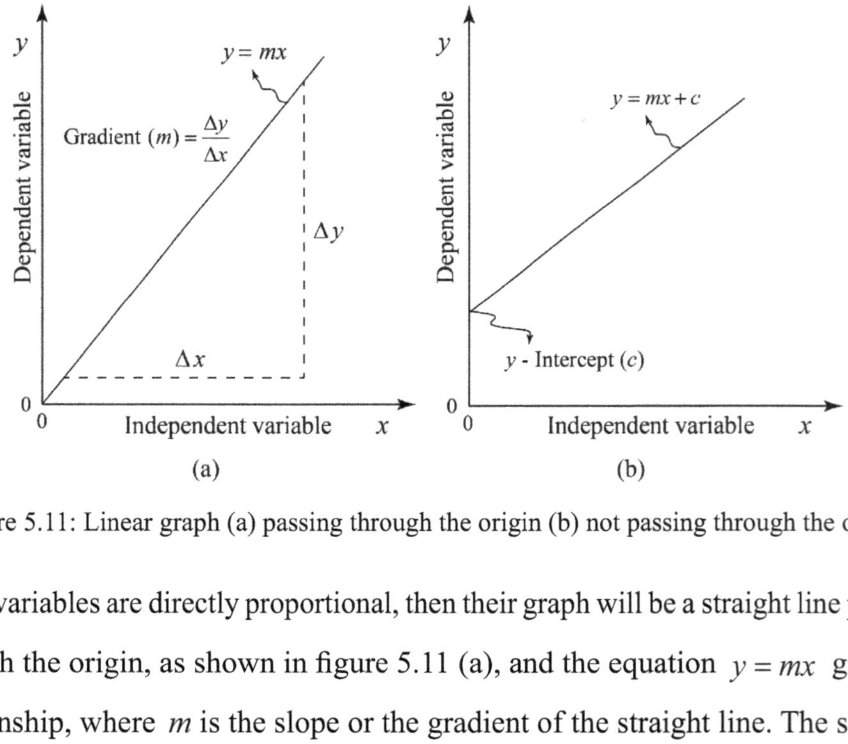

Figure 5.11: Linear graph (a) passing through the origin (b) not passing through the origin

If two variables are directly proportional, then their graph will be a straight line passing through the origin, as shown in figure 5.11 (a), and the equation $y = mx$ gives its relationship, where m is the slope or the gradient of the straight line. The slope or gradient of the straight line is measured by measuring the ratio $\frac{\Delta y}{\Delta x}$. When determining this slope or gradient, use as large a range of the graph as possible. Figure 5.11 (b) shows the linear relationship that is not proportional and does not pass through the origin, and the equation $y = mx + c$ gives its relationship, where m is the slope or the gradient of the straight line and c is the intercept on the y-axis.

When analysing graphs, straight lines are much easier to understand and analyse than curved lines. Usually, when directly measured experimental data are plotted against each other, the lines are often curved rather than linear. Figure 5. 12 shows few graphical relationships, how curves can be replotted to produce straight lines.

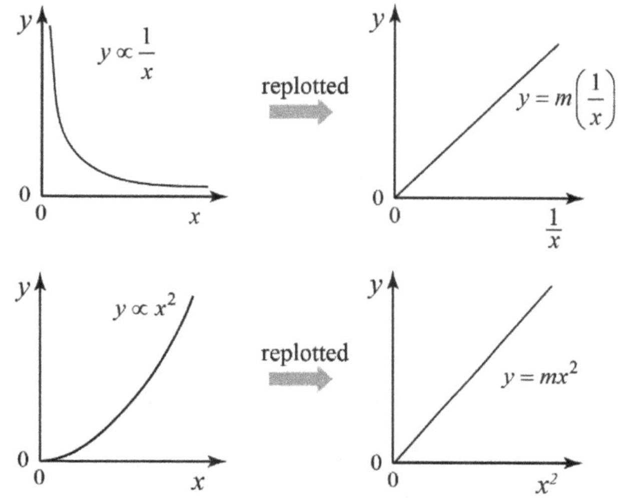

Figure 5.12: Few graphical relationships showing how curves can be replotted to produce straight lines

Estimating the Uncertainty on Slope and Intercept

Most of the experimental data we obtain in a laboratory can be plotted as a straight-line (linear) graph of the form $y = mx + c$, where the m is the slope and c is the y – axis intercept. When data points are plotted, the uncertainties associated with the data are also transferred to the graph. The size of the error (in one or both axes) is represented by error bars that define the uncertainty range.

There are various methods by which the equation of the best straight line may be found. Currently, several data analysis software packages are available, which could be used easily to determine the slope and the intercept using the least squares method. However, the maximum/minimum method is a popular method that could be used in a laboratory when analysing the data by hand.

You may judge the best-fit line as the straight line that passes through or is as close to all your data points as possible. When the best-fit line does not go through the data point itself, it should be as close to the region defined by the error bars as possible. If it is not possible to find a straight line that passes close by all your error bars (compared with the length of the error bars), then your two variables are not linearly related.

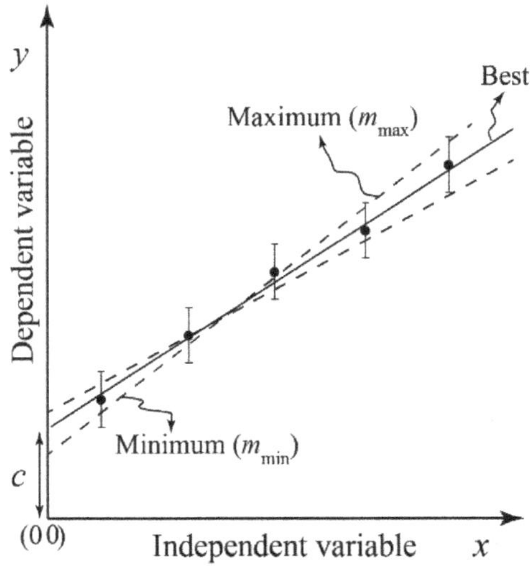

Figure 5.13: Estimation of the line of best-fit and its uncertainty. The slope of all these lines depends on your guess based on the error bars

Figure 5.13 shows the best line (solid line) or the most representative straight line that fits the data points and two other lines (dash lines). Approximately the same number of points lie above and below the best line, and the best line is used to find the slope and the intercept.

The two dashed lines represent the data nearly as well as the best line. One dashed line has the largest plausible or maximum slope $\left(m_{\text{max}}\right)$ and the maximum intercept $\left(c_{\text{max}}\right)$, and the other has the smallest or minimum slope $\left(m_{\text{min}}\right)$ and the minimum intercept $\left(c_{\text{min}}\right)$. The largest slope line can be constructed by drawing a line that goes between points below the best-fit line on the left side of the graph and points above the best-fit line on the right side of the graph. The smallest slope line can be constructed by drawing a line that goes between points above the best-fit line on the left side of the graph and between points below the best-fit line on the right side of the graph. The differences between the slopes and intercepts of these lines yield uncertainties in the slope and intercept, as described below.

The uncertainty in the slope Δm is estimated using the equation

$$\Delta m = \frac{m_{\text{max}} - m_{\text{min}}}{2}$$

and the uncertainty in the intercept Δc is estimated using the equation

$$\Delta c = \frac{c_{\max} - c_{\min}}{2}$$

In some situations, the best straight line may appear to pass through every point (or error bars may be too small to plot), and you may conclude that the results are highly accurate. If the errors are too small to draw the minimum and maximum slope lines, you can consider using the least squares method to estimate the slope and intercept uncertainties. The least squares method is a very popular technique used to compute the slope, intercepts and their respective uncertainty estimations and fit data. It is one of the oldest techniques of modern statistics. This method is not described, as it is beyond the scope of this book.

5.8 Measuring Instruments

Instruments designed to measure fundamental quantities, such as length, mass, time, etc., are called measuring instruments. This section describes standard measuring instruments that we commonly use in laboratories, engineering workshops, metalworking, woodworking and our daily technical use.

Vernier Calliper

The vernier calliper is an instrument that allows one to measure lengths much more precisely than the metric ruler. It measures linear dimensions, such as length, internal and external diameters, and depth of physical objects. Figure 5.14 shows a typical vernier calliper used in laboratories and mechanical engineering workshops.

It consists of a fixed jaw attached to the main scale calibrated in millimetres and a movable jaw attached to a vernier scale which can slide along the main scale. The least count of the vernier calliper shown in figure 5.14 is 0.05 mm or 0.005 cm. Outside jaws (or external jaws) can be used to measure the external diameter of physical objects such as a cylinder. Inside jaws (or internal jaws) can be used to measure the internal diameter of physical objects such as a hollow cylinder, and the depth probe can be used to measure the depth of a hole of physical objects.

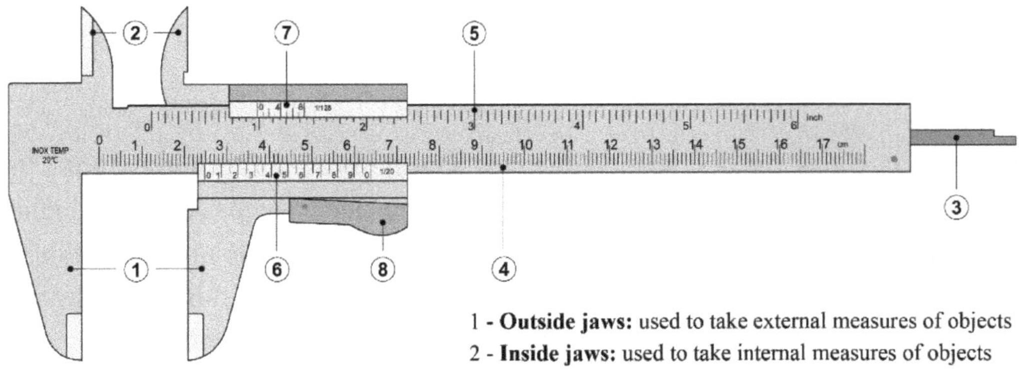

1 - **Outside jaws:** used to take external measures of objects
2 - **Inside jaws:** used to take internal measures of objects
3 - **Depth probe:** used to measure the depth of objects
4 - **Main scale** (cm)
5 - **Main scale** (inch)
6 - **Vernier** (cm)
7 - **Vernier** (inch)
8 - **Retainer:** used to block movable part

Figure 5.14: Vernier Calliper

When measuring a length of an object, the vernier scale can be moved along the main scale, and then by pressing the retainer (or, in some vernier callipers, by tightening the screw), the vernier scale can be kept in a fixed position before taking a reading.

On bringing the jaws of the vernier callipers into contact with each other, the zero of the main scale should coincide with the zero of the vernier scale. If it does not coincide, the vernier is said to have zero error, which could either have a positive zero error or a negative zero error (see the previous subsection). The zero error can appear in some vernier callipers of the above type shown in figure 5.14 due to some reasons, such as corrosion or wasted jaws.

Travelling Microscope

The travelling microscope is extensively used in research, university, school and industrial laboratories. Figure 5.15 shows the standard travelling microscope used in laboratories and its main parts. It consists of an ordinary compound microscope that slides along a graduated vertical pillar and is attached to a horizontal base resting on the levelling screws. The main scale divisions and vernier scale divisions are marked on the horizontal base and the vertical pillar, as shown in figure 5.15.

The microscope can be moved up and down in the vertical pillar and in to and fro direction over the horizontal base. Thus, the microscope can be moved both in the vertical and horizontal directions. Two fine adjustment screws are provided for horizontal and vertical movements. Before using the instrument, the base should be levelled horizontally using the levelling screws. The image of the object can be focused by adjusting the side screw attached to the microscope. The eyepiece of the microscope is provided with a cross wire.

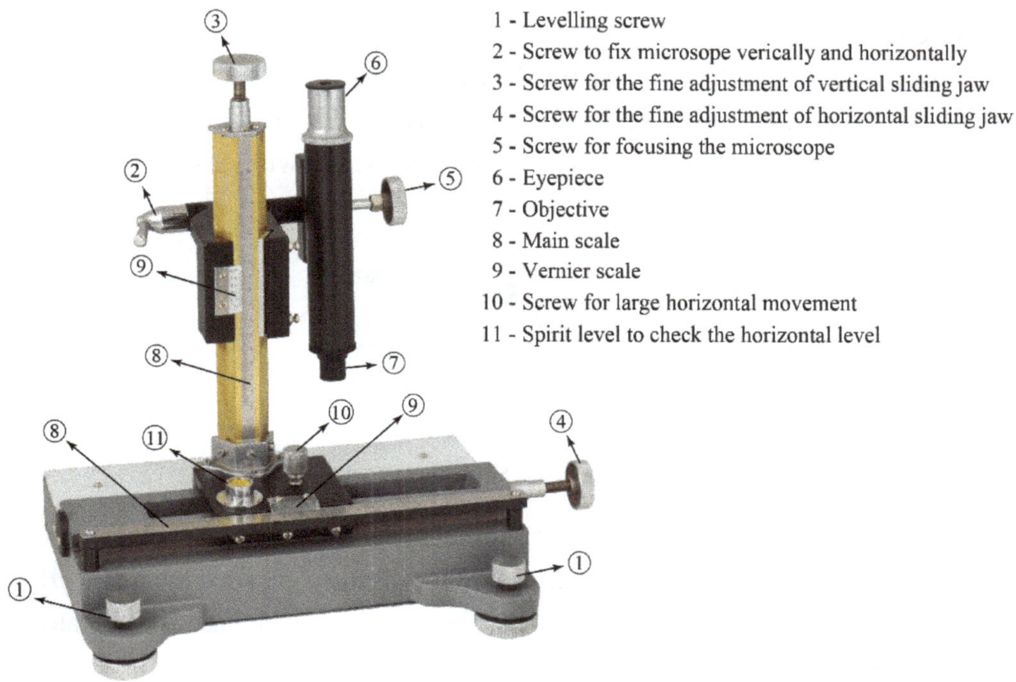

Figure 5.15: Travelling microscope

The least count of the vertical and horizontal scale in a standard travelling microscope is 0.01 mm. The vernier division, which coincides with a scale division, cannot be read with the naked eye. In that case, a magnifying glass can be used. In some travelling microscopes magnifying glass is fixed onto the base close to the horizontal vernier scale. In schools or college laboratories, a travelling microscope is commonly used to obtain measurements such as the length of a mercury thread inserted in a capillary tube, the horizontal and vertical diameters of a capillary tube base, etc.

Micrometer Screw Gauge

A micrometer screw gauge is used to measure the dimensions of physical objects with an accuracy of 0.01 mm or 0.005 mm. These types of screw gages are commonly used in laboratories to obtain measurements such as the diameter of a thin wire, the thickness of a blade, the diameter of a ball bearing, etc.

Intrinsically, a micrometer screw gauge is an instrument of higher precision than vernier callipers, which has been discussed earlier. It is designed with the principle of a screw. A screw has threads and moves linearly by some distance when rotating the screw head. Figure 5.16 shows the diagram of a screw. There are threads on a screw, and the separation between two consecutive threads is the same. The screw can be moved backward or forward by rotating it anticlockwise or clockwise.

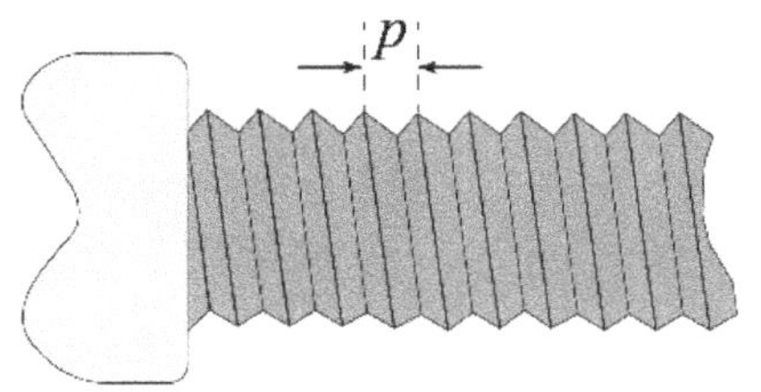

Figure 5.16: A screw showing the pitch (p)

The distance moved by the screw when it makes its one complete rotation is the separation between two consecutive threads. This distance is called the pitch (p) or screw gap of the screw (figure 5.16). Figure 5.17 shows the standard used in laboratories and mechanical engineering workshops.

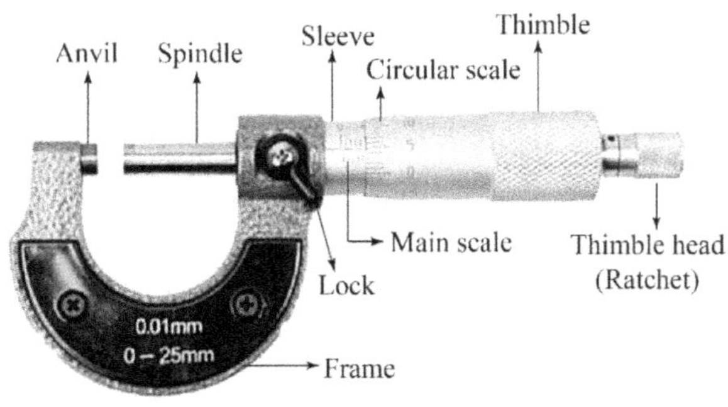

Figure 5.17: Micrometer screw gauge

The micrometer screw gauge consists of a U-shaped piece of solid steel known as a frame. The anvil is attached to one end of the frame, and the other has the spindle connected to a fine thread screw. The main linear scale is calibrated on the sleeve and the screw can move on the sleeve while rotating. The thimble head is connected to the screw and, by rotating the thimble head, the thimble can move along the main linear scale. On the bevelled edge of the thimble, there is a circular scale that is usually divided into 50 or 100 equal parts. The linear distance by which the screw moves during its one complete rotation is called the pitch of the screw. The pitch of the screw is usually 0.5 or 1 mm. Thus, the least count (LC) of the screw gauge, which is the smallest distance that this screw can measure, is

$$LC = \frac{\text{Pitch}}{\text{Total number of division on the circular scale}}$$

Therefore, if there are a total of 50 divisions on its circular scale and the value of one division on the main scale is $0.5\ \text{mm}$ (i.e. the pitch of the micrometer screw gauge is $0.5\ \text{mm}$), then the least count will be $0.01\ \text{mm}$.

When taking measurements, after placing the object between the spindle and the anvil, rotate the thimble until the spindle touches the object slightly. Subsequently, the thimble head is rotated slowly (ratchet) until the thimble head freely rotates, emitting a sound (because of a mechanism inside it). This will prevent pressing the measuring object with the spindle and anvil.

When the surface of the spindle is in contact with the surface of the anvil (touches slightly) and after the thimble head freely rotates, emitting a sound, the zero of the circular scale should coincide with the main linear scale line, as shown in figure 5.18 (a) and is said to have no zero error.

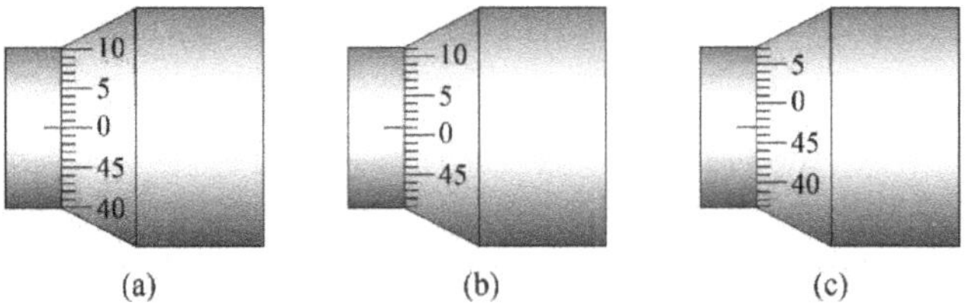

Figure 5.18: Micrometer screw gauge with (a) no zero error, (b) positive zero error and (c) negative zero error

But in some micrometer screw gauges, due to various reasons, such as corrosion or waste of the spindle and anvil, when the surface of the spindle is in contact with the surface of the anvil (touches slightly) after the thimble head freely rotates, emitting a sound, the zero of the circular scale might be in a position as illustrated in figure 5.18 (b) and figure 5.18 (c) and is said to have zero error. Assume that the micrometer screw gauge scale shown in figure 5.18 has the least count of 0.01 mm.

When the circular scale reading across the main linear scale is less than zero (or negative), the micrometer screw gauge is said to have a negative zero error, as shown in figure 5.18 (b). Accordingly, in figure 5.18 (b), the first division of the circular scale reading coincides with the main linear scale line. Therefore, in this case, the positive zero error is 0.01 mm, and for correction, this value should be deducted from the relevant reading.

When the circular scale reading across the main linear scale is more than zero (or positive), the micrometer screw gauge is said to have a positive zero error, as shown in figure 5.18 (c). Accordingly, in figure 5.18 (c), the third division of the circular scale reading coincides with the main linear scale line (i.e. three divisions of the circular scale have been rotated). Therefore, in this case, the negative zero error is 0.03 mm, and for correction, this value should be added from the relevant reading.

Activity 5.5

Write down the micrometer screw gauge reading shown below.

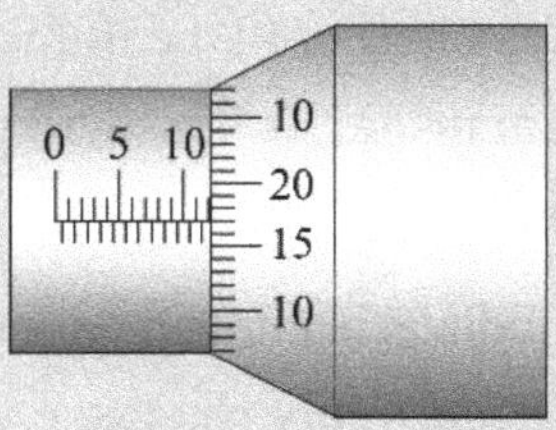

Assume that the least count of the micrometer screw gauge is 0.01 mm and has a positive zero error of 0.02 mm.

Spherometer

Figure 5.19 shows the standard spherometer used in laboratories which uses the screw principle. It is commonly used to take measurements to calculate the radius of curvature of spherical surfaces (convex and concave) and to measure the cover slip or a microscopic slide.

As shown in Figure 5.19, a spherometer consists of three equal legs with the lower tips of the legs (marked A, B, and C) located in the vertices of an equilateral triangle. The spherometer also consists of a central leg with a fine threaded screw that has the lower tip point O, which can be raised or lowered through a threaded hole or nut in the centre of the frame. The lower tip of the central screw, when lowered to the plane (formed by the tips of legs A, B, and C), touches the centre of triangle ABC. The central screw also carries a circular disc at its top, having a circular scale divided into 100 or 200 equal parts. A small vertical scale marked in millimetres or half-millimetres, called the main scale, is also fixed parallel to the central screw at the end of the frame, as shown in figure 5.19.

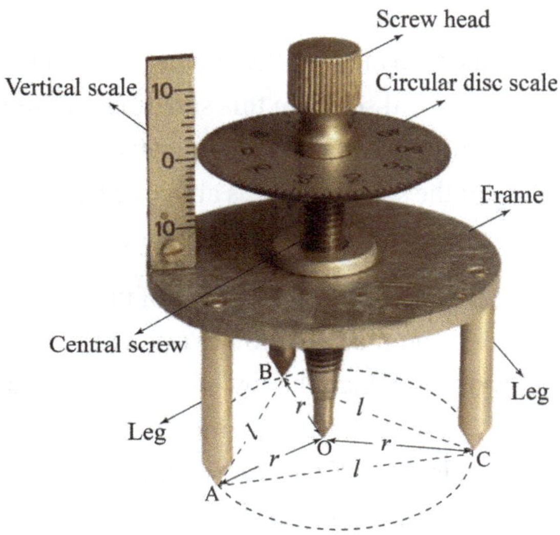

Figure 5.19: Spherometer

This vertical scale is kept very close to the rim of the circular disc but does not touch the circular disc scale. This scale reads the vertical distance from which the central leg moves through the hole or nut. This scale is also known as the pitch scale. The pitch is the vertical distance moved by the central screw in one complete rotation of the circular disc scale.

The least count (LC) of a spherometer is the distance moved by the spherometer screw when it is turned through one division on the circular scale, i.e

$$\text{LC} = \frac{\text{Pitch of the spherometer screw}}{\text{Total number of division on the circular scale}}$$

The spherometer, shown in figure 5.19, has graduations in millimetres on the vertical scale and has 100 equal divisions on the circular disc scale. Therefore, in one rotation of the circular scale, the central screw advances or recedes by 1 mm. Thus, the pitch of the screw is 1 mm. Thus, the least count of the spherometer shown in figure 5.19 is 0.01 mm.

Before using the spherometer for measurements, place the lower tips of the three legs ABC on an optically flat glass surface and rotate the central screw head so that the lower tip of the screw O just touches the glass surface. At this point, the zero mark on the circular disc scale should lie along the zero mark on the vertical scale and is said to have no zero error.

However, in some spherometers, for various reasons, such as corrosion or thread waste on the central screw, the zero mark on the circular disc scale will not lie precisely along the zero mark on the vertical scale. In this situation, the spherometer is said to have zero error. The zero error correction can be positive or negative depending on whether the initial reading of the spherometer is above or below the zero mark on the vertical scale and whether the reading is taken using a convex or concave surface.

Consider a situation where you are in the process of finding the radius of curvature of a spherical surface, and if you place the lower tips of the three legs ABC on the spherical surface, the distance in which the lower tip of the circular screw has to be raised up (or moved down) from the plane. Figure 5.20 shows a schematic diagram of such a situation. The circle AOBXZY shown in figure 5.20 (a) represents the vertical section of the sphere of the radius R with E as its centre (the given spherical surface is part of this sphere).

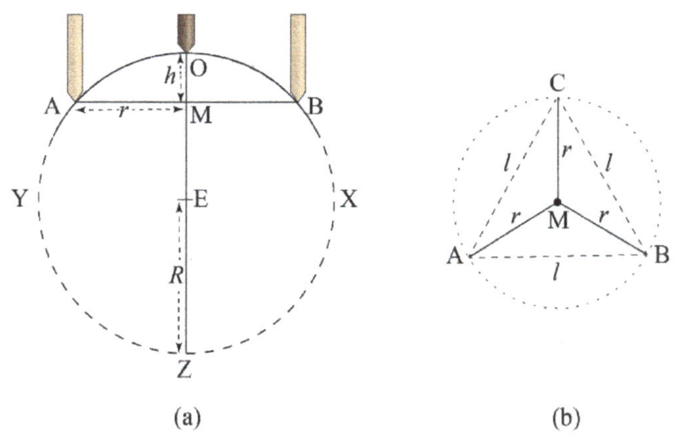

Figure 5.20: (a) Measurement of the radius of curvature of a spherical surface
(b) Base circle of the spherometer

As shown in figure 5.20 (a) and (b), let h be the distance raised up from the plane, which is the distance OM the height of the central screw above the plane of the circular section ABC when its lower tip just touches the spherical surface. Let the radius of the base circle of the spherometer, which passes through the lower tips of three legs, be r, and let the distance between any two legs of the spherometer or the side of the equilateral triangle ABC be l.

The two chords of the circle, such as AB and OZ, intersect at a point M, then if we consider the ΔAMZ and ΔBMO, we have

$$AM \times MB = OM \times MZ$$

But $AM = MB$ and $MZ = OZ - OM$. Also, we have $AM = r$ and $OZ = 2R$. Thus,

$$r^2 = h(2R - h)$$

Hence.

$$R = \frac{r^2}{2h} + \frac{h}{2}$$

But l is the distance between any two legs of the spherometer or the side of the equilateral triangle ABC (figure 5.20(b)), then from geometry, we have

$$r = \frac{l}{\sqrt{3}}$$

Therefore,

$$R = \frac{l^2}{6h} + \frac{h}{2}$$

By measuring h and l, the radius of the given spherical surface R can be calculated.

Measurement of time

Figure 5.21 shows the standard devices that are used in laboratories to measure time. Measurements such as the period of oscillation (e.g. the simple pendulum), the time spent for an object to move a short distance, etc., could be measured using these devices. The stop clock used to measure short time range is shown in figure 5.21 (a). It is a durable, easy-to-use, spring-wound device, which is useful for taking time measurements in laboratories. These types of stop clocks consist of a large dial with big pointer hands that provide easy reading even from a distance. The bigger main dial has a larger sweep hand reading $0 - 60 \times 1$ second and the smaller dial reading $0 - 60 \times 1$ minute. It includes a start/stop lever on the left side and a zero reset lever on the right side and can run for about 30 hours on one complete winding.

(a) (b) (c)

Figure 5.21: (a) Stop clock (b) Mechanical stop watch (c) Digital stop watch

Figure 5.21 (b) and (c) show the stop watches used to measure time with highly accurate precision. They are commonly used in laboratories and sports events such as sprints. Stop watches can be classified into two categories, mechanical stop watches, as shown in figure 2.21(b) and digital stop watches, as shown in figure 2.21 (c). Mechanical stop watches are powered by springs, and two buttons typically control their operation. The first button starts the timer, the second stops the measurement, and the elapsed time is displayed. The time can be restarted by pressing the first button again, and the time is reset to zero by pressing the second button again. Digital stop watches have crystal oscillator timing elements that make them much more accurate than mechanical versions and can measure a time interval of up to 0.01 seconds. The time interval between the start and stop of an event is recorded. When taking measurements using these devices, the accuracy depends on the reaction time of the person who operates the device. A person's reaction time is the time range between observing and responding to an occurrence.

Measurement of mass

A triple beam balance provides greater accuracy than a spring scale in determining the mass of objects in grams. Figure 5.22 (a) shows the standard triple beam balance used in laboratories. It functions by balancing an object with three counterweights that move on the riders attached to three beams to find the object's mass. As shown in figure 5.22(a), three different beams on the triple beam balance function independently assess the mass of an object. A weighted rider is attached to each beam, which you can slide along the beam to determine the object's mass. The back beam has a $100\ \text{g}$ scale with a $10\ \text{g}$ rider, the middle beam with a $500\ \text{g}$ scale with a $100\ \text{g}$ rider, and the front beam with a $10\ \text{g}$ scale with a $0.1\ \text{g}$ rider. When all three riders are fully positioned to the right, their weights increase by $500+100+10=610\ \text{g}$. Thus, the balance can measure the mass of objects up to 610 grams in weight.

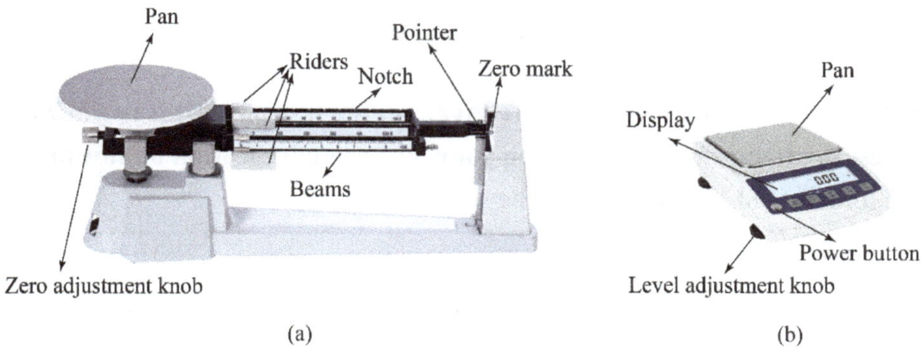

Figure 5.22: (a) Triple beam balance (b) Laboratory electronc blance

Before starting to use the triple balance, make sure that the balance is calibrated. With nothing on the pan, calibrate the triple balance by sliding the three riders to their leftmost positions. Then twist the zero adjustment knob located below the pan until the balance pointer lines up with the fixed zero mark. To measure the object's mass, place it in the centre of the pan. Slide the rider right one notch at a time. When the pointer drops below the zero mark, move the rider backward one notch. Repeat this in all three beams until the pointer aligns with the zero mark. Read the corresponding measurement on each rider and add all values to determine the mass of the object.

Electronic balance is an important instrument for laboratories to precisely measure the mass of substances used in various experiments. Figure 5.22 (b) shows the standard electronic balance used in laboratories. Laboratory electronic balances are commonly used in pharmaceutical, scientific, industrial, and food research, etc. When making measurements, the electronic balance should be placed on a flat, stable surface. Wind, shaky surfaces or any kind of force or vibrations can affect the reading accuracy of the balance. Electronic balances are available in different types, such as top loading balance and analytical balance. A top loading balance will work just fine in measuring the accuracy of 0.001g, whereas the accuracy of an analytical balance is 0.0001 to 0.00001g. If extreme precision and accuracy are vital in an experiment, an analytical balance will be a better choice.

Questions

NOTE: Although the solutions are given on the last page, it is in your own interest to attempt the questions on your own as an indication of how far you have understood the material of the section.

1. Explain the terms *independent variable and dependent variable.*

2. How do you determine the uncertainty of a mean value calculated from a set of repeated measurements of a quantity?

3. What does it mean if experimental results are said to be *repeatable* and *reproducible* in scientific research.

4. A student repeats the experiment in a research laboratory using a digital thermometer that automatically measures the temperature to the nearest $0.01^{\circ}\mathrm{C}$ every $20\ \mathrm{s}$. Explain in which way this method will reduce the effect of random error on the results.

5. A floor tile installer wants to calculate the area of a square tile. He uses a measuring tape to measure the two adjacent sides of a square tile and obtains the following measurements.

   Length of one side $= 74\ \mathrm{mm} \pm 0.5\ \mathrm{mm}$.

   Length of perpendicular side $= 74\ \mathrm{mm} \pm 0.5\ \mathrm{mm}$

   Determine the percentage uncertainty in the measurement of the length of each side of this square tile.

6. In a laboratory, a student made three pairs of perpendicular measurements to determine the diameter of a metal cylinder. Measurements in mm, are 31.8, 31.6, 32.1, 31.7, 31.9, and 32.1. What should be the reported value of the cylinder's diameter?

7. In a measuring instrument, the smallest division of the main scale is $1\ \mathrm{mm}$. The vernier scale of that measuring instrument is constructed by dividing the $19\ \mathrm{mm}$ range from the main scale into 20 equal parts. Calculate the least count of the vernier.

8. In a laboratory, a student used the vernier calliper to measure the diameter of a steel ball. The figure below shows the main and the vernier scale readings that the student has obtained.

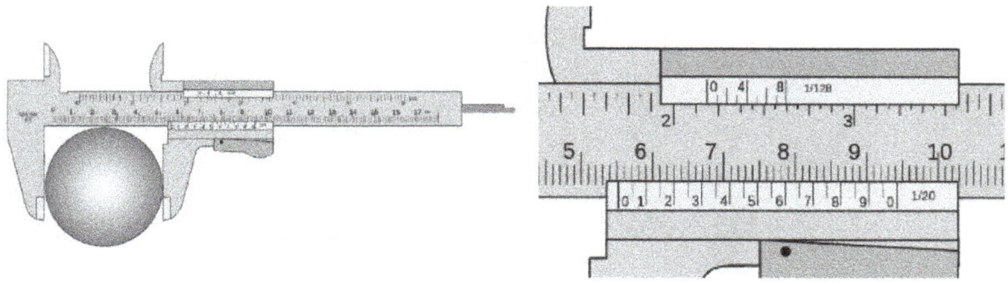

What is the final measurement of the diameter of the steel ball that the student would record? Assume that the vernier calliper does not have zero error and has the least count 0.05 mm.

9. In a metal workshop, a technician used the micrometer screw gauge to measure the width of a metal block. The figure below shows the main and circular scale readings that the technician has obtained.

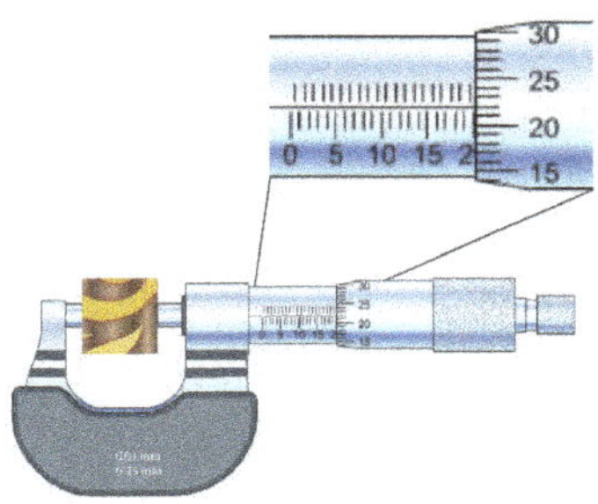

What is the final measurement of the width of the metal block that the technician would record? Assume that the micrometer screw gauge has a negative zero error of 0.01 mm and has the least count of 0.01 mm.

10. In an experiment, a student was asked to investigate the deflection of a given wooden beam against its length when it is fixed only at one end (a cantilever) and a mass (load) hanging from somewhere on the part that extended from the bench top, as shown in the figure below.

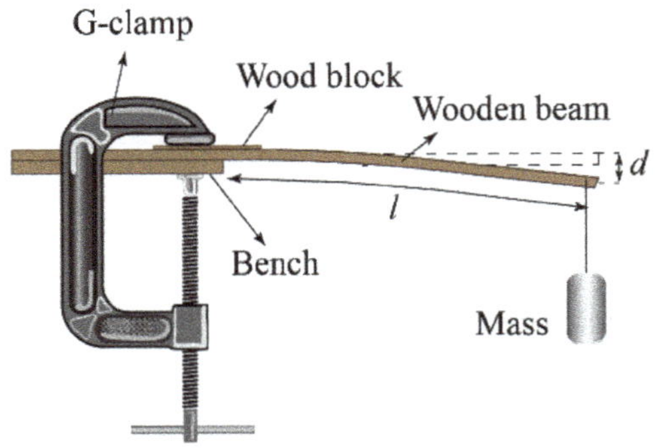

(i) Write down the variables that the student may record before starting the experiment.

(ii) As requested, the student decided to investigate how the deflection, d, depends on the length , keeping all the other variables constant. The student recorded the deflection d of the beam while changing the length l, and they were tabulated as shown in the table below.

Length (l) in cm	Deflection (d) in cm	$\frac{l^2}{d}$	$\frac{l^3}{d}$
30.0	0.4		
40.0	1.0		
50.0	1.9		
60.0	3.3		
70.0	5.4		
80.0	7.9		
90.0	11.1		
95.0	15.26		

(a) Fill the values for l^2, l^3, $\frac{l^2}{d}$ and $\frac{l^3}{d}$ with the appropriate significant figure.

(b) From the analysis you have performed in (ii)(a), predict the possible relationship between the length and the deflection of the beam. Give a reason for your response.

(c) Plot a suitable graph to confirm the correct relationship that you have predicted in (ii) (b).

(d) To improve the graph drawn in (ii)(c), what kind of adjustment should the student make while collecting the data in (ii)?

11. A nurse recorded a patient's body temperature in a hospital in half-hour intervals starting from 8:30 am to 11:00 am. The table below shows the recorded data.

Time	Body temperature in °C
8:30 am	37.7
9:00 am	37.8
9:30 am	38.1
10:00 am	38.3
10:30 am	38.5
12:00 am	38.9

(i) Plot a graph of temperature *versus* elapsed time.

(ii) Using the graph you obtain in part (i), estimate the temperature at 11:00 am.

(iii) Determine the slope of the graph by choosing appropriate points on the graph.

(iv) Would you expect the graph to follow the same trend over the next 12 hours? Explain your answer.

Assume that the maximum body temperature a human can survive is 42.3° C.

12. In a laboratory, students were interested in investigating how the length of a spring varies with the weight (applied force F) hanging from it. They designed the standard experiment and, for various weights, measured the length L of the spring with a metre stick, as shown in the figure below.

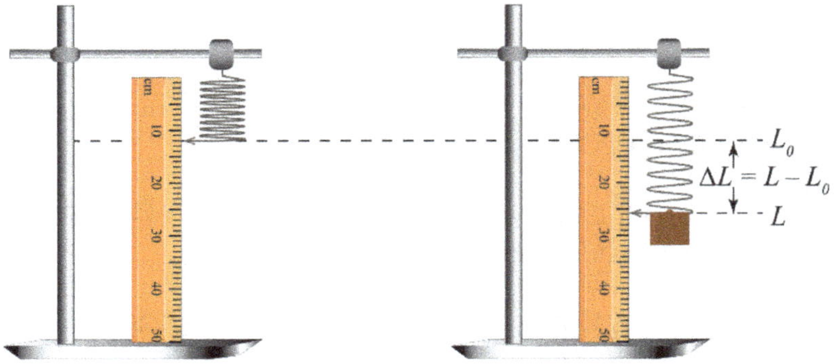

The maximum weight (applied force) that the spring can withstand is 7.00 N. Students obtain the following results shown in the table below.

F(N)	0	0.50	1.00	1.50	2.00	2.50	3.00	3.50	4.00	4.50	5.00	6.00
L(cm)	10.2	12.0	13.5	15.2	16.9	18.3	20.2	21.9	23.3	24.2	26.9	30.2

The relationship between the magnitute of the force (in this case, weight) applied to the spring F and the extension in length ΔL is givn by

$$F = k\Delta L$$

where $\Delta L = (L_0 - L)$ and L_0 is the length of the spring when no weight is hanging from it and k is called the spring constant of the spring.

(i) Identify the independent and dependent variables in this experiment.

(ii) Rewrite the equation $F = k\Delta L$ in a suitable form to plot the graph.

(iii) Plot a suitable graph using the data given and draw the best-fit line.

(iv) In the graph obtained in part (iii), select two suitable points to calculate the slope (or gradient) of the straight-line graph.

(v) Estimate the slope (or gradient) of the straight-line graph obtained in part (iii).

(vi) Hence, determine the value for the spring constant k of the spring used in this experiment.

Chapter 6

Vectors and Scalars

6.1 Introduction

The physical quantities could be classified into two, one having a size (a magnitude) only, and the other having both size (a magnitude) and direction. These two types of distinct quantities have a special mathematical language to differentiate them, and they are called scalars and vectors. In this chapter, we define these two types of specific quantities and learn how they are denoted mathematically by their geometrical representation and mathematical operations, such as addition and subtraction. These two types of quantities are commonly used in physics, engineering, other sciences, and even in everyday speech. Vector quantities are used throughout this text. Therefore, it is important to master the techniques discussed in this chapter.

6.2 Coordinate Systems

Many aspects of physics we deal with use locations in space in some form or another. For example, describing the motion of an object requires a method for describing its position at various times. This description could be accomplished by using coordinates, in which horizontal and vertical axes intersect at a point taken to be the origin. The coordinate systems are used to describe the positions of particles or points at which quantities are to be defined or measured. They are often used as reference for specifying directions. Most commonly used orthogonal coordinate systems are (a) a Cartesian (or rectangular) coordinate system, (b) a spherical polar coordinate system, and (c) a cylindrical coordinate system.

Cartesian coordinate system

Figure 6.1 shows the simplest set of axes known as the Cartesian axes, the x-axis, the y-axis, and the z-axis, where the perpendicular axes x, y, and z intersect at a point defined as the origin O and called the three-dimensional Cartesian coordinate system. Cartesian coordinates are also called rectangular coordinates. In this three-dimensional Cartesian coordinate system, a point is represented by its coordinates (x, y, z), as shown in figure 6.1.

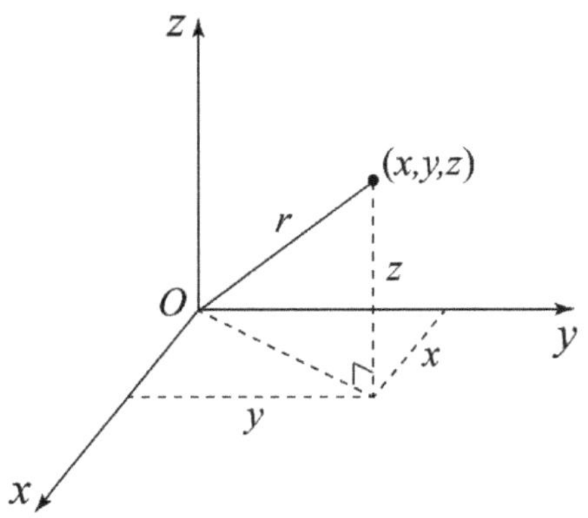

Figure 6.1: Cartesian coordinate system

The distance between the origin O of the Cartesian coordinate system and the point represented by its coordinates (x, y, z) is r (figure 6.1). The value r could be written using the familiar Pythagorean theorem, as

$$r = \sqrt{x^2 + y^2 + z^2} \tag{6.1}$$

Very often, you may find it more convenient to represent a point (say P) in a plane by its plane polar coordinates (r, θ), as shown in figure 6.2. In Cartesian coordinates, this point P is defined by the coordinates (x, y).

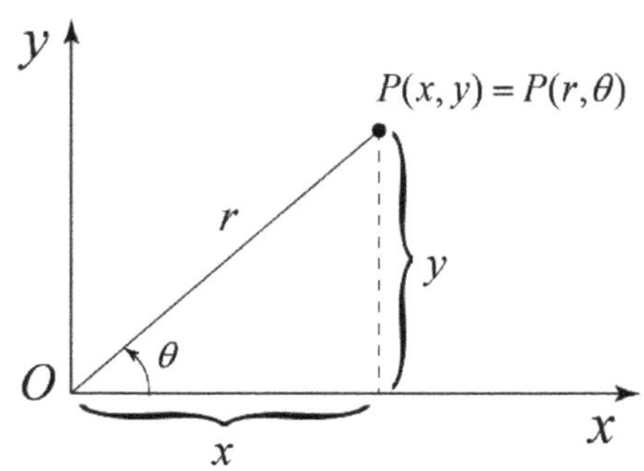

Figure 6.2: The plane polar coordinates of a point are represented by the distance r and the angle θ, where θ is measured counterclockwise from the positive x-axis.

In this polar coordinate system, r is the distance from the origin O to the point P having Cartesian coordinates (x, y) and θ is the angle between the positive x-axis and a line drawn from the origin to the point $P(x, y)$, and is usually measured counterclockwise from the positive x-axis.

From the right angled triangle shown in figure 6.2, we have $\cos\theta = \frac{x}{r}$ and $\sin\theta = \frac{y}{r}$ (refer Chapter 2, section 2.7). Therefore, the Cartesian coordinates (x, y) in terms of plane polar coordinates (r, θ) could be written as

$$x = r\cos\theta \tag{6.2}$$

$$y = r\sin\theta \tag{6.3}$$

$$\tan\theta = \frac{y}{x} \tag{6.4}$$

It should be noted that, for the above three equations (6.2, 6.3, and 6.4) the θ is measured counterclockwise from the positive x-axis.

From the familiar Pythagorean theorem we have

$$r = \sqrt{x^2 + y^2} \tag{6.5}$$

Activity 6.1

Consider the Cartesian coordinates of points $(2,2)$ m and $(-3,4)$ m in the xy plane.

(i) Plot the Cartesian coordinates of the points in the xy plane.

(ii) Find the polar coordinates of these points.

Spherical polar coordinate system

The spherical polar coordinate system is one of the three-dimensional coordinate system relatives of the two-dimensional plane polar coordinate system. Although this coordinate system is not the focus of this book, basics are given for the completion of the content of this section and future references.

Spherical polar coordinates define the position of a point by three coordinates r, θ and ϕ as illustrated in figure 6.3.

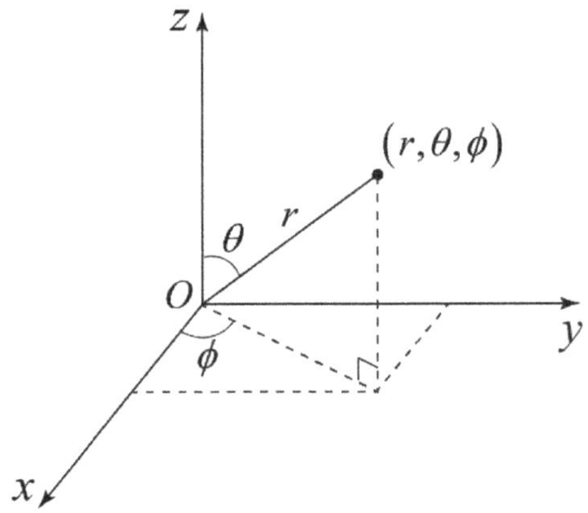

Figure 6.3: The spherical coordinates of a point (r,θ,ϕ).

Here, r is the distance of a point from the origin O, the angle θ (z-inclination angle) is called the polar angle coordinate and the angle ϕ (the angle in the xy-plane) is called the azimuthal angle coordinate.

The following is the relation between the Cartesian and spherical coordinates.

$$r^2 = x^2 + y^2 + z^2\ ,\ \tan\theta = \frac{\sqrt{x^2+y^2}}{z} \quad \text{and} \quad \tan\phi = \frac{y}{x}$$

$$x = r\sin\theta\cos\phi\ ,\quad y = r\sin\theta\sin\phi \quad \text{and} \quad z = r\cos\theta$$

Cylindrical coordinate system

The cylindrical coordinate system is another three-dimensional coordinate system relatives of the two-dimensional plane polar coordinate system. Although this coordinate system is not the focus of this book, basics are given for the completion of the content of this section and future references.

Cylindrical coordinates define the position of a point by three coordinates r, ϕ and z as illustrated in figure 6.4. In this coordinate system the axis of the circular cylinder taken as the z-axis, the perpendicular distance from the cylinder axis is designated by r and the azimuthal angle taken to be ϕ.

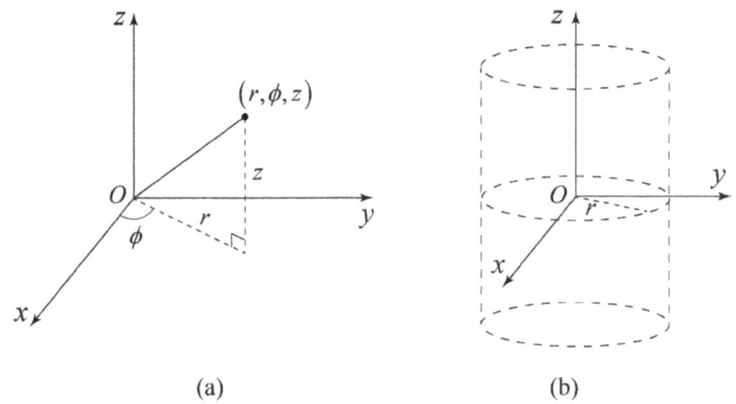

Figure 6.4: (a) The cylindrical coordinates of a point (r,ϕ,z) (b) shows the axis of the circular cylinder taken as the z-axis and the perpendicular distance r from the cylinder axis.

The following is the relation between the Cartesian and cylindrical coordinates.

$$r^2 = x^2 + y^2, \quad \tan\phi = \frac{y}{x} \quad \text{and} \quad z = z$$
$$x = r\cos\phi \quad \text{and} \quad y = r\sin\phi$$

6.3 Scalars and Vectors

Some physical quantities, such as time, mass, length, volume, density, temperature, energy, distance, speed, etc. can be completely described by a single real number associated with a suitable unit without any mention of direction in space. The number describing the physical quantity is known as its magnitude. Because these quantities are describable by giving only a magnitude, they are called scalars. The scalars can be added, subtracted, multiplied, and divided by the usual mathematical operations.

On the other hand, physical quantities, such as displacement, velocity, force, acceleration, torque, momentum, electric field, etc. require both a magnitude and a direction to describe them completely. Such quantities are called vectors, which are extremely useful in physics.

6.4 Notation of Vectors

Lines with arrowheads are used to describe a vector (figure 6.5).

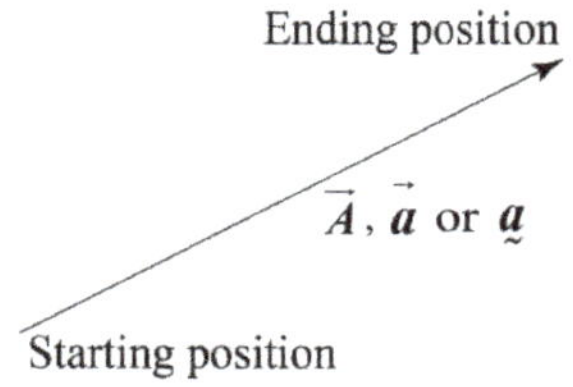

Figure 6.5: Representation of a vector quantity

The tail of the arrow is called the starting position (initial point) of the vector, while the tip of the arrow is called the ending position (or terminal point) of the vector. The length of the line gives the magnitude or size of the vector, and the direction in which the arrow points gives the direction of the vector. We usually use a boldface letter with an arrow over the letter, such as $\vec{\boldsymbol{A}}$, $\vec{\boldsymbol{a}}$, or a boldface letter with the symbol tilde (~) below the letter, such as $\underset{\sim}{\boldsymbol{a}}$ (figure 6.5). In this book, the vector symbols are denoted in boldface to remind us that vector quantities have properties different from those of scalar quantities.

The magnitude of the vector $\vec{\boldsymbol{A}}$ is written as either A or $|\vec{\boldsymbol{A}}|$. The magnitude of a vector quantity is a scalar quantity and is always positive. The magnitude of a vector has physical units, such as metres for displacement or metres per second for velocity.

For example, consider the simplest of all vector quantities, called displacement. When an object changes position, it is said to have undergone a displacement.

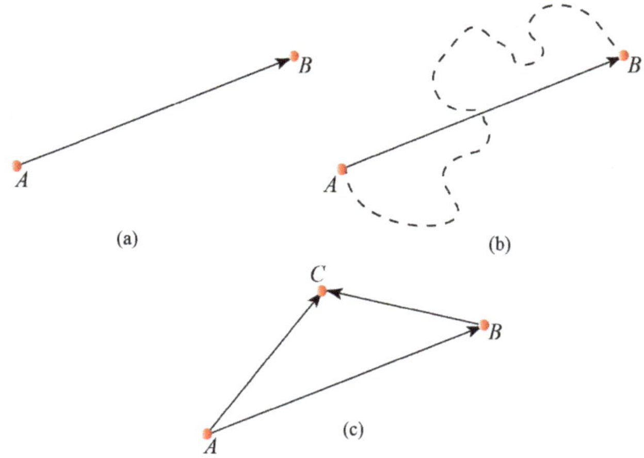

Figure 6.6: Object changes its position and undergoes a displacement (a) displacement is represented by a straight line (b) dotted line shows the actual path of the particle from A to B (c) when particle undergoes a subsequent displacement from B to C, the resultant displacement is $A \rightarrow C$

If it moves from point A to point B, its displacement is represented by a straight line drawn from A to B, as shown in figure 6.6 (a). Here, the arrowhead in B gives the direction of displacement from A to B. This straight line does not necessarily represent the actual motion of the particle from A to B and only shows the net effect of the motion. The dotted line shown in figure 6.6 (b) is the actual path of the particle from A to B. Thus, the actual path is not the same as the displacement and the displacement could be written as $\overrightarrow{AB}$.

Suppose that the object undergoes a subsequent displacement from B to C as shown in figure 6.6 (c), the effect of the displacements $A \rightarrow B$, and $B \rightarrow C$ would be equivalent to a displacement $A \rightarrow C$. Thus AC is known as the sum or resultant of the displacements AB and BC. When representing displacements in the above way, the lengths of the straight lines should be proportional to the magnitudes of the displacements, and their directions would indicate the directions of the displacements. Physical quantities like displacement cannot be described by magnitude alone. To describe them, we must specify both a magnitude and a direction. Thus, in this case, the resultant displacement could be written as $\overrightarrow{AC}$.

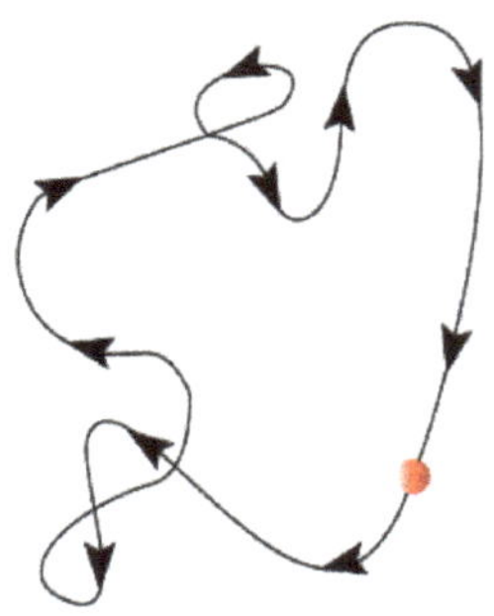

Figure 6.7: The total displacement is zero when the object makes a round-trip

If the object makes a round trip, the total displacement is zero, regardless of the distance travelled (figure 6.7).

6.5 Properties of Vectors

In this section, we shall learn about the general properties of vectors that represent physical quantities. We also investigate adding and subtracting vectors using algebraic and geometric methods.

Equality of two Vectors

Two vectors $\vec{A}$ and $\vec{B}$ are said to be equal if they have the same magnitude and if they point in the same direction (figure 6.8). Thus $\vec{A}=\vec{B}$ only if the magnitude of vectors $\vec{A}$ and $\vec{B}$ are equal (i.e. $|\vec{A}|=|\vec{B}|$), and both vectors point in the same direction along a parallel line.

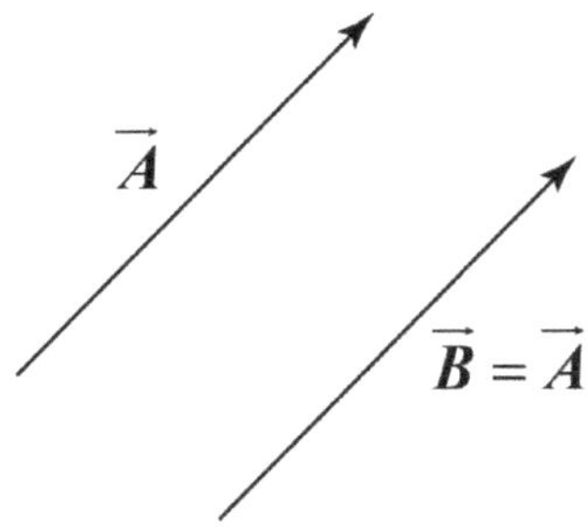

Figure 6.8: Vectors $\vec{A}$ and $\vec{B}$ are equal if they have the same magnitude and if they point in the same direction

As seen in figure 6.8, these two vectors $\vec{A}$ and $\vec{B}$ are equal even though they have different starting points. This property allows us to move a vector to a position parallel to itself in a diagram without affecting the property of the vector.

Negative of a Vector

The negative of the vector $\vec{A}$ is a vector which has the same magnitude but points in the opposite direction, as shown in figure 6.9. Thus, $\vec{B}=-\vec{A}$.

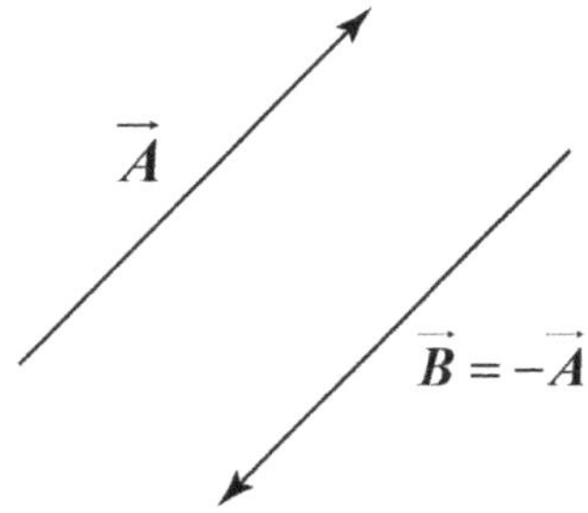

Figure 6.9: Negative vector is a vector which has the same magnitude but points in opposite directions

Adding Vectors

Since vectors involve a direction and a magnitude, they cannot be simply added by the algebraic rules. Conveniently, the geometric method is used to add vectors. If we are to add a vector $\vec{B}$ to a vector $\vec{A}$, first we need to draw a vector $\vec{A}$ on a paper, with its magnitude represented by a convenient length scale, and then draw a vector $\vec{B}$ to the same scale, with its tail starting from the tip of $\vec{A}$ as shown in figure 6.10. This is called the vector trianglar method for adding vectors.

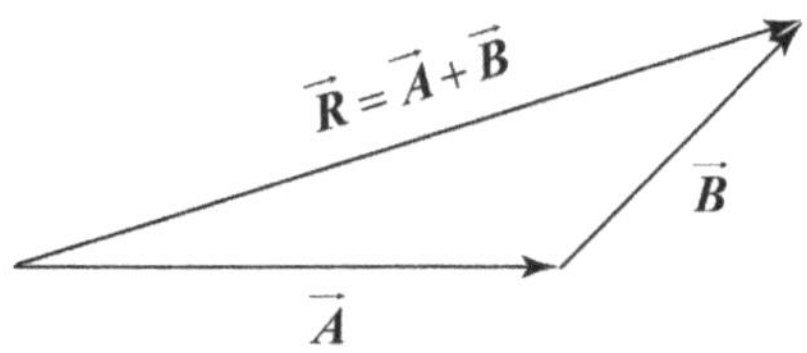

Figure 6.10: The resultant vector $\vec{R}\left(=\vec{A}+\vec{B}\right)$ is the vector sum of vector $\vec{A}$ and vector $\vec{B}$, which runs from the tail of $\vec{A}$ to the tip of $\vec{B}$.

The resultant vector $\vec{R}\left(=\vec{A}+\vec{B}\right)$ is the vector drawn from the tail of the vector $\vec{A}$ to the tip of the vector $\vec{B}$. Further, if two vectors are added, the sum is independent of the order of the addition. This means $\vec{A}+\vec{B}=\vec{B}+\vec{A}$ and is known as the commutative law of addition.

If we have more than two vectors, the same geometric construction can be used to add those vectors. Suppose if we have five vectors $\vec{A},\vec{B},\vec{C},\vec{D}$, and $\vec{E}$ the resultant vector $\vec{R}$ of these five vectors is the vector sum $\vec{A}+\vec{B}+\vec{C}+\vec{D}+\vec{E}$ that completes the polygon, as shown in figure 6.11.

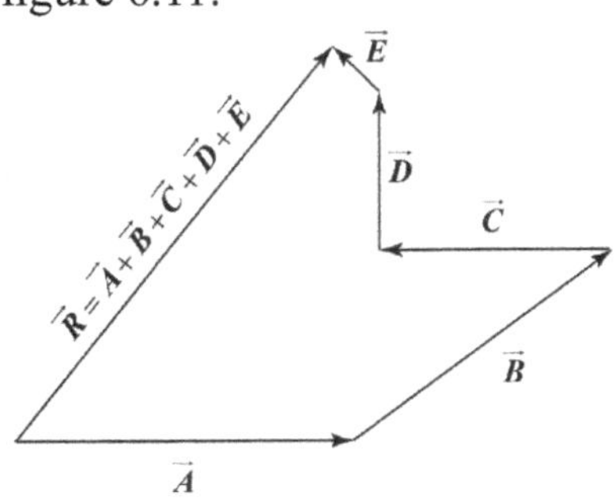

Figure 6.11: Geometric construction for adding more than two vectors: vector addition of five vectors, $\vec{A},\vec{B},\vec{C},\vec{D}$ and $\vec{E}$.

There is another way of adding two vectors. Instead of making the second vector start where the first one finishes, we make them both start at the same place and complete a parallelogram. As shown in figure 6.12, if we draw vectors $\vec{A}$ and $\vec{B}$ with their tails at the same point, the vector $\vec{C}$ is the diagonal of a parallelogram constructed with vectors $\vec{A}$ and $\vec{B}$ as two adjacent sides. This method is accomplished using the property of equality of two vectors and the properties of a parallelogram in which opposite sides are equal and in the same direction.

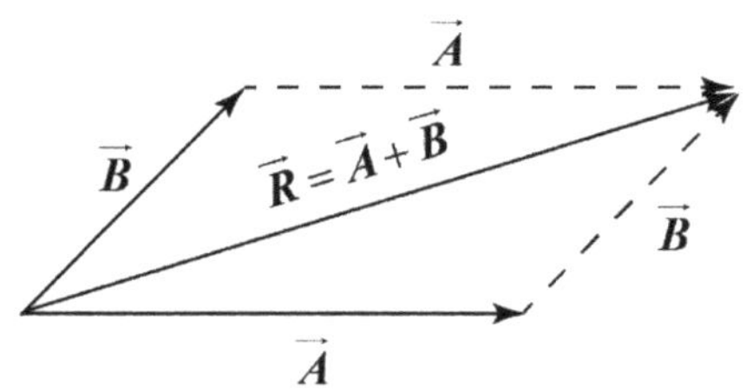

Figure 6.12: Adding two vectors by placing them tail to tail and constructing a parallelogram

This is called the parallelogram law for adding vectors. It gives the same result as the vector trianglar method discussed earlier.

Activity 6.2

A passenger vehicle travels 10 km north and then 25 km east.

(i) How far and in what direction are you from your starting point?

(ii) What are the magnitude and direction of your resultant displacement?.

Activity 6.3

If 3 N and 4 N forces act on a body shown below, determine its resultant force acting on the body.

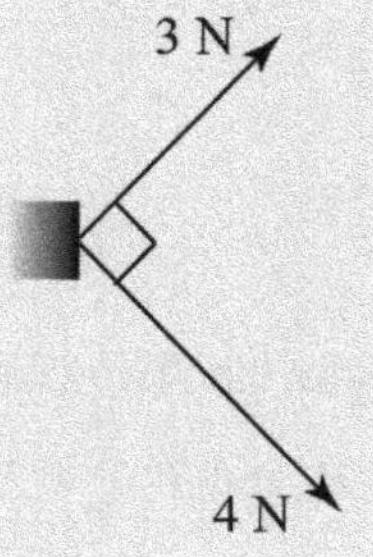

Subtracting Vectors

Using the definitions of the operation of vector addition and the negative of a vector, we could define the operation of vector subtraction. Accordingly, we define the operation $\vec{A}-\vec{B}$ as vector $-\vec{B}$ added to vector $\vec{A}$. That is $\vec{A}+\left(-\vec{B}\right)=\vec{A}-\vec{B}$. Figure 6.13 shows the geometric construction for subtracting vector $\vec{B}$ from vector $\vec{A}$.

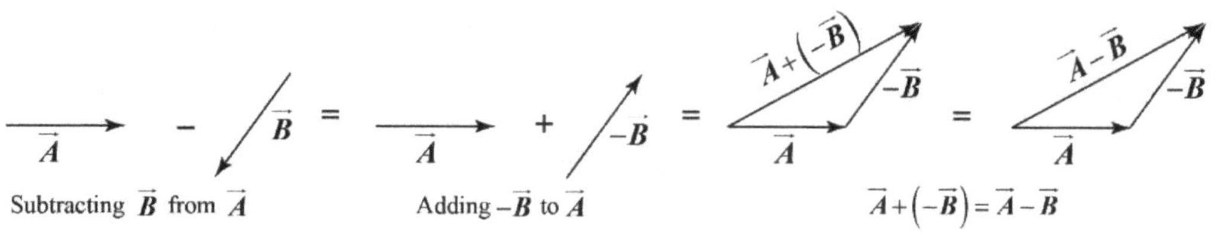

Figure 6.13: Geometric construction for subtracting vector $\vec{B}$ from vector $\vec{A}$:

$$\vec{A}+\left(-\vec{B}\right)=\vec{A}-\vec{B}.$$

Multiplying a Vector by a Scalar quantity

Suppose vector $\vec{A}$ is multiplied by a positive scalar quantity (say n), the product $n\vec{A}$ is a vector that has the same direction as vector $\vec{A}$ with the magnitude of $\left|n\vec{A}\right|=nA$. On the other hand, if vector $\vec{A}$ is multiplied by a negative scalar quantity (say $-n$), the product $-n\vec{A}$ is directed opposite to the vector $\vec{A}$ with the magnitude of $\left|-n\vec{A}\right|=nA$.

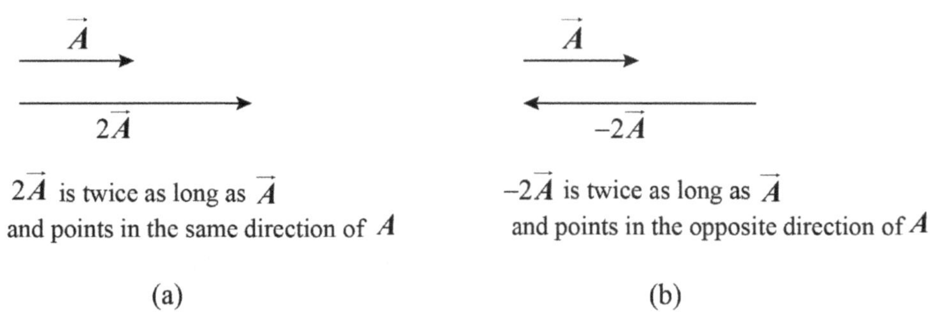

Figure 6.14: (a) Multiplying a vector by a positive scalar quantity (say 2) changes the magnitude (twice its length) of the vector and points in the same direction of the vector (b) Multiplying a vector by a negative scalar (say - 2) changes its (twice its length) magnitude and points in the opposite direction of the vector

Figure 6.14 (a) shows the geometric construction for multiplying the vecor $\vec{A}$ by a positive scalar quantity 2 and figure 6.14 (b) shows the geometric construction for multiplying the vector $\vec{A}$ by a negative scalar quantity -2.

6.6 Components of Vectors

Up to now, we have seen that when adding vectors, the vector triangular method (graphical method) is used to find the resultant vector. This method can be tedious and is not recommended when high accuracy is required or in three-dimensional problems.

So we need a simple but more general method for adding vectors. The usual method that we use in physics is to resolve the vector into two perpendicular components, called components of vectors or rectangular components. These components of the vectors are the projections of the vector along the coordinate axes. Any vector can be completely described by its components.

To define the components of vectors, let's start with a Cartesian coordinate system. Consider a vector $\vec{A}$ lying in the xy plane and making an arbitrary angle θ with the positive x axis as shown in figure 6.15.

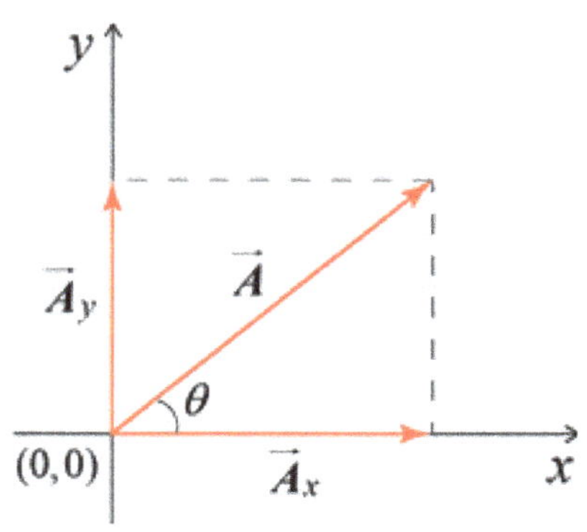

Figure 6.15: A vector $\vec{A}$ lying in the xy plane can be represented by its component vectors $\vec{A}_x$ and $\vec{A}_y$.

To find the projection of the vector $\vec{A}$ along an axis, draw perpendicular lines from the two ends of the vector to the axis, as shown in figure 6.15. The vector $\vec{A}_x$ is the component of the vector along the x axis, and the vector $\vec{A}_y$ is the component along the y axis. The vector $\vec{A}$ can be expressed as the sum of two component vectors $\vec{A}_x$ which is parallel to the x axis and $\vec{A}$ which is parallel to the y axis (figure 6.16). That is $\vec{A} = \vec{A}_x + \vec{A}_y$, these three vectors form a right angled triangle.

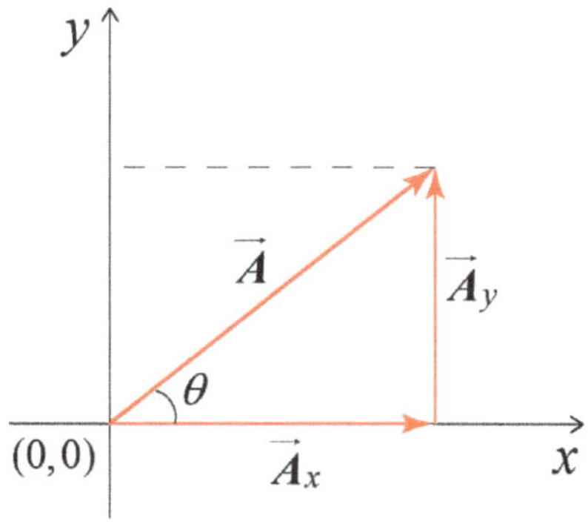

Figure 6.16: The y component vector $\vec{A}_y$ can be moved to the right and added to $\vec{A}_x$: $\vec{A} = \vec{A}_x + \vec{A}_y$.

From figure 6.16 and using the trigonometry, we have $\cos\theta = \frac{A_x}{A}$ and $\sin\theta = \frac{A_y}{A}$ where $|\vec{A}_x| = A_x$, $|\vec{A}_y| = A_y$ and $|\vec{A}| = A$.

Hence, magnitudes of the components of vector $\vec{A}$ are

$$A_x = A\cos\theta \tag{6.6}$$

$$A_y = A\sin\theta \tag{6.7}$$

The magnitudes of these components (A_x and A_y) are the lengths of the two sides of a right triangle with a hypotenuse of length A. Therefore, the magnitude and direction of vector $\vec{A}$ are related to its components through the expressions

$$A = \sqrt{A_x^2 + A_Y^2} \tag{6.8}$$

$$\theta = \tan^{-1}\left(\frac{A_y}{A_x}\right) \tag{6.9}$$

From the equation 6.9 it is evident that the signs of the components A_x and A_y depend on the angle θ. Figure 6.17 shows the signs of the components when vector $\vec{A}$ lies in the various quadrants.

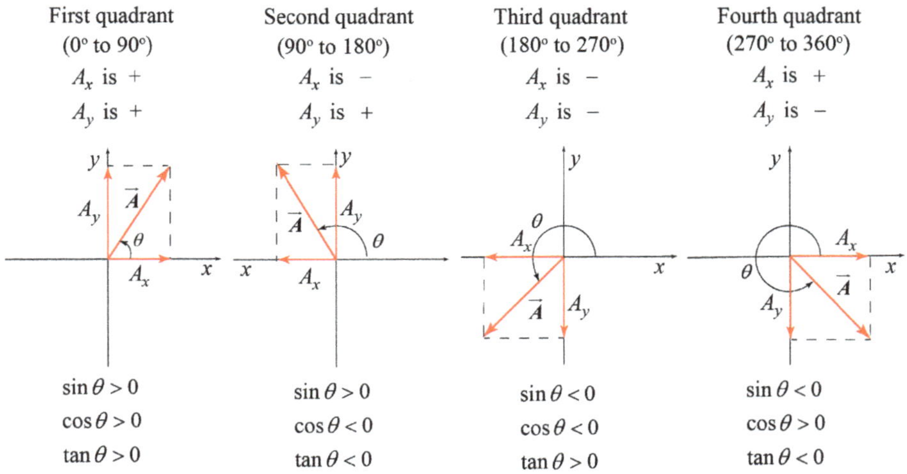

Figure 6.17: The signs of the components of a vector $\vec{}$ in different quadrants

In the first quadrant, A_x is positive and A_y is positive; in the second quadrant, A_x is negative and A_y is positive; in the third quadrant, A_x is negative and A_y is negative and in the fourth quadrant, A_x is positive and A_y is negative.

When solving problems in physics, most of the time, one may require resolving a vector into its components on appropriate perpendicular axes (not necessarily that the axes should be horizontal and vertical, but the axes should be perpendicular to each other). For example, when considering the motion of objects sliding down inclined planes, it is often convenient to orient the x axis parallel to the plane and the y axis perpendicular to the plane.

Activity 6.4

A displacement vector $\underset{\sim}{r}$ in the xy plane is 20.0 m long and directed at an angle $\theta = 30°$, as shown in the figure below. Determine the x component and the y component of the displacement vector.

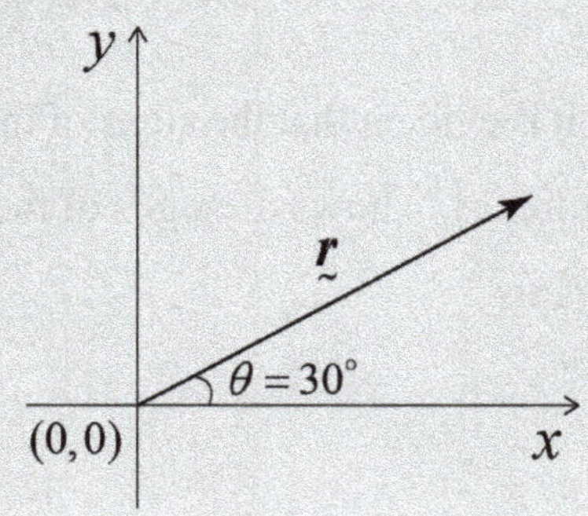

Activity 6.5

If the x and y components of the vector $\vec{P}$ are $P_x = 30.0$ m and $P_y = 50.0$ m respectively, determine

(i) the magnitude of the vector $\vec{P}$, and

(ii) the angle θ between the direction of the vector $\vec{P}$ and the positive direction of the x axis.

Activity 6.6

Obtain expressions for the x and y components for the position vector having the polar coordinates $\left(15.0\ \text{m}, 150^\circ\right)$.

Activity 6.7

A worker pulls the heavy box forward, using a rope attached to it with force 30.5 N at an angle 38° above the horizontal floor.

(i) How much force is given horizontally to pull the box forward?

(ii) How much force is given vertically upward?

Using components of vectors to add two or more vectors

We could use vector components to add two or more vectors. Figure 6.18 shows two vectors $\vec{A}$ and $\vec{B}$, and their resultant vector $\vec{R}$ (vector sum) along with the x and components of all three vectors, A_x, B_x, R_x and A_y, B_y, R_y respectively.

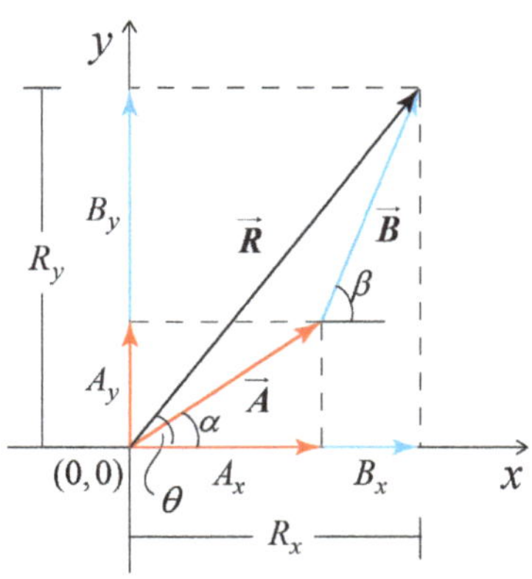

Figure 6.18: Resultant vector $\vec{R}$ is a sum of vectors $\vec{A}$ and $\vec{B}$: the x and y components of all three vectors are A_x, B_x, R_x and A_y, B_y, R_y respectively.

From figure 6.18, we can see that the magnitude of the x and y components of the resultant vector $\vec{R}$ (i.e., R_x and R_y) is the sum of the magnitude of the x and y components of the vector (A_x and A_y) and $\vec{B}$ (B_x and B_y), and is given by

$$R_x = A_x + B_x$$

$$R_y = A_y + B_y \tag{6.12}$$

The magnitudes of the x and y components of vectors $\vec{A}$ and $\vec{B}$ are

$$A_x = A\cos\alpha \quad \text{and} \quad A_y = A\sin\alpha$$

$$B_x = B\cos\beta \quad \text{and} \quad B_y = B\sin\beta \tag{6.10}$$

where, $|\vec{A}| = A$, $|\vec{A}_x| = A_x$, $|\vec{A}_y| = A_y$, $|\vec{B}| = B$, $|\vec{B}_x| = B_x$ and $|\vec{B}_y| = B_y$.

The magnitude and direction of the resultant vector $\vec{R}$ is

$$R = \sqrt{R_x^2 + R_y^2}$$

$$\tan\theta = \frac{R_y}{R_x} \tag{6.11}$$

where, $|\vec{R}| = R$, $|\vec{R}_x| = R_x$ and $|\vec{R}_y| = R_y$.

Activity 6.8

Two vectors $\underset{\sim}{a}$ and $\underset{\sim}{b}$ make an angle θ with each other when placed tail to tail as shown below.

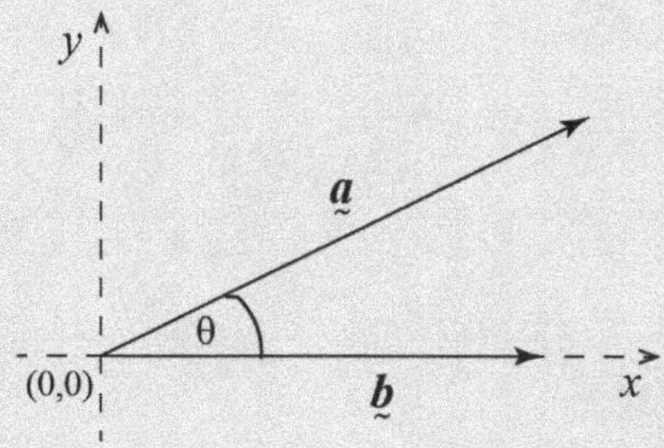

Prove, by taking components along two perpendicular axes that the length of the resultant vector $\underset{\sim}{r}$ is $r=\sqrt{a^2+b^2+2ab\cos\theta}$, where a,b and r are the lengths of the vectors $\underset{\sim}{a}$, $\underset{\sim}{b}$ and $\underset{\sim}{r}$ respectively.

Activity 6.9

A student walks 30° north of east for 4.0 km. How far would the student have to walk due north and due east to arrive at the same location?

Activity 6.10

Consider three vectors $\underset{\sim}{a},\underset{\sim}{b},\underset{\sim}{c}$ and $\underset{\sim}{d}$ shown in figure below, where, $|\underset{\sim}{a}|=20.0$ unit, $|\underset{\sim}{b}|=35.0$ unit, $|\underset{\sim}{c}|$ 15.0 unit and $|\underset{\sim}{d}|=20.0$ unit. 20°

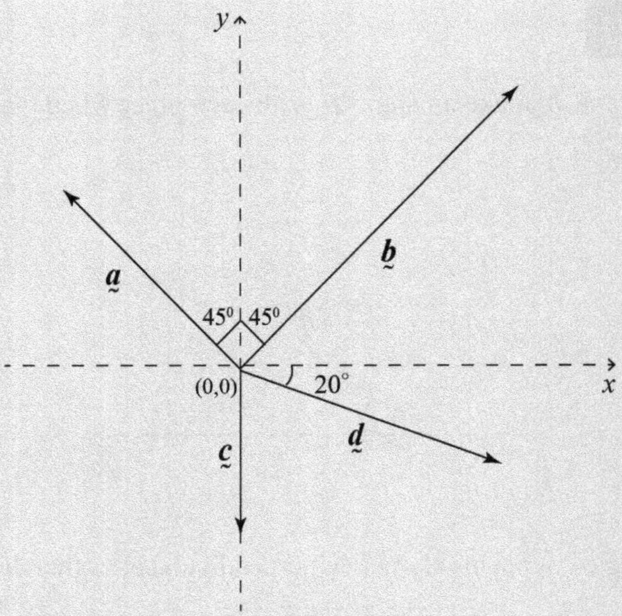

Determine

(a) the resultant vector in x, y direction and

(b) the magnitude and direction of the resultant vector

Questions

NOTE: Although the solutions are given on the last page, it is in your own interest to attempt the questions on your own as an indication of how far you have understood the material of the section.

1. Convert the following plane polar coordinates to Cartesian coordinates.

   (i) $\left(2, \dfrac{\pi}{4}\right)$ (ii) $\left(3, -\dfrac{\pi}{6}\right)$

2. Convert the following Cartesian coordinates to plane polar coordinates.

   (i) $\left(1, -\sqrt{3}\right)$ (ii) $\left(-1, \sqrt{3}\right)$

3. The rectangular coordinates of a point are given by $(3, y)$, and its polar coordinates are $(r, 60^\circ)$. Determine the value of y and the value of r.

4. A force of magnitude 5 N acts on an object at the origin in a direction 30° above the positive horizontal axis, as shown in the figure below. A second force of magnitude 4 N acts on the object in the direction of the positive vertical axis, as shown. Determine the magnitude and direction of the resultant force.

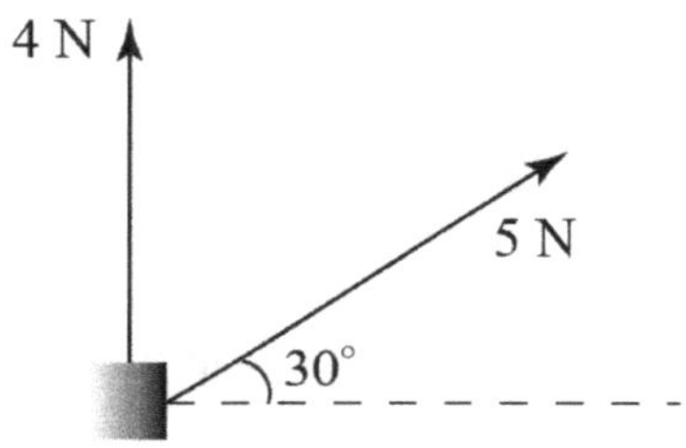

5. Vector $\vec{P}$ has a magnitude of 15 units and points in the positive y axis. When the vector $\vec{Q}$ is added to $\vec{P}$ the resultant vector $\vec{R} = \vec{P} + \vec{Q}$ points in the negative y axis having a magnitude of 5 units, find the magnitude and direction of the vector $\vec{Q}$.

6. For the vectors $\vec{P}$ and $\vec{Q}$ shown in the figure below,

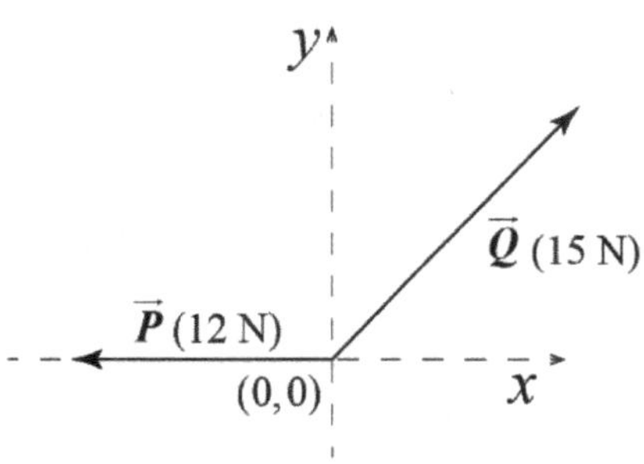

Sketch the vector diagram for

(i) $\vec{P} + \vec{Q}$ (ii) $\vec{P} - \vec{Q}$ (iii) $-\vec{P} - \vec{Q}$ (iv) $\vec{Q} - \vec{P}$

7. An eagle flies at a velocity of 30.0 m s^{-1} at an angle 60.0° up from the horizontal, as shown in the figure below.

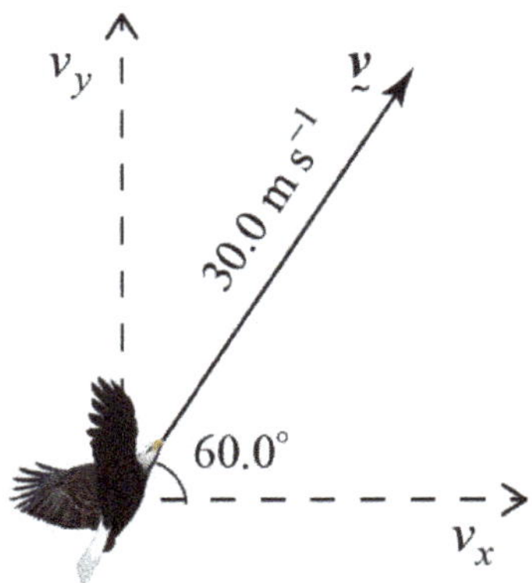

Determine the vertical and horizontal components of the eagle's velocity.

8. A wooden box with weight 100 N is at rest on a plank on a slope inclined at an angle 25° to the horizontal, as shown in the figure below.

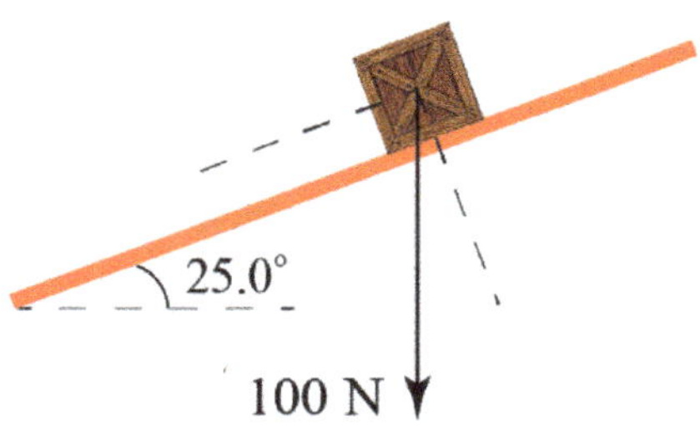

Determine the component of its weight that acts along the slope and perpendicular to the slope.

9. A milk man delivers milk covering his route by travelling 1.0 km west, 2.0 km north, and 1.5 km east.

   (i) What is his resultant displacement?

   (ii) What is the total distance the milk man travels?

10. On the construction side, a crane is used to raise one end of the steel girder off the ground, as shown in the figure below. When the cable attached to the end of the steel girder is at $\theta = 25.5^\circ$ to the vertical, the force of the cable on the steel

girder is $F = 8.0$ kN . Calculate the horizontal and vertical components of the force.

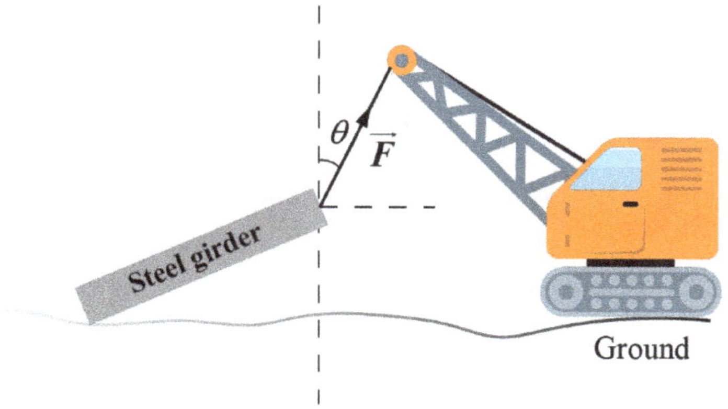

11. A courier service employee drives a delivery truck in the following distances in a horizontal plane. He drives 3.5 km north, 6.0 km east and 4.2 km at an angle 30.0° north of east.
    (i) Draw a suitable vector diagram to show the travel route.
    (ii) Determine the northward and eastward components of the travel.
    (iii) Determine the magnitude and direction of the truck's resultant displacement.

12. A sailor on a small sailboat encounters shifting winds and has to reach his destination, which is 6.0 km directly east of the starting point. From the starting point, the sailor sails 2.0 km east, then 4.5 km southeast, and then an additional distance in an unknown direction, as shown in the figure below. Determine the magnitude and direction of the third segment of the journey.

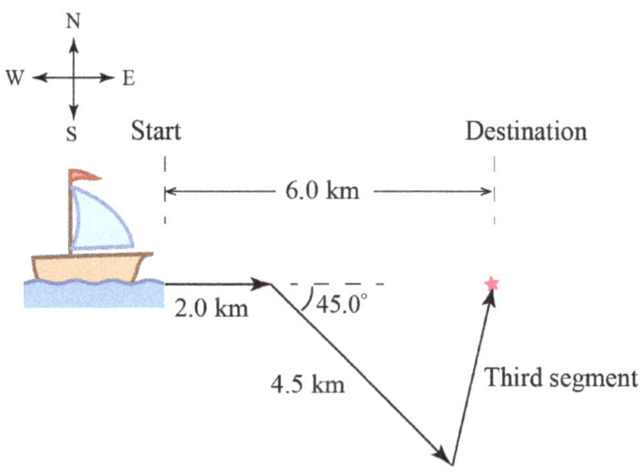

Chapter 7

Motion Along a Straight Line

7.1 Introduction

Studying the motion of objects or systems of groups of objects is an important area in physics called kinematics and is one of the disciplines in mechanics. Kinematics in mechanics can be characterised as the description of motion without considering its causes (i.e. forces). Even though we live in a three-dimensional world, where we observe three-dimensional motions, it is always useful to understand the motion of objects in just one-dimension since many physical problems can be reduced to this simpler case. The one-dimensional motion describes objects moving in straight lines. For example, the motion of a falling object is in the fundamental nature of a one-dimensional problem where the object moves in only one direction (down).

Motion is a continuous change of position with time. In this chapter, we will limit our discussion to the simplest possible situation, a single object moving along a straight line, and we will assume that the object can be treated as a particle. Further, the straight line where the object moves can be one of the axes of a coordinate system and could describe the position of the object by its displacement from the origin of coordinates along this line.

7.2 Displacement, Velocity and Acceleration

In this section, we develop mathematical methods to describe motion. Motion is a continuous change of position with time, and to describe motion explicitly, we need a way to say where an object is located. To describe the location of an object, we use a point of reference called the origin and give its distance from the origin and direction. For motion along a straight line, we can choose the line of motion to be the x axis of a coordinate system. The origin can be the point at $x = 0$, and the position of an object can be described by its x coordinate, which tells us both how far the object is from the origin and on which side (gives the direction).

Displacement

Displacement of an object is defined as the change of position (the final position minus the initial position). The displacement along the x axis is written as Δx, where the symbol Δ (the upper case Greek letter delta) means the change in the quantity that follows.

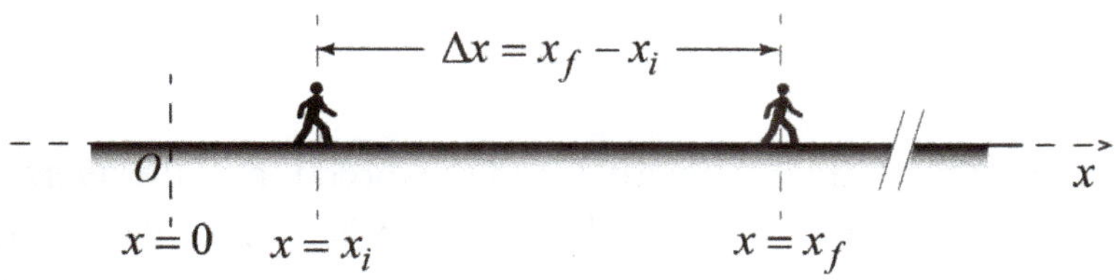

Figure 7.1: Initial position x_i and final position x_f of a walking person

When the object moves from an initial position x_i to a final position x_f as shown in figure 7.1, its displacement is given by

$$\Delta x = x_f - x_i \tag{7.1}$$

Remember that displacement Δx is a vector that has magnitude and direction.

Consider the motion of a passenger car moving on a straight horizontal road, as shown in figure 7.2. Place the origin O at the starting point and the horizontal line along the road as the x axis of the coordinate system. The distance moved has to be measured when the front of the car passes the starting line (lamp post) of the origin and is given by the coordinate x, which varies with time t.

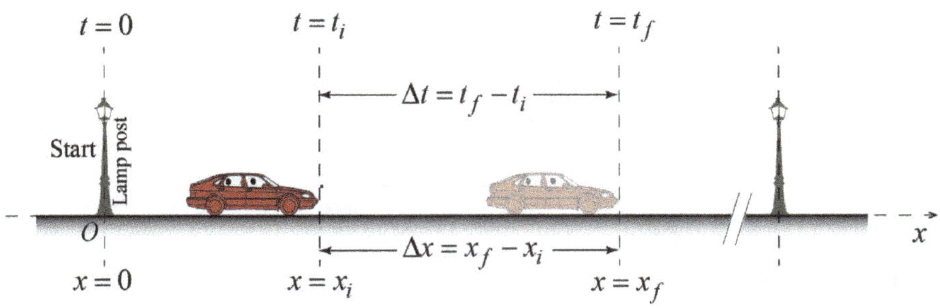

Figure 7.2: Positions of a passenger car moving on a straight horizontal road

At time $t = t_i$ a point at the front of the car is at position $x = x_i$ and at time $t = t_f$ it is at position $x = x_f$. In figure 7.2, the car has moved a displacement $\Delta x = x_f - x_i$ along the positive x direction, during the time interval $\Delta t = t_f - t_i$.

Average velocity

The average velocity ($v_{avg,x}$) of the car (the object) along the x axis is defined as the car's displacement $\Delta x = x_f - x_i$ divided by the time interval $\Delta t = t_f - t_i$ during which that displacement occurs.

$$v_{avg,x} = \frac{x_f - x_i}{t_f - t_i} = \frac{\Delta x}{\Delta t} \tag{7.2}$$

The time interval Δt is always positive, and the direction of the average velocity $v_{avg,x}$ is the same as the direction of the displacement. Therefore, the average velocity $v_{avg,x}$ is a vector quantity. If $x_f > x_i$, Δx is positive and $v_{avg,x}$ is positive and the motion is in the positive x direction. If $x_f < x_i$, Δx is negative and $v_{avg,x}$ is negative and the motion is in the negative x direction. The SI unit of average velocity $v_{avg,x}$ is $\mathrm{m\,s^{-1}}$.

It is often useful to represent the relationship of x and t by means of a graph. Let us consider the figure 7.3 which shows the position of a moving object on a straight line as a function of time.

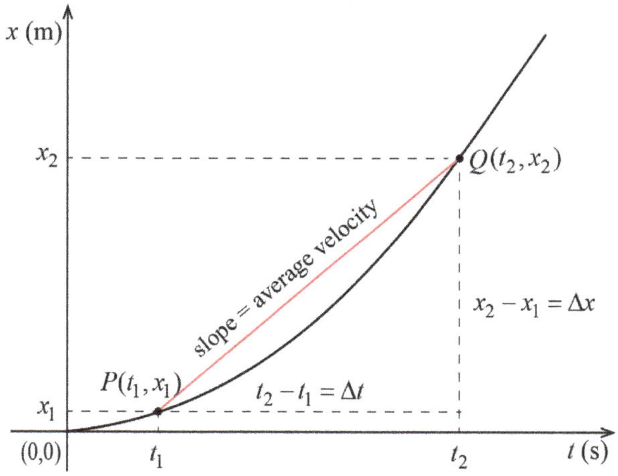

Figure 7.3: Graph of the position of the moving object as a function of time

The x axis represents time t in seconds, and the y axis represents the position of the object x in metre. Note that the curve in the figure does not represent the actual path of the moving object. The points P and Q in figure 7.3 represent the corresponding positions and time coordinates (x_1, t_1) and (x_2, t_2) respectively. The slope of the line joining P and Q (vertical side of the triangle Δx divided by the horizontal side of the triangle Δt) is the average velocity $v_{avg,x}\left(=\frac{\Delta x}{\Delta t}\right)$.

The term speed refers to how far an object moves in a given time interval regardless of direction. Often we use the terms speed and velocity interchangeably and incorrectly, but there is a very important distinction between them. Speed is a scalar quantity and simply a positive number with a unit. On the other hand, velocity is a vector quantity with a unit and has a magnitude and direction.

The average speed (v_{avg}) of an object is defined as the total distance (say d) travelled along its path divided by the total time interval (say Δt) required to travel that distance.

$$v_{avg} = \frac{d}{\Delta t} \tag{7.3}$$

The SI unit of average speed is the same as the unit of average velocity $\mathrm{m\ s^{-1}}$. Average speed is a scalar quantity and is always expressed as a positive number with a unit.

Instantaneous velocity

When an object speeds up or slows down, we always want to know its velocity at any specific instant of time or at one specific point in the path. Such a velocity is called the object's instantaneous velocity to that particular instant or at that specific point in the path. More precisely, the instantaneous velocity at any moment is defined as the average velocity during an infinitesimal short interval. Thus, this could be described as the limit of the average velocity over an infinitesimal time interval as the time interval approaches zero.

This means, mathematically, the instantaneous velocity could be defined as the limit of $\frac{\Delta x}{\Delta t}$ as Δt approaches zero. Thus the x component of instantaneous velocity could be written as

$$v_x = \lim_{\Delta t \to 0} \frac{\Delta x}{\Delta t} \tag{7.4}$$

In the language of calculus, it is represented as the derivative of the position function with respect to time, i.e. $v_x = \frac{dx}{dt}$.

To visualise the concept of instantaneous velocity, let us consider the graph of position *versus* time shown in figure 7.4, which shows the position of a moving object on a straight line as a function of time. The x axis represents time t in seconds, and the y axis represents the position of the object x in metres. Note that the curve in the figure does not represent the path of the moving object.

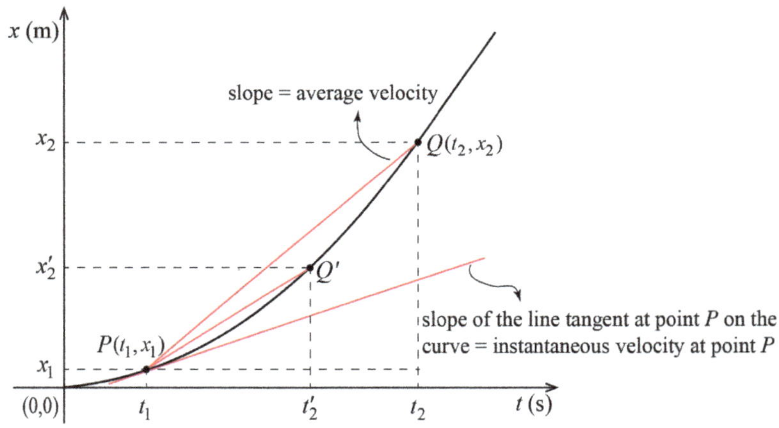

Figure 7.4: The slope of the line tangent at a point on the curve is equal to the instantaneous velocity at that point

The points P and Q in figure 7.4 represent the corresponding positions and time coordinates (x_1, t_1) and (x_2, t_2) respectively. The line between points P and Q becomes shallower and shallower as point Q becomes extremely close to point P (line joining P and Q' is shallower than line joining P and Q), and eventually, the line becomes a tangent line to the curve at point P (on the other hand, the line between points P and Q becomes steeper and steeper, as points P become extremely close to point Q, and eventually the line becomes a tangent line to the curve at point Q – not shown in figure 7.4). Thus, the instantaneous velocity at any point is equal to the slope of the tangent line to the curve at that point.

In particular, the instantaneous velocity is a vector quantity, which means that it has both magnitude and direction. The instantaneous speed of a moving object is defined as the magnitude of its instantaneous velocity.

Activity 7.1

An object moves according to the equation $x = 10t^2$, where x is measured in metres and t is measured in seconds. Determine the average velocity for the time interval from $2.0\ \text{s}$ to $5.0\ \text{s}$.

Activity 7.2

A runner runs on a flat horizontal road in the eastward direction. The runner's position as a function of time is plotted along the x axis of the coordinate system. As shown in the figure below, the runner starts to run eastwards from the position $x = 100\ \text{m}$, and at $x = 400\ \text{m}$ and then the runner changes his direction and completes his running westwards to a position $x = 300$ m.

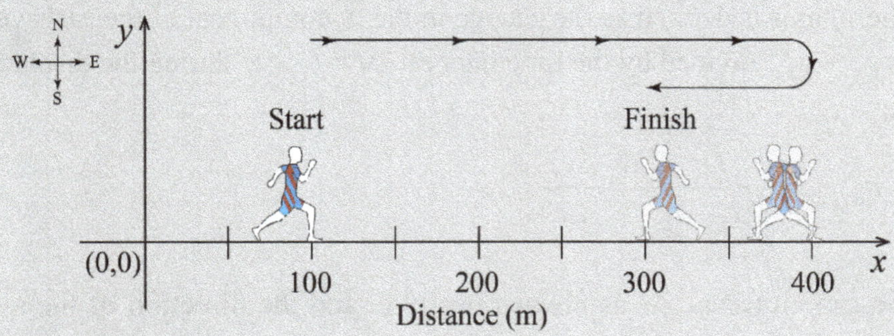

If the runner has taken $320\ \text{s}$ to complete his running, determine

(i) the total distance covered by the runner.

(ii) the total displacement of the runner.

(iii) the runner's average velocity.

(iv) the runner's average speed.

Acceleration

When the velocity of a moving object changes with time, the object is said to undergo acceleration. To understand the concept of acceleration, consider the motion of a passenger car (an object) moving on a straight horizontal road, as shown in figure 7.5. Place the origin O at the starting point and the horizontal line along the road as the x axis of the coordinate system.

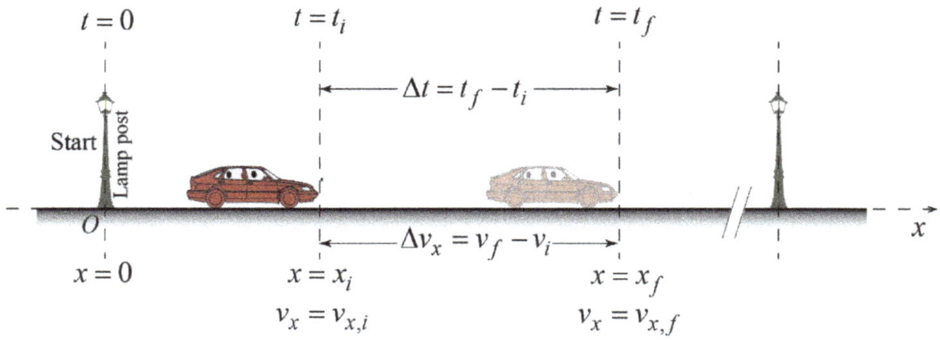

Figure 7.5: Change of velocity of a passenger car moving on a straight horizontal road

As shown in figure 7.5, let at $t=t_i$, the car is at the position $x=x_i$, its velocity along the x axis $v_x = v_{x,i}$, and at $t=t_f$, the car is at the position $x=x_f$, its velocity along the x axis $v_x = v_{x,f}$. The average acceleration ($a_{avg,x}$) of the car (the object) along the x axis is defined as the change in the x component of the car's velocity $\Delta v_x = v_{x,f} - v_{x,i}$ divided by the time interval $\Delta t = t_f - t_i$ during the displacement $\Delta x = x_f - x_i$.

$$a_{avg,x} = \frac{v_{x,f} - x_{x,i}}{t_f - t_i} = \frac{\Delta v_x}{\Delta t} \tag{7.5}$$

The time interval Δt is always positive, and the direction of the average acceleration $a_{avg,x}$ is the same as the direction of the change in velocity Δv_x. Therefore, the average acceleration $a_{avg,x}$ is a vector quantity. If the velocity is increasing, i.e. $v_{x,f} > v_{x,i}$, the acceleration is positive, and if the velocity is decreasing, i.e. $v_{x,f} < v_{x,i}$, the acceleration is negative. When an object is slowing down, we sometimes refer to it as decelerating. Deceleration does not mean that the acceleration is necessarily negative.

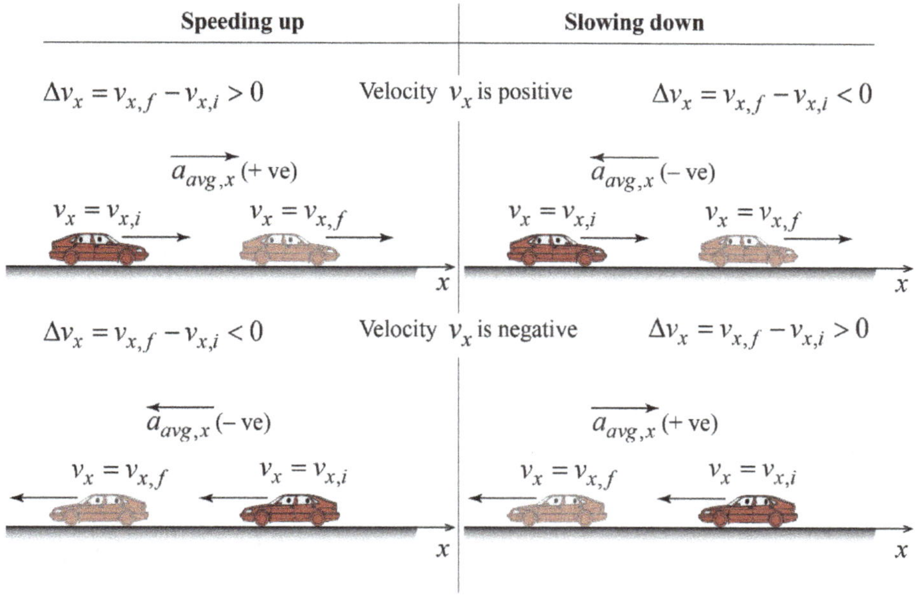

Figure 7.6: The relationship between the signs of acceleration

Figure 7.6 shows the possible instances where acceleration describes how an object's motion changes with time. The velocity and acceleration of an object do not necessarily have the same direction or sign. For example, an object that is moving in a positive direction but slowing down has positive velocity and negative acceleration.

Activity 7.3

A car is travelling on a flat horizontal road in the eastward direction. If the car accelerates uniformly starting with the initial velocity of 20 m s^{-1} for 5 seconds and reaches its final velocity of 40 m s^{-1}, determine the car's average acceleration during this time period. Take the eastward direction as the +ve x direction.

Instantaneous acceleration

acceleration is a measure of how quickly an object's velocity is changing at a specific point in time. Such an acceleration is called the object's instantaneous acceleration to that particular instant or at that specific point in the path. More precisely, the instantaneous acceleration at any moment is defined as the average acceleration during an infinitesimal short interval. Thus, this could be described as the limit of the average acceleration over an infinitesimal time interval as the time interval approaches zero.

This means, mathematically, the instantaneous acceleration could be defined as the limit of $\frac{\Delta v_x}{\Delta t}$ as Δt approaches zero, where Δv_x is the change of velocity in direction of +ve x axis. Thus the x component of instantaneous acceleration could be written as

$$a_x = \lim_{\Delta t \to 0} \frac{\Delta v_x}{\Delta t} \tag{7.6}$$

In the language of calculus, it is represented as the derivative of the velocity function with respect to time, i.e. $v_x = \frac{dx}{dt}$.

To visualise the concept of instantaneous acceleration, let us consider the graph of velocity *versus* time shown in figure 7.7, which shows the horizontal component of the velocity of a moving object on a horizontal straight line as a function of time. The x axis represents time t in seconds, and the y axis represents the horizontal component of the velocity of the object v_x in metre per second.

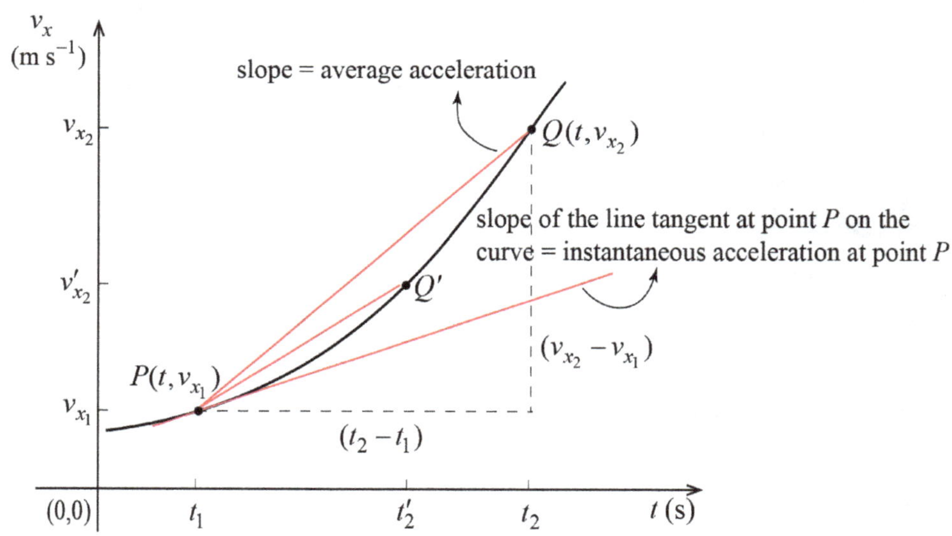

Figure 7.7: The slope of the line tangent at a point on the curve is equal to the instantaneous acceleration at that point

The points P and Q in figure 7.7 represent the corresponding velocity and time coordinates (v_{x_1}, t_1) and (v_{x_2}, t_2) respectively. The line between points P and Q becomes shallower and shallower as point Q becomes extremely close to point P

(line joining P and Q' is shallower than line joining P and Q), and eventually, the line becomes a tangent line to the curve at point P (on the other hand, the line between points P and Q becomes steeper and steeper, as points P become extremely close to point Q, and eventually the line becomes a tangent line to the curve at point Q – not shown in figure 7.7). Thus, the instantaneous acceleration at any point is equal to the slope of the tangent line to the curve at that point. The instantaneous acceleration is a vector quantity, which means that it has both magnitude and direction.

Activity 7.4

Each figure below illustrates the qualitative behaviour of the velocity of the x component of an object moving on a straight line in two different instances. For these, each figure sketches a qualitative graph of the behaviour of the acceleration as a function of time.

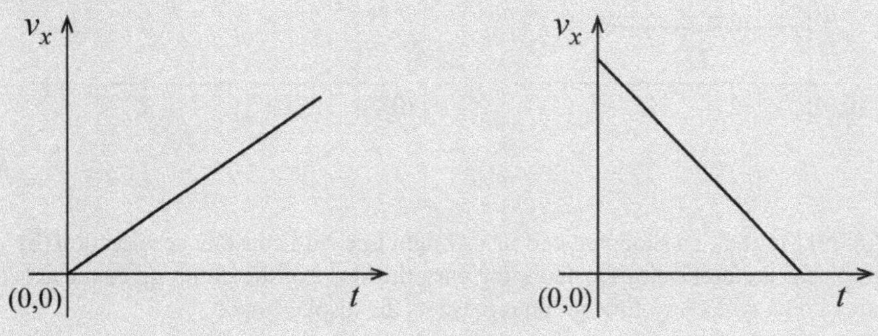

(a) (b)

7.3 Straight line motion under Constant Acceleration

Straight line motion under constant acceleration refers to the movement of an object along a straight line path, where the acceleration of the object remains constant. In this type of motion, the velocity of the object changes at a constant rate, and the acceleration remains constant throughout the motion. As a result, the velocity-time graph is a straight line. The slope of the line represents the acceleration of the object. This type of motion can be described using the equations of motion, which describe the position, velocity, and acceleration of the object at any given time.

Imagine that you are pushing a shopping cart in a straight line and pushing it harder to make it move faster (figure 7.8). Assume that when pushing the cart, its velocity changes at a constant rate. This means that the cart is moving with constant acceleration. Take that the motion of the cart is in the +ve x direction.

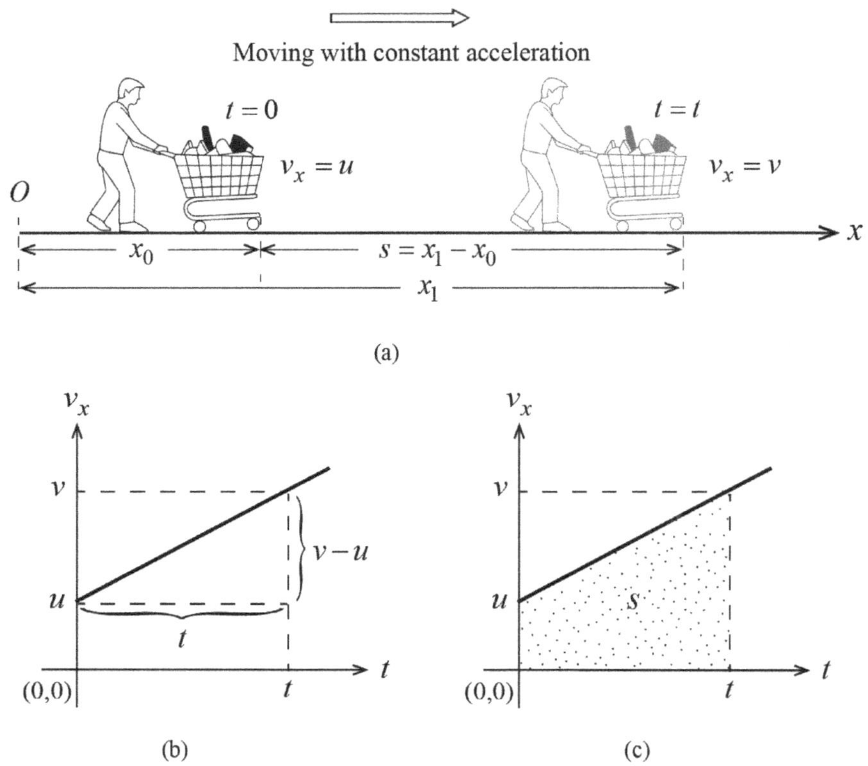

Figure 7.8: (a) Motion of a shopping cart in a straight line with constant acceleration (b) Velocity – time graph for the motion of the shopping cart, the slope of the graph gives the acceleration (c) The area under a velocity-time graph represents the displacement

Suppose at $t = 0$, the cart's position (front segment) is at a distance x_0 from the origin O, its velocity $v_x = u$ and at time $t = t$ its position is at a distance x_1 from the origin O, its velocity $v_x = v$, as illustrated in figure 7.8(a). Thus the displacement of the cart would be

$$s = x_1 - x_0 \tag{7.7}$$

The velocity–time graph for the above motion is illustrated in figure 7.8(b) and (c). The slope of the graph gives the constant acceleration a of the moving shopping cart along the straight line toward the +ve x direction. Thus,

$$a = \frac{v-u}{t-0} = \frac{v-u}{t} \tag{7.8}$$

Therefore,

$$v = u + at \tag{7.9}$$

This equation tells us the velocity of an object at any given time, given its initial velocity, acceleration, and time.

The area under a velocity-time graph represents the displacement of an object (in this case, the shopping cart) over a given time interval as shown in figure 7.8(c). In other words, it means the distance that an object has travelled from its starting point. Thus, displacement (s) = area under the velocity-time graph over a given time interval. Therefore,

$$s = \left(\frac{v+u}{2}\right)t \tag{7.10}$$

Hence, using the equation 7.8, we get

$$s = \left(\frac{u+at+u}{2}\right)t = ut + \frac{1}{2}at^2$$

Thus,

$$s = ut + \frac{1}{2}at^2 \tag{7.11}$$

This equation tells us the distance that an object (in this case, the shopping cart) has travelled from its starting point at any given time, given its initial velocity, acceleration, and time.

By using the equations 7.8 and 7.10, we have

$$s = \left(\frac{v+u}{2}\right)t = \left(\frac{v+u}{2}\right)\left(\frac{v-u}{a}\right) = \frac{v^2-u^2}{2a}$$

Thus,

$$v^2 = u^2 + 2as \tag{7.12}$$

This equation gives us the velocity of the object (in this case, the velocity of the shopping cart) at any position without knowing the time when it is at that position.

In summary, the equation of motion for an object with constant acceleration can be derived from the velocity-time graph. These equations are commonly known as kinematic equations and describe the relationship between displacement, velocity, acceleration, and time. Further, these kinematic equations may be used to solve any problem involving an object under constant acceleration moving in a straight line (in one dimension). They are

$$\boxed{\begin{aligned} v &= u + at \\ s &= ut + \frac{1}{2}at^2 \\ v^2 &= u^2 + 2as \end{aligned}} \tag{7.13}$$

where v is the final velocity, u is the initial velocity, a is the acceleration, s is the displacement and t is the time.

Activity 7.5

A car initially travelling along a straight stretch of highway at $52.2\ \text{km h}^{-1}$ accelerates with a constant acceleration of $2.0\ \text{m s}^{-}$ to overtake a motor vehicle moving in front. Assume that the car's velocity and acceleration are in the same direction.

(i) Determine the velocity of the car after $6.0\ \text{s}$.

(ii) Determine the distance the car has travelled during the time interval $6.0\ \text{s}$.

Activity 7.6

A fishing boat initially at rest moves away from the jetty along the eastward direction with constant acceleration of $2.50\ \text{m s}^{-2}$.

(i) Determine the time taken for the boat to reach a velocity $15.0\ \text{m s}^{-1}$ along the eastward direction.

(ii) Determine the distance travelled when it reaches its velocity $15.0\ \text{m s}^{-1}$ in the eastward direction.

Activity 7.7

A jetliner travelling northward is landing with a velocity of $70.0\ \text{m s}^{-1}$. Once the jet touches down, it has $1\ \text{km}$ of runway in which to reduce its velocity to $7.0\ \text{m s}^{-1}$along the northwad direction. Determine the acceleration of the jetliner during its travel along the runway. Assume that the jetliner is having constant acceleration along the runway.

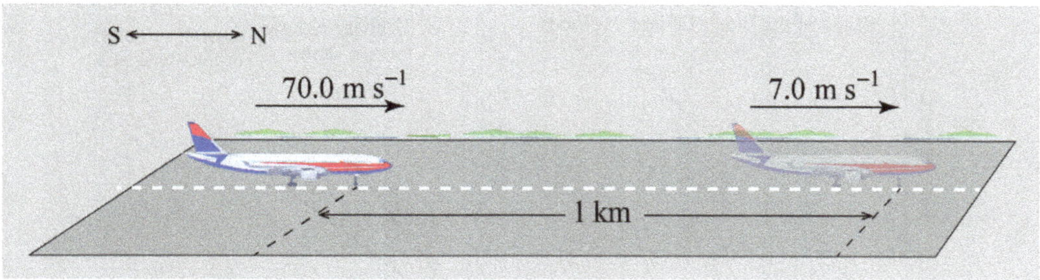

7.4 Graphical representation of Displacement, Velocity, and Acceleration

The graphical representation of displacement, velocity, and acceleration is a powerful tool to visualise and understand the motion of an object. It provides a visual representation of the motion of an object, making it easier to understand the relationships between displacement, velocity, and acceleration and to analyse the motion of an object in a more intuitive way.

Displacement *vs* time graph

The displacement *vs* time graph represents the change in the position of an object relative to a reference point as a function of time. It shows how the displacement of an object changes over time, taking into account the direction of motion. The x axis represents time, and the y axis represents the displacement of the object. The shape of the curve on the graph can be used to determine the type of motion (uniform or nonuniform) of the object.

For uniform motion, the displacement *vs* time graph is a straight line with a constant slope, as shown in figure 7.9 (a). This means that the object is moving at a constant speed in a single direction. The slope of the line represents the velocity of the object; so, as ilustrated in figure 7.9 (b) a steeper slope in graph A means a higher velocity, and a shallower slope in graph B means a lower velocity.

Figure 7.9 (c) shows the displacement *vs* time graph when the object's displacement is not changing. Thus, the slope of the line is zero. Thus, the velocity of the object is zero. Figure 7.9 (d) illustrates the displacement *vs* time graph when the slope suddenly becomes negative. This indicates that the object's velocity changes to negative at $t = t_1$, which means that at $t = t_1$, the object is changing to the opposite direction to the previous direction.

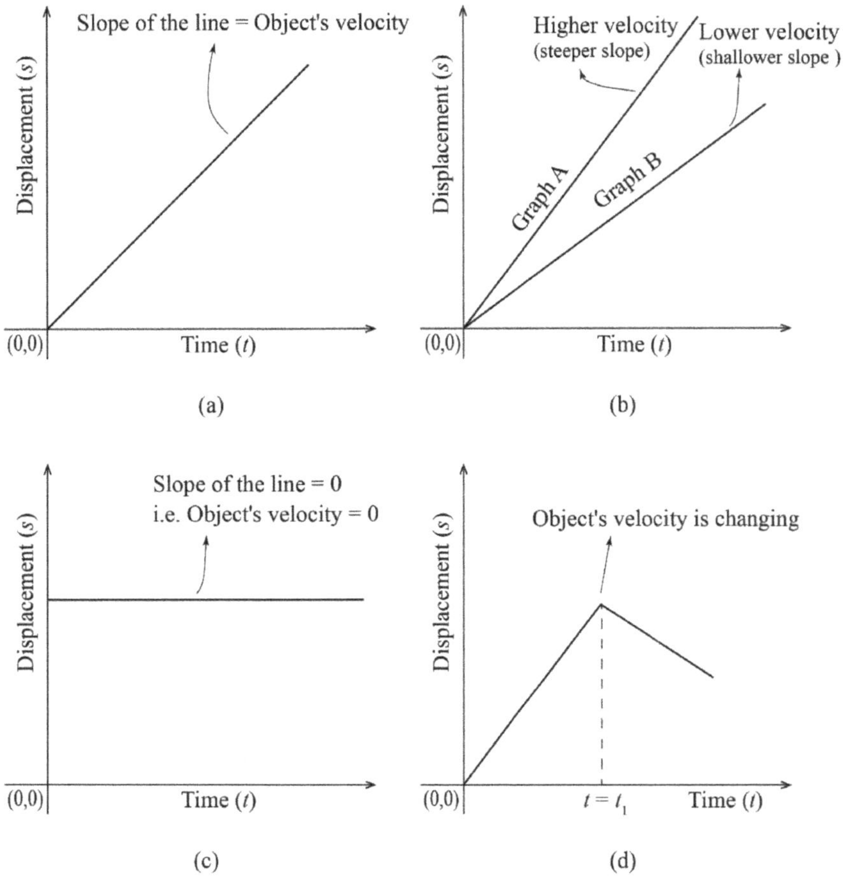

Figure 7.9: Displacement *vs* time graph for the uniform motion of an object (a) the straight line shows that the object's velocity is constant (b) the slope shows which is moving faster° a steeper slope in graph A means a higher velocity, and a shallower slope in graph B means a lower velocity (c) displacement is not changing, the slope of the line is zero, and the velocity is zero (d) the slope of the graph suddenly becomes negative, velocity is negative, and the object is moving in the opposite direction

For nonuniform motion, the displacement *vs* time graph is a curve, as illustrated in figure 7.10. The shape of the curve can be used to determine the type of motion, such as acceleration or deceleration (acceleration is $-\text{ve}$). The slope of the curve on the displacement *vs* time graph represents the velocity of the object, and the rate of change of the slope represents the acceleration.

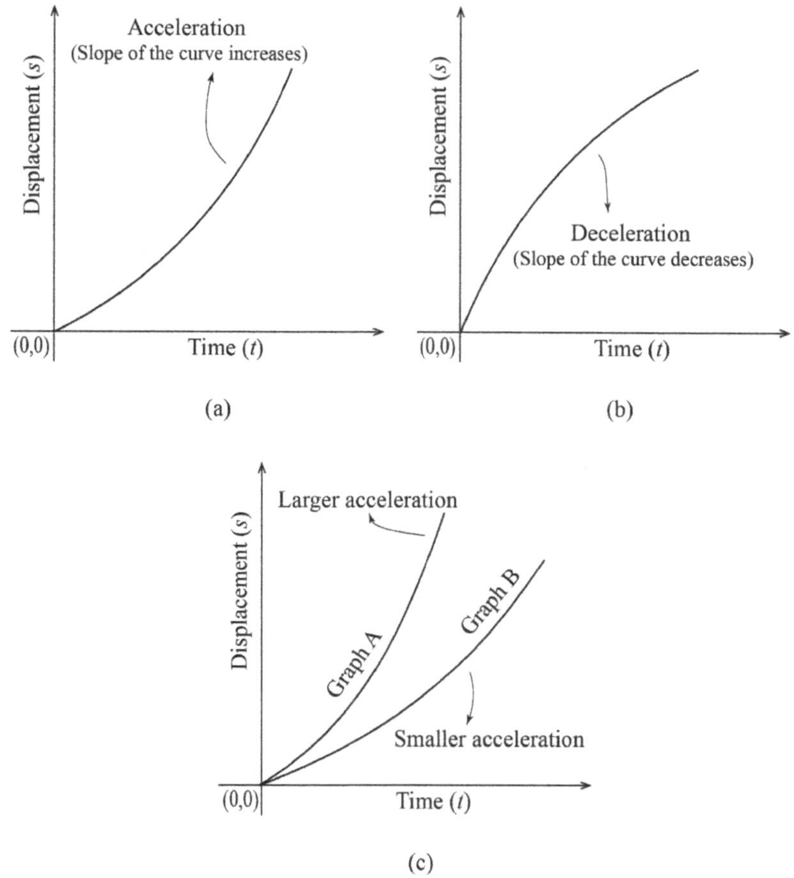

Figure 7.10: Displacement *vs* time graph for the nonuniform motion of an object (a) where the slope of the curve is increasing (b) where the slope of the curve is decreasing (c) the effect of changing the acceleration° graph A has a larger acceleration and graph B has a smaller acceleration

For example, a concave upward curve represents an object that is accelerating, where the slope or gradient of the curve increases, as shown in figure 7.10 (a), while a concave downward curve represents an object that is decelerating, where the slope or gradient of the curve decreases as shown in figure 7.10 (b). In figure 7.10 (c), a steeper slope in graph A represents a greater velocity than in graph B and has a larger acceleration, while a shallower slope in graph B represents a smaller velocity than in graph A and has a smaller acceleration.

Velocity *vs* time graph

The velocity *vs* time graph represents the rate of change of displacement with respect to time.

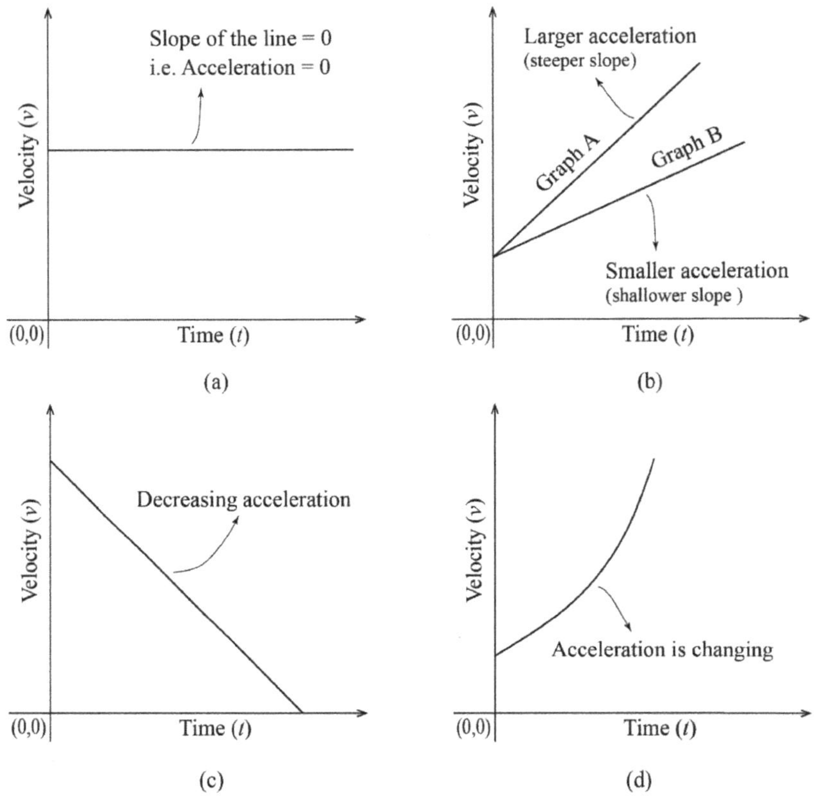

Figure 7.11: The velocity *vs* time graph (a) motion with constant velocity (b) steeper slope in graph A means a larger acceleration and shallower slope in graph B means a smaller acceleration (c) negative slope represents deceleration i.e. acceleration is decreasing (d) slope of the line at any point represents the acceleration at that point

On this graph, the x axis represents time, and the y axis represents the velocity. The slope of the curve on the graph at any point represents the acceleration of the object at that point in time.

For the uniform motion, the velocity *vs* time graph is a horizontal line with a constant velocity, as illustrated in figure 7.11 (a). This means that the object is moving at a constant speed in a single direction, thus the acceleration of the object is zero.

For uniform acceleration, the velocity *vs* time graph is a straight line with a non-zero slope. The slope of the line represents the acceleration of the object. As illustrated in figure 7.11 (b), a steeper slope in graph A means a larger acceleration, and a shallower slope in graph B means a smaller acceleration. Figure 7.11(c) shows a negative slope of the velocity *vs* time graph, which represents an object that is decelerating, i.e. the object's acceleration is decreasing.

For the nonuniform motion, the velocity *vs* time graph is a line with a non-zero slope, as illustrated in figure 7.11 (d). The slope of the line at any point represents the acceleration of the object at that point in time, and a positive slope means that the object is accelerating.

Acceleration *vs* time graph

The acceleration *vs* time graph represents the rate of change of velocity with respect to time.

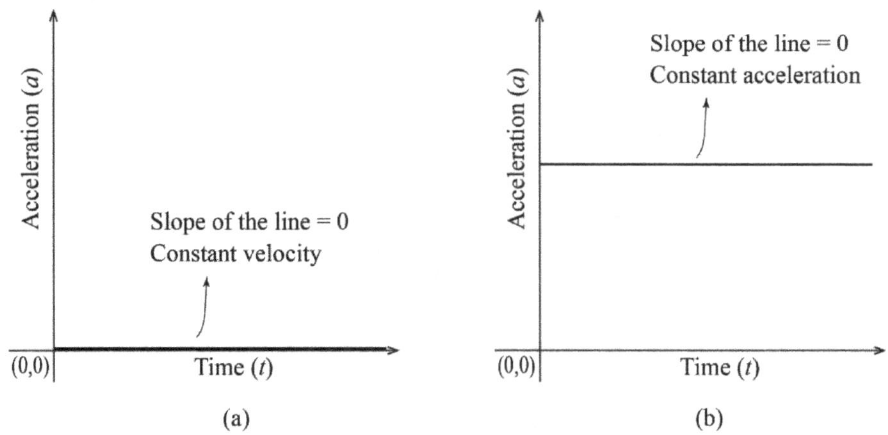

Figure 7.12: The acceleration *vs* time graph (a) uniform motion (b) constant acceleration

For uniform motion, the velocity is constant, and thus the acceleration of an object is zero. Therefore, for uniform motion, the acceleration *vs* time graph is a horizontal line with a value of zero, as shown in figure 7.12 (a). For an object undergoing constant acceleration, the acceleration of the object remains the same with time, as shown in figure 7.12 (b).

Activity 7.8

The graph below illustrates the velocity *vs* time graph of a moving car on a straight line road.

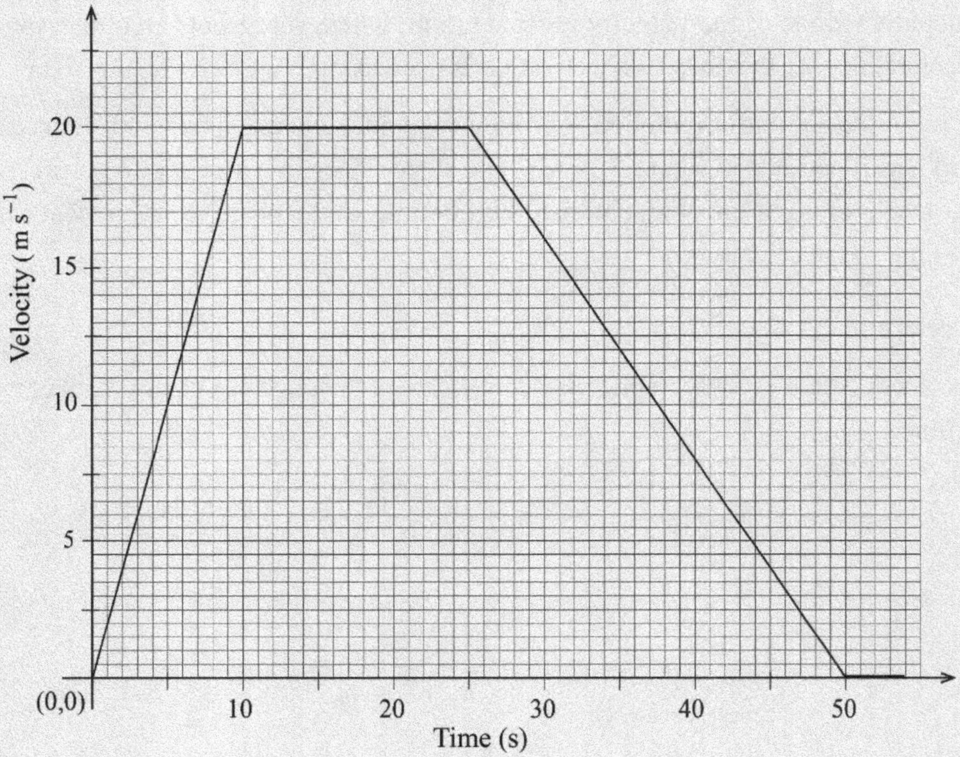

(i) Describe the motion of the car for the 50 s period and determine the relevant accelerations during the motion.

(ii) Determine the total distance travelled by the car during the 50 s period.

(iii) Draw the acceleration *vs* time graph for the moving car for the 50 s period.

7.5 Acceleration due to Gravity

The concept of acceleration due to gravity is a significant part of physics that represents the rate of change in velocity of an object as it falls towards the centre of the Earth during a free fall. A free fall is a simplified form of motion where there is no air resistance or other external forces present. The constant acceleration experienced by an object in free fall is referred to as the acceleration due to gravity, which is represented by the letter *"g"* and has a constant value 9.8 m s^{-2}. (Note: In this book we take $g = 9.8 \text{ m s}^{-2}$. However for approximate calculations we could take $g = 10 \text{ m s}^{-2}$.)

Acceleration due to gravity is significant in various fields of physics and engineering, including the examination of projectile movement, the calculation of weight, and the development of falling objects such as parachutes. For example, it is used to determine the speed of an object at the highest point of its trajectory, the range of a projectile, and the time it takes for an object to fall a specific distance.

In addition, acceleration due to gravity is utilised in computations related to celestial mechanics, like determining escape velocity for planets or satellites, computing orbits of celestial bodies and planets, and examining tides.

Galileo's study on free fall motion

Galileo's investigation into free fall was a crucial event in physics history and is considered one of the initial systematic studies of motion. Galileo performed a series of experiments to examine the motion of objects under the influence of gravity, and his findings challenged the prevailing Aristotelian view that heavier objects fall faster than lighter ones.

Figure 7.13 (a) illustrates Galileo's experiments of dropping objects of different weights from the Leaning Tower of Pisa and observing their motion. He noted that objects of different weights fell at the same rate, reaching the ground at the same time. This was a significant finding as it contradicted the Aristotelian view that heavier objects fall faster than lighter ones.

Galileo's investigation of free fall laid the foundation for the scientific study of motion and provided a basis for the development of Newton's laws of motion. It was one of the first systematic experiments in the history of science, and its results helped to establish the empirical method as a powerful tool for scientific investigation.

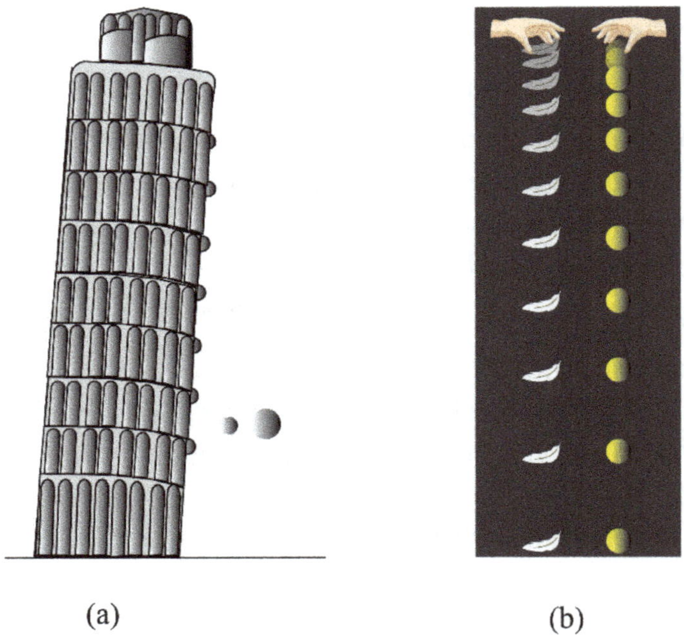

Figure 7.13: (a) Illustration of Galileo's experiments performed at the Leaning Tower of Pisa. (b) The illustration depicts a ball and a feather in free fall inside a vacuum chamber

In addition to its scientific importance, Galileo's investigation of free fall was also significant for its philosophical implications. His findings challenged the prevailing Aristotelian view of motion and paved the way for the development of the scientific revolution, which transformed our understanding of the natural world.

Falling of objects at the same pace

The study of falling bodies has been extensively researched with great accuracy. When there are no external forces, such as air resistance, as in a vacuum chamber, all objects fall at the same acceleration, regardless of their size or weight. This observation has been confirmed through careful experimentation. The illustration in figure 7.13 (b) shows a ball and a feather falling inside a vacuum chamber. Upon release simultaneously, they both fall at the same rate, unaffected by differences in size or weight.

Activity 7.9

A tennis ball is dropped from the edge of the rooftop of a tall building. If it starts from rest and falls freely, determine its position and velocity after 1, 2, 3 and 4 s.

Activity 7.10

As shown in the figure, a person throws a ball upward into the air with an initial velocity 14.0 m s^{-1}. Neglect the air resistance or other external forces acting on the ball during its motion.

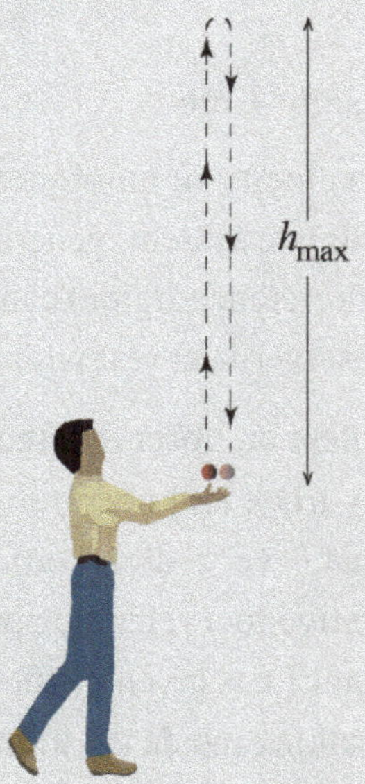

(i) Determine the maximum height the ball can reach above the person's hand.
(ii) Determine the time taken for the ball to reach the maximum height h_{max} above the person's hand.
(iii) Determine the time taken for the ball to return to the person's hand.
(iv) Determine the velocity of the ball when it reaches the person's hand.
(v) Determine the time taken for the ball to pass the position 7.0 m above the person's hand.
(vi) Sketch the displacement *vs* time and velocity *vs* time graph for the entire motion.

Activity 7.11

Two coconuts fall freely from rest simultaneously, with one being dropped from a tree that is twice as tall as the other.

(i) If the coconut from the taller tree reaches the ground with speed v, show that the coconut from the shorter tree will reach the ground with speed $\frac{v}{\sqrt{2}}$.

(ii) If the coconut from the shorter tree takes time t to reach the ground, show that the coconut from the taller tree will reach the ground in time $\sqrt{2}t$.

7.6 Relative Velocity along a Straight Line

When expressing a position and velocity of an object, we always describe them with reference to a particular coordinate system, generally referred to as a frame of reference. A frame of reference (or reference frame) can be described by a Cartesian coordinate system for which an observer is at rest with respect to the origin.

Let us consider a situation where an observer standing at rest (Earth frame of reference) on the side of a railway track and observing a walking person inside a train that is moving towards the east (+ve x direction), as shown in figure 7.3. The person's position at time $t = t$, relative to a reference point O_T in the moving train (i.e. in the reference frame of the train T), is given by the distance $x_{P,T}$. At time $t = t$ the position of the person who is walking inside the train relative to a reference point O_E of the stationary observer on the ground (i.e. in the Earth frame of reference E) is given by the distance $x_{P,E}$ and at time $t = t$ the position of the train's reference point O_T relative to O_E is given by the distance $x_{T,E}$. Assume, at $t = 0$, the positions of O_E coincide O_T and is at $x = 0$.

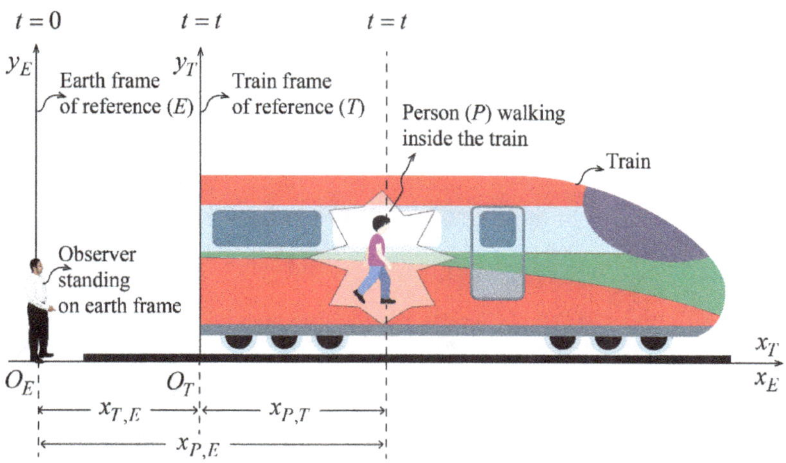

Figure 7.14: A person walks inside a moving train while an observer at rest on the earth frame standing by the side of a railway track

Therefore, from figure 7.14, we can see that during the time interval $\Delta t = t - 0 = t$ the displacement made by the person who is walking inside the train relative to a reference point O_T could be written as

$$x_{P,T} = x_{P,E} - x_{T,E} \tag{7.14}$$

If we divide the equation by the time interval $\Delta t = t$, we get

$$\frac{x_{P,T}}{t} = \frac{x_{P,E}}{t} - \frac{x_{T,E}}{t}$$

Thus, we arrive at the velocity relation given by

$$v_{P,T} = v_{P,E} - v_{T,E} \tag{7.15}$$

The notation $v_{T,E}$ means the velocity of the train T as measured by the observer on the Earth frame of reference E in the direction towards east (+ve x direction). The notation $v_{E,T}$ means the velocity of the observer on the Earth frame of reference E as measured by moving train T in the opposite direction. The magnitudes of $v_{T,E}$ and $v_{E,T}$ are equal and their directions are opposite to each other.

Using the properties of vectors, we have $v_{T,E} = -v_{E,T}$, thus

$$v_{P,T} = v_{P,E} + v_{E,T} \tag{7.16}$$

In general, if A and B are two objects moving on a straight line with a uniform velocity $v_{A,E}$ and $v_{B,E}$ respectively relative to Earth frame E, heading in the same direction, then the velocity of object A relative to B, (i.e., $v_{A,B}$) is given by

$$v_{A,B} = v_{A,E} - v_{B,E} \tag{7.17}$$

But, $v_{B,E} = -v_{E,B}$, thus

$$v_{A,B} = v_{A,E} + v_{E,B} \tag{7.18}$$

Example: Two cars A and B a distance apart, moving towards each other with a constant velocity $36\ \text{km h}^{-1}$ and $54\ \text{km h}^{-1}$ respectively relative to Earth frame E along a straight road as shown in the figure below. Take the eastward direction as +ve x direction. Determine the velocity of car A with respect to car B.

Velocity of the car A respect to Earth E, $v_{A,E} = 36\ \text{km h}^{-1} \rightarrow$

Velocity of the car B respect to Earth E, $v_{B,E} = 54\ \text{km h}^{-1} \leftarrow$

Let the velocity of the car A respect to the car B be $v_{A,B}$

We have $v_{A,B} = v_{A,E} + v_{E,B} = v_{A,E} - v_{B,E}$

Therefore $v_{A,B} = \left(36\ \text{km h}^{-1} \rightarrow\right) - \left(54\ \text{km h}^{-1} \leftarrow\right) = (36+54)\ \text{km h}^{-1} \rightarrow$

$v_{A,B} = 90\ \text{km h}^{-1} \rightarrow$ i.e. in eastward direction.

Activity 7.12

The figure below shows that a motorcycle rider *M* travelling in a motorcycle moves with a constant velocity 65 km h^{-1} towards eastward. A steady wind *W* blows from east to west at a velocity 20 km h^{-1}. Determine the velocity of the wind felt by the motorcycle rider.

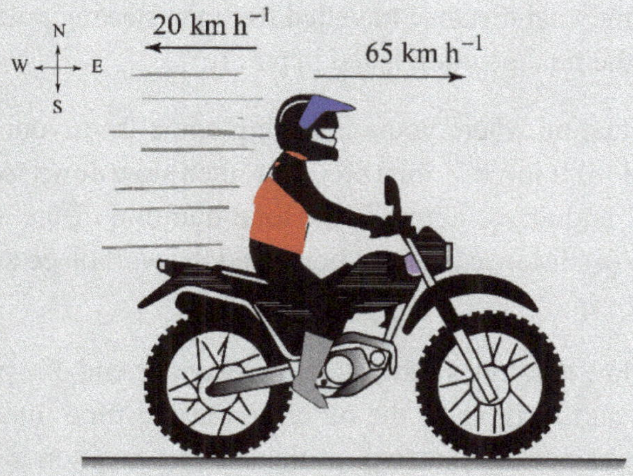

Questions

NOTE: Although the solutions are given on the last page, it is in your own interest to attempt the questions on your own as an indication of how far you have understood the material of the section.

1. Explain the difference between the distance travelled and the displacement.

2. Explain the difference between speed and velocity.

3. Does the speedometer of a car measure speed or velocity?

4. Can an object have a varying speed if its velocity is constant? Explain.

5. If the average velocity of an object is zero for a certain time interval, what can you say about the displacement of the object during that interval?

6. A motor car, running along a horizontal straight line road, starts at a position 60 m east of a landmark (milestone marker) and heads west. After a short time interval, the motor car is 800 m west of the landmark. Choose east to be the positive x direction.

(i) What is the displacement of the motor car from the starting point?

(ii) What is his displacement from the landmark?

(iii) The motor car then turns around and heads east. If at a later time the motor car is 860 m east of the landmark, what is his displacement from the starting point at this time?

(iv) What is the total distance travelled from the starting point if the motor car stops at the final position listed in part (c)?

7. Consider a situation where you are driving home from your workplace at the average speed $80.0\ \text{km h}^{-1}$ for $160.0\ \text{km}$, then slow down to an average speed $50.0\ \text{km h}^{-1}$ and arrive home. If the total duration of the travel is $3\ \text{hours}$, determine the (a) distance between home and the workplace and (b) the average speed of your trip.

8. The figure below illustrates a swimmer leaving one end of a pool of length l at a time $t = 0$ and arriving at the other end at the time interval $\Delta t = t_1$. The swimmer swims back and arrives at the starting position at the time interval $\Delta t = t_2$.

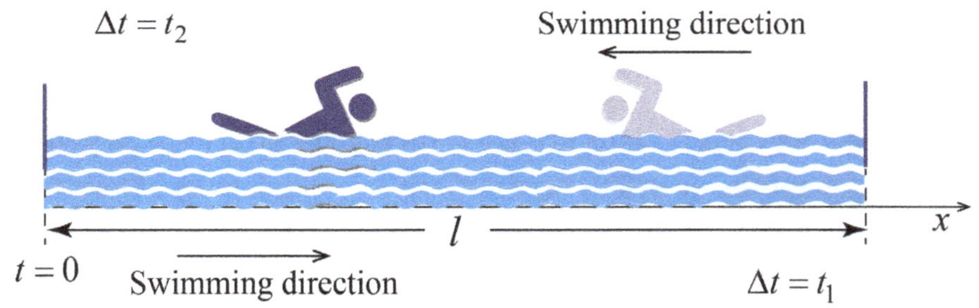

If the swimmer is initially in the positive x direction, determine the swimmer's average velocities in

(i) (a) the first half of the swim.

(b) the second half of the swim, and

(c) the round trip.

(ii) What is the swimmer's average speed for the round trip?

9. Each figure below illustrates the qualitative behaviour of the velocity of the x component of an object moving on a straight line in two different instances. For these, each figure sketches a qualitative graph of the behaviour of the acceleration as a function of time.

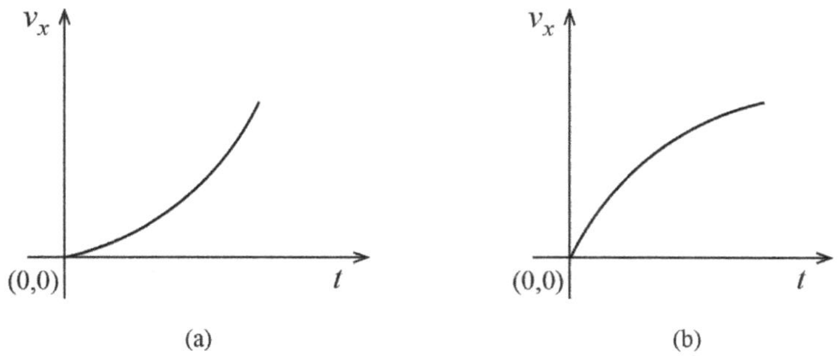

10. A position–time graph for an object moving along the x axis is shown in the figure below.

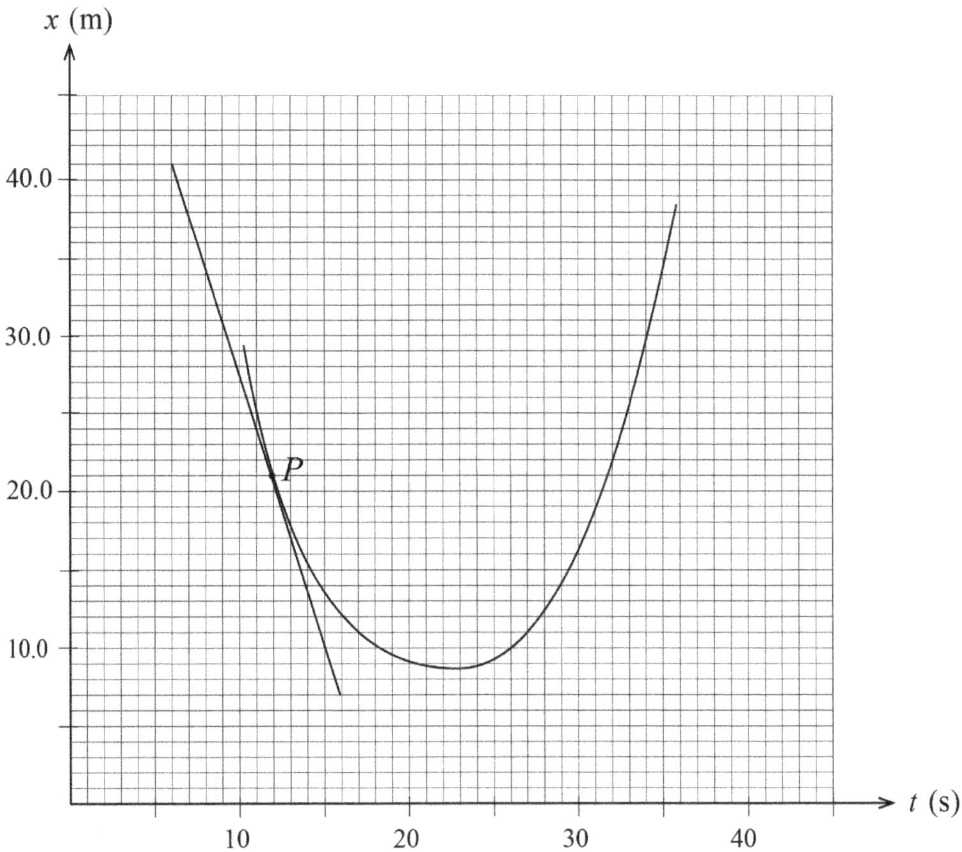

(i) Determine the average velocity in the time interval $t = 11$ s to $t = 17$ s.

(ii) Determine the instantaneous velocity at point P ($t = 12$ s) by using the graph.

11. An observer observes a lioness in its ambush of an antelope. The lioness reaches $72.0\ \text{km h}^{-1}$ in $2.5\ \text{s}$ starting from rest. Assume that the lioness chases the antelope with a constant acceleration and runs on a flat horizontal surface towards northward on a straight line. Determine

    (i) its average acceleration.

    (ii) the distance travelled by the lioness during that time.

12. As shown in the figure below, a car travels east on a straight line road at a constant velocity $54.0\ \text{km h}^{-1}$ in a busy city area where the speed limit is $40.0\ \text{km h}^{-1}$. The traffic police on a motorcycle parked on the side of a lamp post noticed that the car was passing the lamp post. One second after the car passes the lamp post, the traffic police start chasing the car, accelerating at a constant rate $2.0\ \text{m s}^{-2}$.

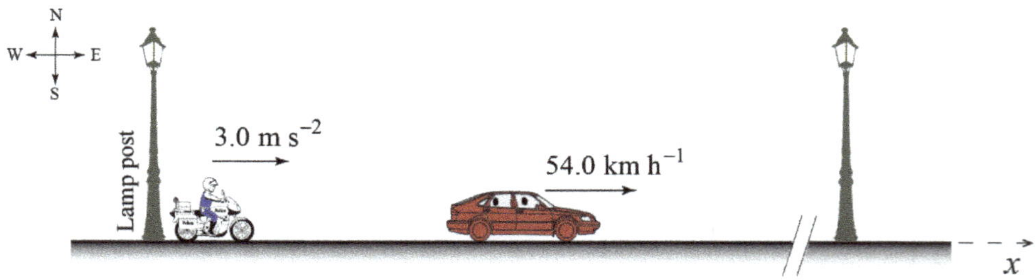

    (i) How long does it take for the traffic police to overtake the car?

    (ii) Determine the speed of the traffic police at that point.

    (iii) Determine the total distance that the traffic police have travelled at that point.

    (iv) Investigate the situation if the traffic police would have had a more powerful motorcycle with higher acceleration.

13. A boat carrying passengers starts from rest at a dock and heads east with a constant acceleration of magnitude $3.0\ \text{m s}^{-2}$. After travelling $150\ \text{m}$, the motor of the boat got choked and forced to slow down the boat at a constant acceleration of magnitude $1.5\ \text{m s}^{-2}$ (while still moving east) until its velocity reached $15\ \text{m s}^{-1}$. Just as the boat attains the speed of $15\ \text{m s}^{-1}$, it passes a buoy due east of the dock.

    (i) Sketch a qualitative velocity *vs* time graph for the boat, taking the +ve axis point east, $t(\text{s})$.

(ii) Determine the velocity of the boat when it reaches the distance 150 m.

(iii) Determine the total time required by the boat to reach the buoy.

(iv) Determine the distance between the dock and the buoy.

14. The displacement *vs* time graph of an object moving along a straight line is shown in the figure below. Draw the velocity *vs* time and the acceleration *vs* time graph.

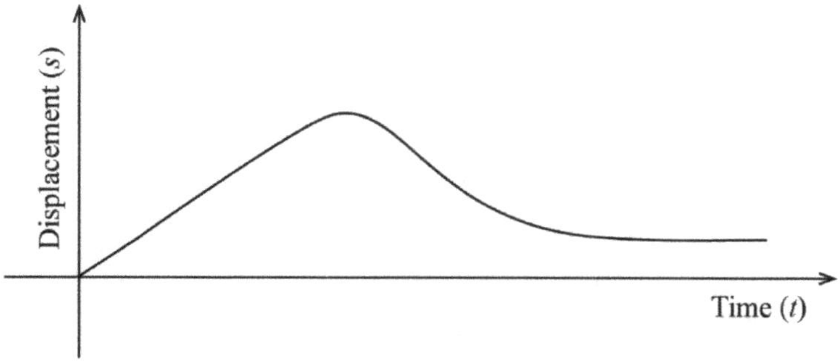

15. The cheetah and antelope are both at rest and 100 m apart. The cheetah starts to chase the antelope from rest and accelerates uniformly, reaches a velocity of magnitude 108 km h^{-1} in 2.0 s, and then maintains this velocity for 16 s. The antelope takes 0.8 s to react, which then accelerates uniformly for 2.0 s, reaches a velocity of magnitude 90 km h^{-1} and then maintains this velocity.

(i) Draw the velocity *vs* time graph for cheetah and antelope on the same display.

(ii) Determine the cheetah's acceleration.

(iii) Determine the distance travelled by the cheetah while accelerating.

(iv) Determine the distance travelled by the cheetah while it is moving at a constant velocity.

(v) Determine the distance covered by the antelope 18 s after the cheetah started to run.

(vi) How far apart are the cheetah and the antelope 18 s after the cheetah starts to run?

16. A person in a hot air balloon travelling vertically at a constant velocity of $8.0\ \text{m s}^{-1}$ dropped a sand bag when the balloon was $50.0\ \text{m}$ above the ground as shown in figure below. Neglect the air resistance or other external forces acting on the sand bag during its fall.

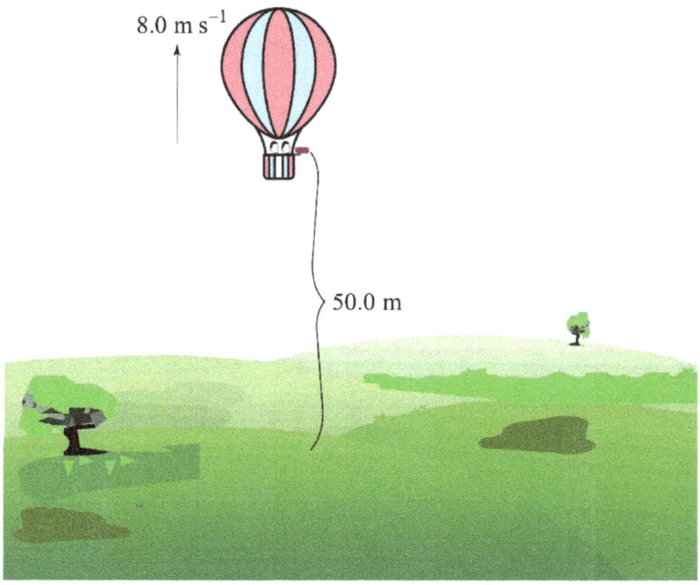

(i) Determine the position and velocity of the sand bag at $0.5\ \text{s}$ and $1.5\ \text{s}$ after its release.

(ii) Determine the time taken to strike the ground after the sand bag is released.

(iii) Determine the velocity of the sand bag upon reaching the ground.

(iv) Determine the highest height above the ground that the sand bag reaches.

(v) Draw graphs for the acceleration, velocity, and vertical position of the sand bag as functions of time.

17. The graph below depicts the relationship between velocity and time for a bouncing rubber ball on the ground.

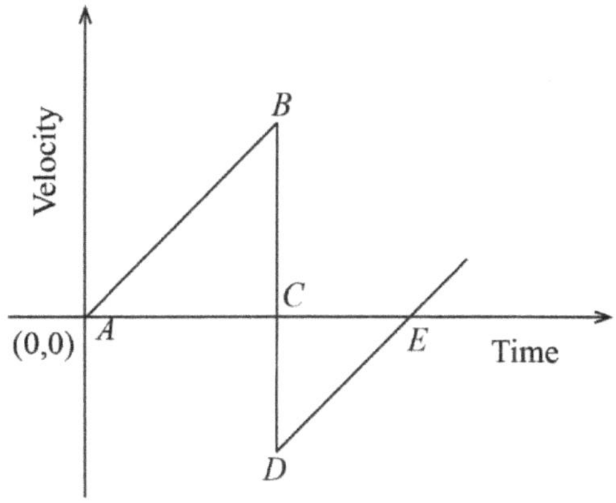

According to the graph, the ball is initially dropped from rest at *A* and strikes the ground at *B*. The ball then bounces off the ground at *D* and reaches its highest at *E*. Assume that the effect of the air resistance is negligible.

(i) Why is the velocity at point *D* negative?

(ii) Do the slopes of lines *AB* and *DE* the same? Explain.

(iii) What does the area between line *AB* and the time axis represent?

(iv) Are the areas of triangles *ABC* and *CDE* equal? Justify your answer.

(v) A ball is dropped from rest at an initial height of 1.5 m. Upon hitting the ground, the ball bounces back to a height of 1.0 m. The ball is in contact with the ground between *B* and *D* for a duration of 0.2 seconds. Using free fall motion, determine

(a) the speed of the ball just before hitting the ground.

(b) the speed of the ball immediately after hitting the ground.

(c) the magnitude and direction of the acceleration of the ball during contact with the ground. Assume that the acceleration during impact is constant.

18. The driver of a car travelling north at a constant velocity v on a straight road encountered a fallen tree ahead. Despite immediately applying the brakes, the car continued to travel $50\ \text{m}$ at a velocity v before the brakes took effect. Subsequently, the car skidded along the road for $144\ \text{m}$, leaving skid marks visible on the road (see the figure below). During skidding, the car decelerated at a constant rate of $2.0\ \text{m s}^{-2}$. The traffic police who started investigating the incident were interested in finding out whether the driver exceeded the speed limit.

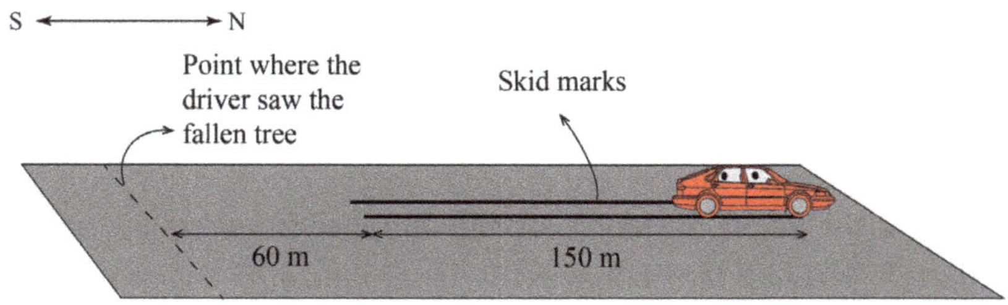

(i) Determine the initial speed of the car before the brakes are applied.

(ii) Detemine elapsed time from the moment the driver spotted the fallen tree until the brakes took effect.

(iii) If the average reaction time for humans is $0.3\ \text{s}$, does the driver alert to the danger.

(iv) If the speed limit is $100\ \text{km h}^{-1}$ does the driver exceed the speed limit.

19. Suppose that you are driving north on a straight two-lane road at a constant velocity $70\ \text{km h}^{-1}$. An ambulance travelling at a constant velocity $100\ \text{km h}^{-1}$ in the same direction approaches you in the other lane, as shown in the figure below. Determine the velocity of the ambulance relative to you. Take the northward direction as the +ve x direction.

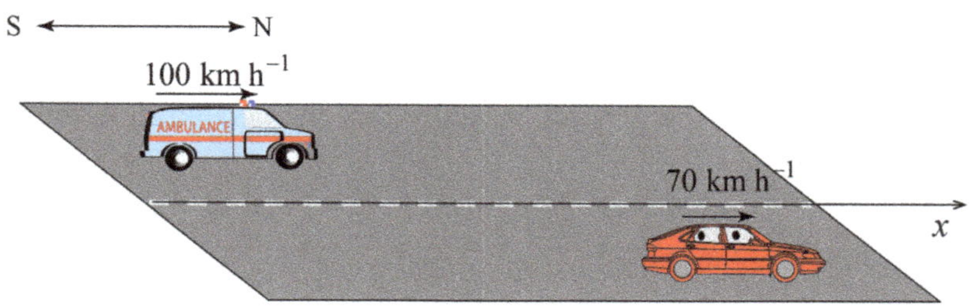

20. A train travels toward the east on a straight rail track at a constant velocity $75\ \text{km h}^{-1}$. As shown in the figure below, a motorcycle rider travels along a straight road close and parallel to the rail track in the same direction at a constant velocity $85\ \text{km h}^{-1}$. If the length of the train is $175\ \text{m}$, how long will it take for the motorcycle rider to cross the entire length of the train?

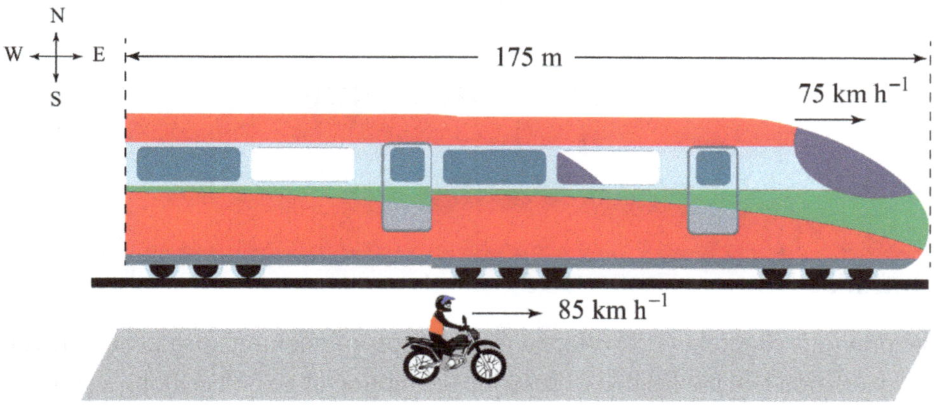

Chapter 8

Motion in a Plane

8.1 Introduction

In the previous chapter, we developed the concepts of position, displacement, velocity and acceleration that are needed to describe the motion of an object along a straight line.

Now, we will explore the topic of motion in a plane, which involves the movement of objects across a two-dimensional or 2D surface. Objects in motion in a plane can move in a straight line or follow a curved path, and their speed and direction can change over time.

To describe this type of motion, we use mathematical equations that express the position of the object at any given time. These equations are based on vectors, mathematical entities that have both magnitude and direction.

As a simple example of motion in a plane, we will examine motion with constant acceleration, including a detailed treatment of projectile motion. Additionally, we will discuss the uniform circular motion, a familiar class of motion that has practical significance in everyday life.

8.2 Position, Displacement, Velocity and Acceleration in a plane

Displacement, velocity, and acceleration are three fundamental concepts describing the motion of an object in a plane.

Position and Displacement in a plane

To describe the motion of an object in a plane, the object's position must be described. The position $\underset{\sim}{r}$ of the object can be represented as a vector from the origin O to the object's location, as shown in figure 8.1(a). Consider the object moving from its initial position with a position vector $\underset{\sim}{r}_i$ to its final position with a position vector $\underset{\sim}{r}_f$ in a xy plane, as shown in figure 8.2(b).

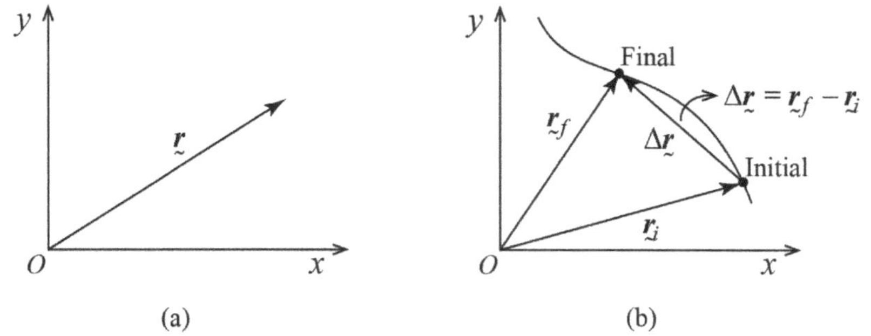

Figure 8.1: (a) A position vector $\underset{\sim}{r}$ (b) The final position vector $\underset{\sim}{r}_f$ minus the initial position vector $\underset{\sim}{r}_i$ is the displacement, i.e. $\Delta\underset{\sim}{r} = \underset{\sim}{r}_f - \underset{\sim}{r}_i$.

The displacement in a plane refers to the change in an object's position $\Delta\underset{\sim}{r}$ from its initial position to its final position and is given by

$$\Delta\underset{\sim}{r} = \underset{\sim}{r}_f - \underset{\sim}{r}_i \tag{8.1}$$

Displacement is a vector quantity that has both magnitude and direction. When considering motion in a plane, displacement is represented by a 2D vector that describes the changes in the x and y coordinates of the object.

8.3 Velocity in a plane

The velocity in a plane describes the rate of change of an object's position. It is a vector quantity, which means it has both magnitude and direction and is often represented as $\underset{\sim}{v}$. Consider an object at point P as it goes around a curved section of a path in a xy plane, and at a slightly later time, say Δt the object is at point Q, as shown in figure 8.2.

Let $\underset{\sim}{r}_i$ be the position vector of the object at P at a time t_i, and $\underset{\sim}{r}_f$ be the position vector of the object at Q at a time t_f. Therefore, the displacement of the object from position P to position Q during the time interval $\Delta t (= t_f - t_i)$ would be $\Delta\underset{\sim}{r} (= \underset{\sim}{r}_f - \underset{\sim}{r}_i)$. Hence, the average velocity of the object between P and Q could be written as

$$\underset{\sim}{v}_{avg} = \frac{\Delta\underset{\sim}{r}}{\Delta t} = \frac{\underset{\sim}{r}_f - \underset{\sim}{r}_i}{\Delta t} \tag{8.2}$$

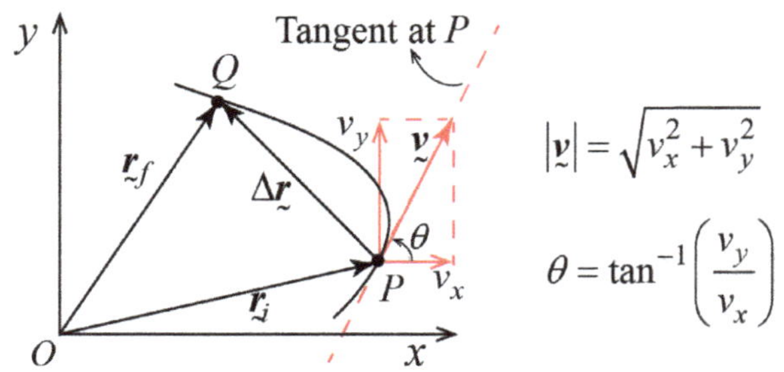

Figure 8.2: Position vectors $\underset{\sim}{r}_i$ and $\underset{\sim}{r}_f$ of the points P and Q, respectively° the displacement $\Delta\underset{\sim}{r}$ during the time interval Δt ° the instantaneous velocity at P becomes tangent to the curve at P.

The direction of the average velocity is the direction of the displacement $\Delta\underset{\sim}{r}$.

The instantaneous velocity at P could be described as the limit of the average velocity over an infinitesimal time interval as the time interval approaches zero. This means, mathematically, that the instantaneous velocity at P could be described as the limit of $\frac{\Delta\underset{\sim}{r}}{\Delta t}$ as Δt approaches zero. Thus the instantaneous velocity at P could be written as

$$\underset{\sim}{v} = \lim_{\Delta t \to 0} \frac{\Delta\underset{\sim}{r}}{\Delta t}$$

The instantaneous velocity at P is tangent to the curve at P, and its x and y components are v_x and v_y respectively, as shown in figure 8.2. The magnitude of the velocity can be calculated from $|\underset{\sim}{v}| = \sqrt{v_x^2 + v_y^2}$. The direction of the velocity at P is given by the angle θ (counterclockwise from the $+x$ axis) and could be determined using $\theta = \tan^{-1}\left(\frac{v_y}{v_x}\right)$. Even if the speed of an object remains constant, the direction of its velocity vector changes over time in two-dimensional motion along a curved path.

8.4 Acceleration in a plane

Acceleration in a plane refers to the rate of change of velocity of an object in a two-dimensional plane. It is a vector quantity that has both magnitude and direction and is often represented as $\underset{\sim}{a}$. Consider an object moving on a curved path in a xy plane.

As shown in figure 8.3, let $\underset{\sim}{\boldsymbol{v}}_i$ be the instantaneous velocity vector of the object at P at a time t_i, and $\underset{\sim}{\boldsymbol{v}}_f$ be the instantaneous velocity vector of the object at Q at a time t_f. These two instantaneous velocities differ in both magnitude and direction. Therefore, the change in velocity of the object from position P to position Q during the time interval $\Delta t(=t_f - t_i)$ would be $\Delta\underset{\sim}{\boldsymbol{v}}(=\underset{\sim}{\boldsymbol{v}}_f - \underset{\sim}{\boldsymbol{v}}_i)$. Hence, the average acceleration of the object between P and Q could be written as

$$\underset{\sim}{\boldsymbol{a}}_{avg} = \frac{\Delta\underset{\sim}{\boldsymbol{v}}}{\Delta t} = \frac{\underset{\sim}{\boldsymbol{v}}_f - \underset{\sim}{\boldsymbol{v}}_i}{\Delta t} \tag{8.4}$$

The direction of the average acceleration is a vector quantity in the same direction as the change in velocity vector $\Delta\underset{\sim}{\boldsymbol{v}}$.

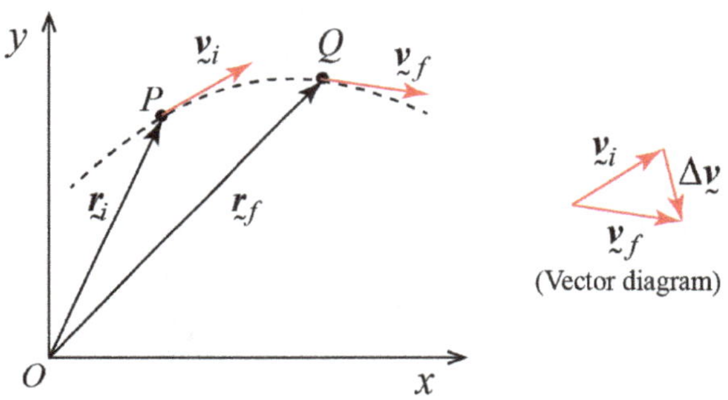

Figure 8.3: An object moves from position *P* to position *Q*, and therefore velocity vector changes from $\underset{\sim}{\boldsymbol{v}}_i$ to $\underset{\sim}{\boldsymbol{v}}_f$. The vector diagram shows the change in velocity $\Delta\underset{\sim}{\boldsymbol{v}}$

The instantaneous acceleration at *P* could be described as the limit of the average acceleration over an infinitesimal time interval as the time interval approaches zero.

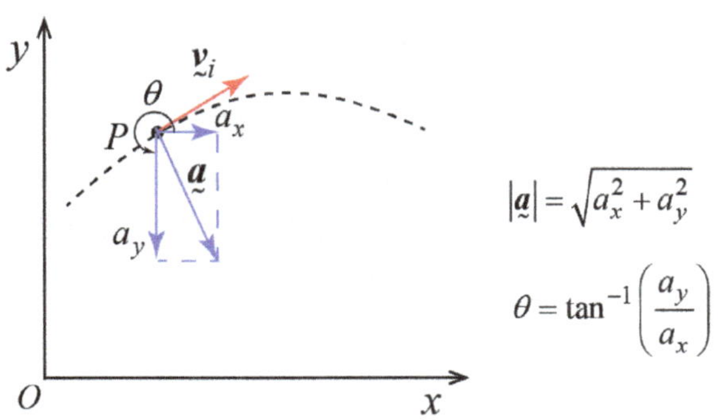

Figure 8.4: The instantaneous acceleration of a position *P* in the $x-y$ plane and its acceleration vector always points toward the concave side of the curved path

This means, mathematically, that the instantaneous acceleration at P could be described as the limit of $\frac{\Delta \underset{\sim}{v}}{\Delta t}$ as Δt approaches zero. Thus the instantaneous velocity at P could be written as

$$\underset{\sim}{a} = \lim_{\Delta t \to 0} \frac{\Delta \underset{\sim}{v}}{\Delta t} \tag{8.5}$$

The instantaneous acceleration at point P does not have the same direction as the instantaneous velocity vector $\underset{\sim}{v}$. As shown in figure 8.4, the acceleration vector always points toward the concave side of the curved path. Furthermore, when an object is in motion along a curved path in a $x-y$ plane, it will experience acceleration, even if its speed remains constant. This is due to the change in the direction of the velocity vector over time.

The magnitude of the acceleration can be calculated from $|\underset{\sim}{a}| = \sqrt{a_x^2 + a_y^2}$. The direction of the acceleration at P is given by the angle θ (counterclockwise from the $+x$ axis) and could be determined using $\theta = \tan^{-1}\left(\frac{a_y}{a_x}\right)$.

8.5 Motion in a plane with Constant Acceleration

The motion of an object in a $x-y$ plane with constant acceleration can be easily modelled and analysed by utilising mathematical equations and vectors. This type of motion permits movement in both x and y directions simultaneously. To analyse the motion, we can utilise the kinematic equations for constant acceleration, which were discussed in the previous chapter, section 7.3, separately for the x components (horizontal) and the y components (vertical).

If we consider the x and y components of the moving object's initial velocity as u_x and u_y respectively, and the object's displacement, final velocity, and acceleration at a time t as s_x, v_x, a_x and s_y, v_y, a_y respectively, the general kinematic equations for motion in the x and y directions can be found in table 8.1.

Table 8.1: General kinematic equations for constant acceleration in a $x-y$ plane.

x components (horizontal)	y components (vertical)
$v_x = u_x + a_x t$	$v_y = u_y + a_y t$
$s_x = u_x t + \frac{1}{2} a_x t^2$	$s_y = u_y t + \frac{1}{2} a_y t^2$
$v_x^2 = u_x^2 + 2a_x s_x$	$v_y^2 = u_y^2 + 2a_y s_y$

8.6 Projectile motion

The projectile motion describes the trajectory of an object that has been launched into the air and is solely influenced by the freefall acceleration g, which acts in a downward direction. This type of motion involves both vertical and horizontal components, the vertical component being affected by gravity and the horizontal component being uniform. The result of the path of a projectile motion under the influence of gravity is a parabolic curve, as shown in figure 8.5 (a).

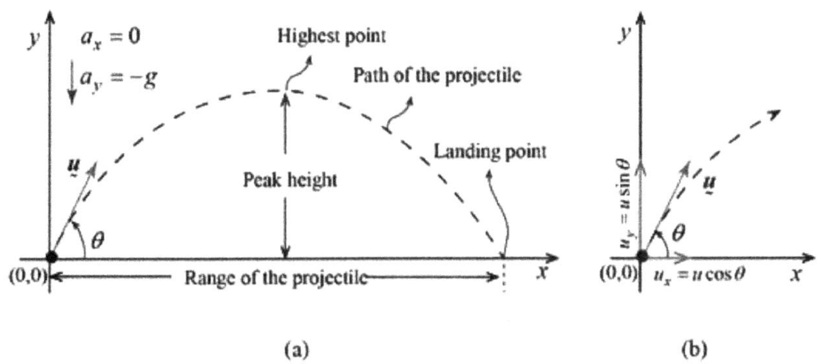

Figure 8.5: The path of a projectile motion and the vertical and horizontal components of the initial velocity

The highest point on this parabolic path is the peak and the point where the object strikes the ground is the landing point. The range of the projectile, which is the horizontal distance covered by the object, and the time of flight, which is the total time taken by the object to reach the ground, can be determined using the equations of motion. Since the projectile motion is influenced only by gravity, the horizontal (x) component of acceleration a_x is zero, and the vertical (y) component of acceleration

a_y is constant and equal in magnitude to the acceleration of free fall, i.e. g , it acts in the downward direction. Thus,

$$a_x = 0 \qquad \text{and} \qquad a_y = -g \tag{8.6}$$

(Note: in this book, we take $g = 9.8 \text{ m s}^{-2}$, however, for approximate calculations we could take $g = 10 \text{ m s}^{-2}$)

Figure 8.5 illustrates the launch of an object with an initial velocity $\underset{\sim}{\boldsymbol{u}}$ and an angle θ relative to the positive x axis. As shown in figure 8.5 (b), the magnitude of the initial velocity, $\underset{\sim}{\boldsymbol{u}}$, can be expressed as u, while the magnitudes of its x and components are

$$u_x = u\cos\theta \qquad \text{and} \qquad u_y = u\sin\theta \tag{8.7}$$

respectively. Due to the fact that the horizontal (x) component of acceleration a_x is zero, the x component of the velocity u_x remains constant throughout the motion. However, as shown in figure 8.6, constant vertical acceleration $(a_y = -g)$ causes the y component of the velocity (u_y) to undergo uniform changes at equal time intervals throughout the motion.

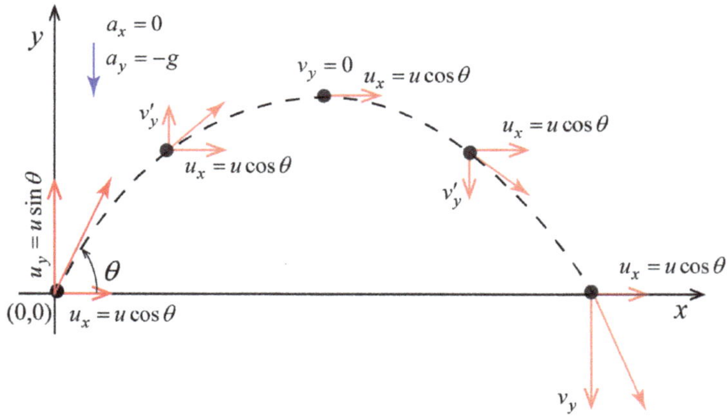

Figure 8.6: Projectile motion showing the variation of vertical and horizontal components of the velocities during the motion

As shown in Figure 8.6, the vertical velocity (v_y) of the projectile decreases initially until it reaches zero at the top of the trajectory. Then, due to the pull of gravity, the projectile falls back downward. During the downward motion, the vertical velocity (v_y) continues to change at the same constant rate as it did on the way up and at the top of the path. The acceleration remains constant, with a magnitude of g and the same direction throughout the path. The horizontal and vertical motions of a projectile are independent of each other and can be analysed separately.

Horizontal Range and Maximum Height of a Projectile

Horizontal range and maximum height are two essential quantities used to describe the motion of a projectile. As shown in figure 8.7, assume a projectile is launched from the origin at a time $t\quad 0$ with a velocity $\underset{\sim}{u}$, angle of launch θ and returns to the same horizontal level after flight time $t = t_f$. We often observe similar situations in sports such as football, where the ball frequently lands at the same level. The horizontal range R is the distance travelled by the projectile in the horizontal direction, while the maximum height h is the highest point reached by the projectile in its trajectory. These quantities could be determined using the initial velocity $\underset{\sim}{u}$, angle of launch θ, and gravitational acceleration g.

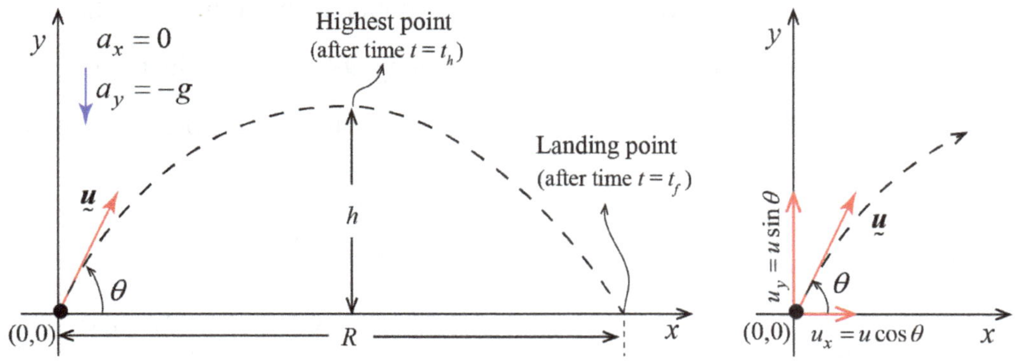

Figure 8.7: A projectile is launched from the origin at a time $t = 0$ with a velocity $\underset{\sim}{u}$, angle of launch θ and returns to the same horizontal level after flight time $t = t_f$.

The projectile motion exhibits symmetry about the highest point. Therefore, the horizontal distance covered by the projectile to reach the maximum height *(h)* is half of the total horizontal distance R covered during the entire motion. Furthermore, the time required for the projectile to reach the maximum height *(h)*, denoted by t_h, is equal to half of the total flight time t_f.

Use the equation $v = u + at$ for the projectile motion in the vertical direction from the initial point to the highest point. Initial velocity $u_y = u\sin\theta$, the velocity at the highest point is zero, and the time taken to reach the highest point is t_h, thus,

$$0 = u\sin\theta + (-g)t_h$$

Therefore,

$$t_h = \frac{u \sin\theta}{g} \tag{8.8}$$

Use the equation $v^2 = u^2 + 2as$ for the projectile motion in the vertical direction from the initial point to the highest point. Initial velocity $u_y = u\sin\theta$, the velocity at the highest point is zero, and the displacement when reaching the highest point is h, thus,

$$0 = (u\sin\theta)^2 + 2(-g)h$$

Therefore,

$$h = \frac{u^2 \sin^2\theta}{2g} \tag{8.9}$$

Use the equation $s = ut + \frac{1}{2}at^2$ for the projectile motion in the vertical direction from the initial point to the landing point. Initial velocity $u_y = u\sin\theta$, the total vertical displacement is zero, and the time taken for the entire flight of the projectile motion is t_f, thus,

$$0 = (u\sin\theta)t_f + \frac{1}{2}(-g)t_f^2$$

The solution will be, $t_f = 0$ or $t_f = \frac{2u\sin\theta}{g}$

But, $t_f = 0$ is not an acceptable solution, and this represents the initial time.

Therefore,

$$t_f = \frac{2u\sin\theta}{g} \tag{8.10}$$

From equations 8.8 and 8.10, we have

$$t_f = 2t_h$$

Use the equation $s = ut + \frac{1}{2}at^2$ for the projectile motion in the horizontal direction from the initial point to the landing point. Initial velocity $u_x = u\cos\theta$, the total horizontal displacement and the time taken for the entire flight are R and t_f, respectively. Thus,

$$R = (u\cos\theta)t_f + \frac{1}{2}\times 0\times t_f^2$$

i.e.

$$R = (u\cos\theta)\left(\frac{2u\sin\theta}{g}\right) \quad \Rightarrow \quad R = \frac{u^2 2\sin\theta\cos\theta}{g}$$

But, we have $\sin 2\theta = 2\sin\theta\cos\theta$, thus

$$= \frac{{}^2\sin 2\theta}{\quad}$$

From equation 8.11, the horizontal displacement (R) will be maximum when $\sin 2\theta = 1$, and it will happen only when $\theta = 45^\circ$ in which $2\theta = 90^\circ$ and $\sin 2\theta = 1$. Thus, the maximum horizontal displacement $(R_{\max})$ will occur when the angle of launch θ is 45° and $R_{\max} = \frac{u^2}{g}$.

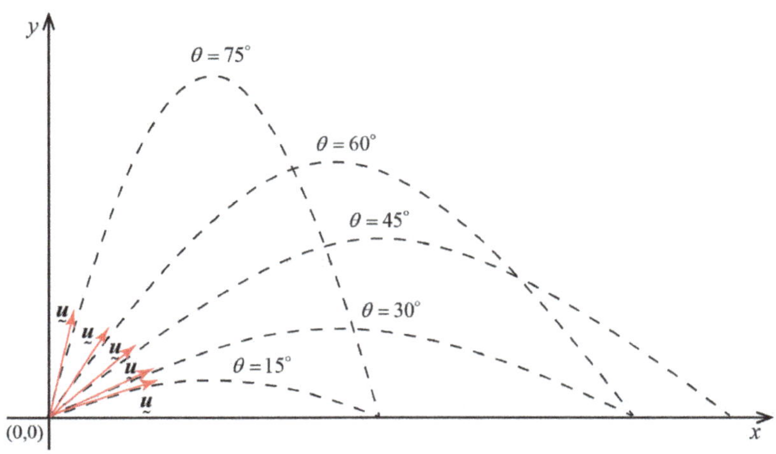

Figure 8.8: A projectile launched over a flat surface from the origin with a specific initial speed at various angles of projection

Figure 8.8 illustrates various trajectories for a projectile having a given initial speed but launched at different angles. As you can see, the horizontal displacement is maximum when $\theta = 45^\circ$. In Figure 8.9, the x and y components of a projectile's velocity are plotted as a function of time. The projectile is launched from ground level at time $t = 0$, and later returns to the same elevation at time $t = t_f$. The y component of velocity decreases linearly as shown in figure 8.9 (a) from its initial value with a slope of $a_y = -g$, while the y component of velocity has the same magnitude as at t 0 but with the opposite sign ($v_{fy} = -v_{iy}$) The horizontal component of velocity v_x is constant, resulting in a flat line as shown in figure 8.9(b).

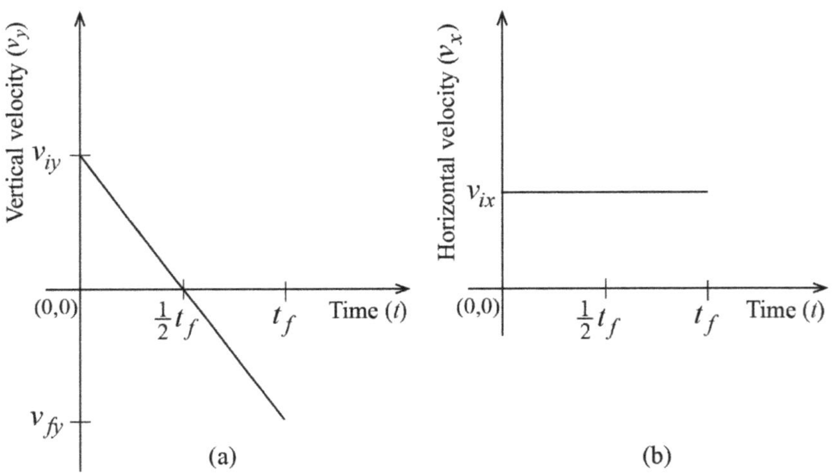

Figure 8.9: (a) Variation of the vertical componet of the velocity *versus* time. (b) Variation of the horizontal componet of the velocity *versus* time.

Activity 8.1

A football is launched from the ground at an angle of 35.0° to the horizontal with an initial velocity of 15.0 m s^{-1}. Calculate the following, assuming there is no air resistance or ball rotation, and taking $g = 10$ m s^{-2}:

(i) The maximum height the football reaches above the ground.

(ii) The time it takes for the football to hit the ground.

(iii) The horizontal distance the football travels before hitting the ground.

(iv) The velocity vector of the football at its maximum height.

(v) The magnitude and direction of the acceleration at the maximum height.

Activity 8.2

A student used a catapult to launch a stone with a velocity of 20.0 m s^{-1} at an angle of 30.0° with respect to the horizontal from a height of 1.65 m above the ground. Calculate the following, neglecting air resistance and taking $g = 10 \text{ m s}^{-2}$:

(i) The maximum height that the stone reaches above the ground.

(ii) The amount of time that the stone is in the air before hitting the ground.

(iii) The horizontal distance that the stone travels before hitting the ground.

8.7 Relative Velocity in a Plane

In the previous chapter, we discussed the concept of relative velocity along a straight line in section 7.6. In this chapter, we will extend our discussion to relative velocity in a plane by using vector addition, which was covered in section 6.5.

If we apply what we learned in section 7.6 to a two-dimensional scenario, we can determine the relative velocity between two objects, *A* and *B*, moving in a plane with uniform velocities, $\underset{\sim}{v}_{A,E}$ and $\underset{\sim}{v}_{B,E}$, respectively, relative to the Earth frame *E*. The velocity of object *A* relative to *B*, represented by $\underset{\sim}{v}_{A,B}$, can be calculated by subtracting the velocity of object *B* from that of object *A* (see section 7.6). Thus,

$$\underset{\sim}{v}_{A,B} = \underset{\sim}{v}_{A,E} - \underset{\sim}{v}_{B,E} \tag{8.12}$$

It is important to note that both $\underset{\sim}{v}_{A,E}$ and $\underset{\sim}{v}_{B,E}$ are vector quantities.

But, $\underset{\sim}{v}_{B,E} = -\underset{\sim}{v}_{E,B}$, thus

$$\underset{\sim}{v}_{A,B} = \underset{\sim}{v}_{A,E} + \underset{\sim}{v}_{E,B} \tag{8.15}$$

By extending our knowledge of relative velocity to a plane, we can solve complex problems involving moving objects in a two-dimensional space. This concept is crucial in various fields, including physics, engineering, and aviation. With an understanding of relative motion and velocity, we can predict collisions, optimize systems, and analyze the behaviour of objects in motion.

Example: Consider a situation in which a boat is being rowed across a wide river by a rower with a velocity of 2.5 m s^{-1} relative to the water. The water in the river is flowing uniformly at a velocity of 1.2 m s^{-1} due east relative to the Earth. If the boat is rowed directly north, what is the boat's velocity relative to an observer standing on the bank?

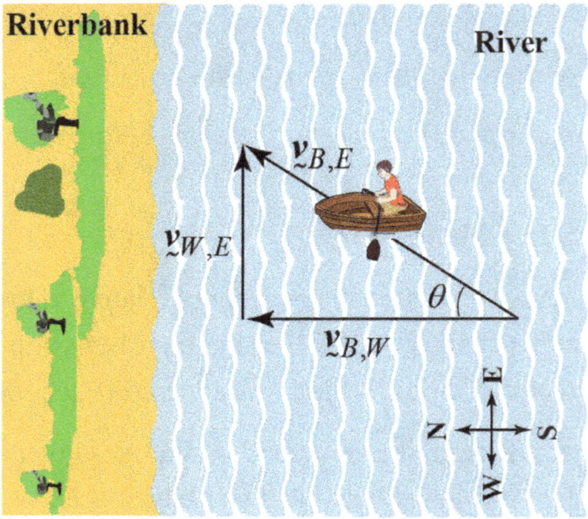

Let, $\underset{\sim}{v}_{B,W}$ be the velocity of the boat relative to water.

$\underset{\sim}{v}_{W,E}$ be the velocity of the water relative to Earth.

$\underset{\sim}{v}_{B,E}$ be the velocity of the boat relative to Earth.

We have $\left|\underset{\sim}{v}_{B,W}\right| = v_{B,W} = 2.5 \text{ m s}^{-1}$, $\left|\underset{\sim}{v}_{W,E}\right| = v_{W,E} = 1.2 \text{ m s}^{-1}$

According to the vector diagram, $\underset{\sim}{v}_{B,E} = \underset{\sim}{v}_{B,W} + \underset{\sim}{v}_{W,E}$

Therefore, using Pythagorean theorem, the magnitide of the velocity of boat relative to the Earth will be $\left|\underset{\sim}{v}_{B,E}\right| = v_{B,E} = \sqrt{v_{B,W}^2 + v_{W,E}^2}$, thus,

$$v_{B,E} = \sqrt{2.5^2 + 1.2^2} = 2.8 \text{ m s}^{-1}, \quad \theta = \tan^{-1}\left(\frac{v_{W,E}}{v_{B,W}}\right) = \tan^{-1}\left(\frac{1.2}{2.5}\right) = 25.6^\circ$$

The boat is moving with the magnitide of the velocity 2.8 m s^{-1} in the direction 25.6° east of north relative to the Earth. The boat will end up towards downstream.

Activity 8.3

Consider a scenario in which a boat is observed travelling directly north across a wide river. The boat's velocity relative to the water is 2.5 m s^{-1}. The river water flows uniformly at a velocity of 1.2 m s^{-1} due east relative to the Earth. Determine the boat's velocity relative to earth and the direction in which the boat should have been rowed.

Questions

NOTE: Although the solutions are given on the last page, it is in your own interest to attempt the questions on your own as an indication of how far you have understood the material of the section.

1. A projectile is launched at an angle θ to the horizontal with an initial speed v_i if the air resistance is negligible.

   (i) Is the projectile a freely falling body?

   (ii) What is its acceleration in the vertical direction?

   (iii) What is its acceleration in the horizontal direction?

2. Can two projectiles, having the same initial speed but launched at different angles of elevation, reach the same landing point? Explain. Ignore air resistance.

3. A javelin is thrown from a height of 1.60 m with an initial velocity of 20.0 m s^{-1} at an angle of 45.0° to the horizontal ground, as illustrated in the figure.

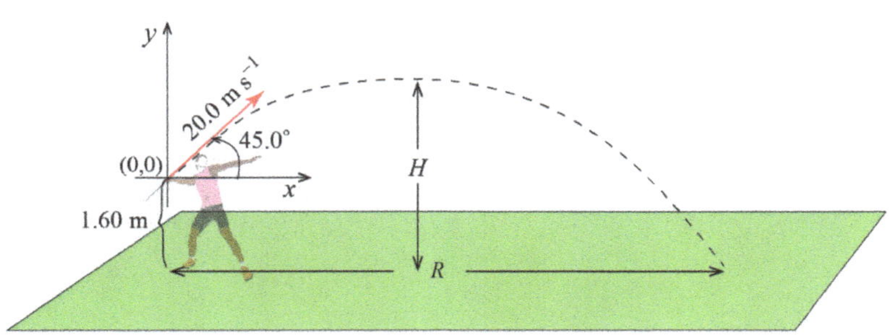

Assuming that the javelin behaves as a particle, neglecting air resistance, and taking $g = 10\ \text{m s}^{-2}$, determine the following:

(i) The maximum height (H) that the javelin reaches above the ground.

(ii) The time the javelin spends in the air before landing on the ground.

(iii) The horizontal distance (R) the javelin travels before landing on the ground.

(iv) The magnitude of the velocity and the direction of the javelin while landing on the ground.

4. As shown in the figure, a tennis player intends to serve the ball horizontally over a net that is $0.91\ \text{m}$ high. The serving point is $11.88\ \text{m}$ away from the net and has a height of $2.50\ \text{m}$.

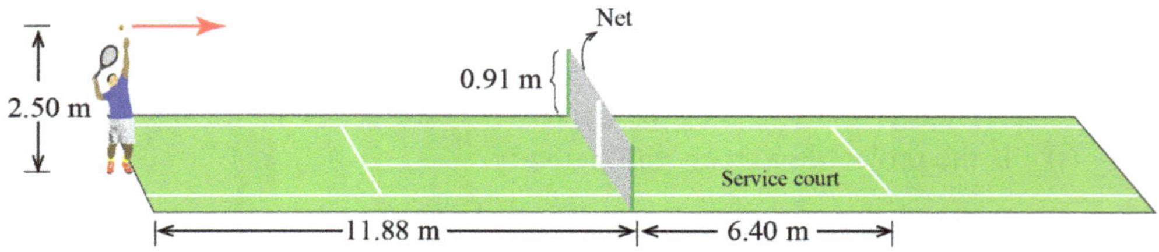

If the ball can be treated as a particle and there is no air resistance, answer the following questions by taking $g = 10\ \text{m s}^{-2}$.

(i) What is the minimum velocity required for the ball to clear the $0.91\ \text{m}$ height net?

(ii) How long will the ball be in the air?

(iii) If the distance from the net to the service line is $6.40\ \text{m}$, will the velocity calculated in part (a) be sufficient to land the ball in the service court?

5. As shown in the figure, a grasshopper jumps off the edge of a table at an angle 48.0° to the horizontal x axis. After reaching a maximum vertical height of 5.0 cm above the tabletop, it lands on the horizontal ground 48.0 cm away.

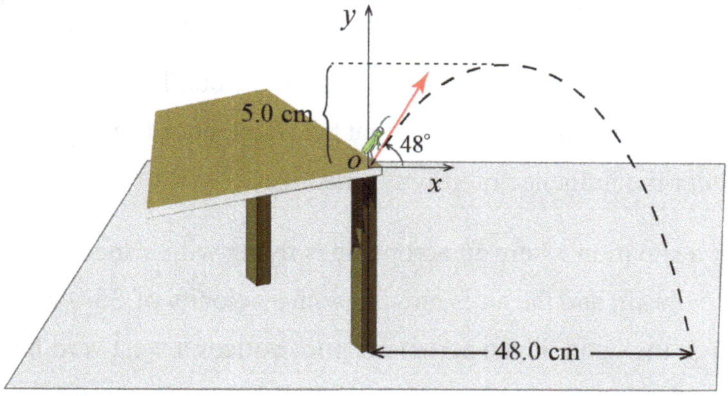

Assuming that the grasshopper can be treated as a particle and that there is no air resistance, answer the following questions by taking $g = 10\ \text{m s}^{-2}$.

(i) What is the initial velocity of the grasshopper?

(ii) How long does it take for the grasshopper to land on the ground?

(iii) What is the height of the table?

6. A soccer player kicks a soccer ball from the ground with an initial velocity of $20.0\ \text{m s}^{-1}$ at an upward angle of 45.0°. At the same instant, a player positioned 50.0 m away in the direction of the kick starts running to meet the ball. Assuming that the ball can be treated as a particle and there is no air resistance, answer the following questions using $g = 10\ \text{m s}^{-2}$.

(i) How long will the soccer ball remain in the air before it hits the ground?

(ii) How far will the soccer ball travel horizontally from the initial kicking point?

(iii) What average speed must the running player maintain to meet the ball just before it lands on the ground?

7. In a firefighting operation, firefighters aim to direct a water cannon towards a blaze located 12.0 m above ground level. The water is shot from the cannon at 30 m s^{-1} at a fixed angle of 45.0° above the horizontal level and the cannon itself is positioned 1.50 m above ground level. To effectively hit the blaze, what are the possible distances between the base of the building and the water cannon that should be maintained? Assume that the water, once leaving the cannon, falls freely under the influence of gravity, and take $g = 10 \text{ m s}^{-2}$.

8. Consider a situation where an aeroplane is flying with a speed of 200 km h^{-1} relative to the air, and the air is moving with a velocity of 35.0 km h^{-1} towards the north in the wind. If the aeroplane pilot notices a westward heading on the compass, what is the aeroplane's velocity relative to the ground?

9. A bird is migrating and flying at a velocity of 115 km h^{-1} relative to the wind in a north-south direction. However, it is facing a wind blowing from west to east at a velocity of 44 km h^{-1}. Determine:

   (i) The angle the bird should head, relative to the north-south direction, in order to travel directly southward relative to the ground.

   (ii) The speed of the bird relative to the ground.

   (iii) The duration the bird will require to cover a ground distance of 500 km from north to south.

10. A cutting tool in an aluminium sheet factory cuts aluminium sheets that move on a conveyor belt at a constant speed of 12 cm s^{-1}. To compensate for the belt's motion, the cutter descends at an angle during fixed intervals. The cutter moves across the width of the 100 cm wide aluminium sheet at a speed of 20 cm s^{-1}. What angle should the cutter be set to maintain the constant speed of the conveyor belt?

Chapter 9

Principles of Force and Motion

9.1 Introduction

In this chapter, we will explore the fundamental principles of force and motion that underlie the behaviour of all objects in the physical world. Understanding these principles is crucial for anyone interested in physics, engineering, or any other field that deals with the behaviour of objects.

The chapter begins with an introduction to the fundamental concepts of force, which are critical to understanding the behaviour of objects in motion. We will explore basic force concepts, including different types of force, such as contact and non-contact forces. Additionally, we will discuss methods for measuring force and determining the resultant force of multiple forces acting on an object. Moreover, we will cover the unit of force and its relation to other fundamental units of physics.

Next, the chapter will focus on Newton's Laws of Motion, which are the foundation of our understanding of force and motion. These laws will be examined in detail, revealing how they define the connection between force and motion and how they can be applied in practical situations.

Furthermore, the concepts of inertia, mass, and weight, which are critical to understanding the behaviour of objects in motion, will be covered in detail. The process of drawing a free-body diagram, a powerful tool for visualising the forces acting on an object, will also be explained.

Finally, the applications of Newton's laws of motion will be discussed, providing insight into how they can be used to solve real-world problems.

9.2 Force

Force is a fundamental concept in physics that describes the influence that an object can exert on another object. It is characterised as any action capable of altering the motion of an object, which can manifest as either a push or a pull. However, it is important to note that forces do not always result in motion. In certain cases, a force may have no net impact on the motion of an object if it is counterbalanced by opposing forces or if the object is stationary. For example, if a book is placed on a table, it does not move because the force of gravity acting on the book is balanced by the normal force of the table pushing up on the book. The two forces cancel each other out, resulting in a net force of zero. Forces can be classified into two main categories: contact forces and non-contact forces (or field force).

Contact Forces

Contact forces (figure 9.1) are forces that occur when two objects come into physical contact with each other. Some common examples of contact forces include friction force – figure 9.1(a), tension – figure 9.1(b), normal force – figure 9.1(c), air resistance or drag force – figure 9.1(d), and applied force – figure 9.1(e).

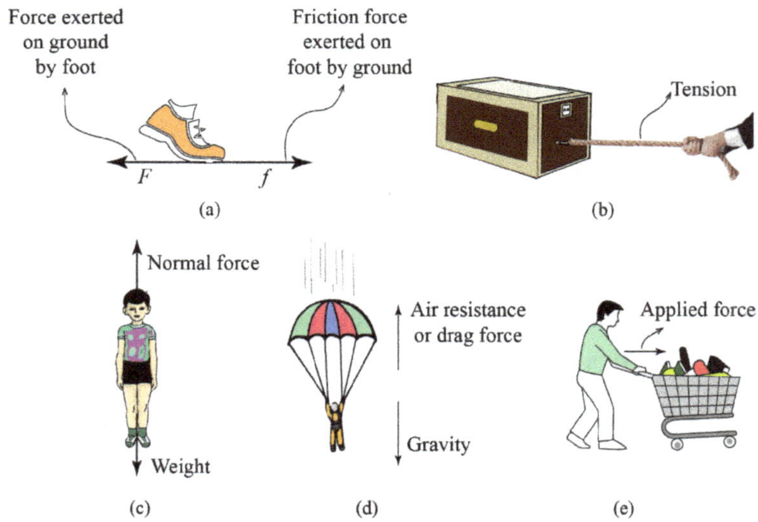

Figure 9.1: Contact forces (a) Frictional force (b) Tension (c) Normal force (d) Air resistance or drag force (e) Applied force

Friction force: This is the force that opposes the motion of an object when it comes into contact with a surface. Friction is what allows you to walk on the ground, stop a car, or write with a pencil.

Tension: This is the force that is transmitted through a string, rope, or cable when it is pulled tight. For example, tension is what keeps a suspension bridge in place or allows a person to pull a heavy object with a rope.

Normal force: This is the force that acts perpendicular to the surface of an object when it is in contact with another surface. For example, when you stand on the ground, the ground exerts an upward normal force on your feet to keep you from sinking into the ground.

Air resistance (Drag force): This is the force that opposes the motion of an object as it moves through the air. Air resistance is what causes a parachute to slow down or force experienced by a cyclist or a runner as they move through the air.

Applied force: This is a force that is applied to an object by another object or person. For example, pushing a shopping cart, throwing a ball, or hitting a nail with a hammer all involve applying a force to an object in contact.

Non-contact Forces

Non-contact forces are forces that occur without any physical contact between objects. Some common examples of non-contact forces include gravitational force – figure 9.2 (a), electric force – figure 9.2 (b), magnetic force – figure 9.2 (c).

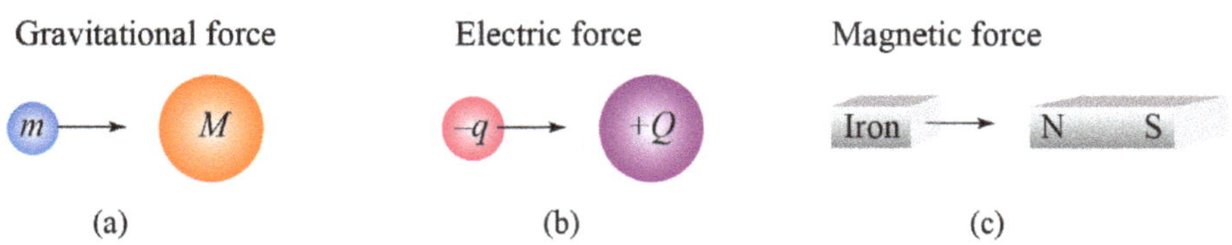

Figure 9.2: Non-contact forces (a) gravitational force (b) electric force (c) magnetic force

Gravitational force: This is the force of attraction between two objects with mass. For example, the force keeps objects bound to the Earth or the force that keeps the planets in orbit around the Sun. Even though the planets and the Sun are not in physical contact with each other, they are still affected by each other's gravitational force.

Electric force: This is the force that one electric charge exerts on another. For example, an electric force between an electron and a proton that forms a hydrogen atom.

Magnetic force: This is the force that exists between magnets or between a magnet and a magnetic material. For example, the force a bar magnet exerts on a piece of iron or the force that holds a refrigerator magnet to the door. Even though the magnet and the door are not in physical contact, they are still affected by the magnetic force between them.

There are only four fundamental forces in nature, all of which are non-contact forces or field forces. These include (i) the gravitational force between objects, (ii) the electromagnetic force between electric charges, (iii) the strong force between subatomic particles, and (iv) the weak force that arises in certain radioactive decay processes. The scope of this book will be limited to the discussion of gravitational and electromagnetic forces.

Measuring Force

Measurement of the magnitude of the force is an important aspect of understanding how objects interact with each other in the physical world. Various instruments are used to measure force, including spring scales, strain gauges, load cells, force plates,

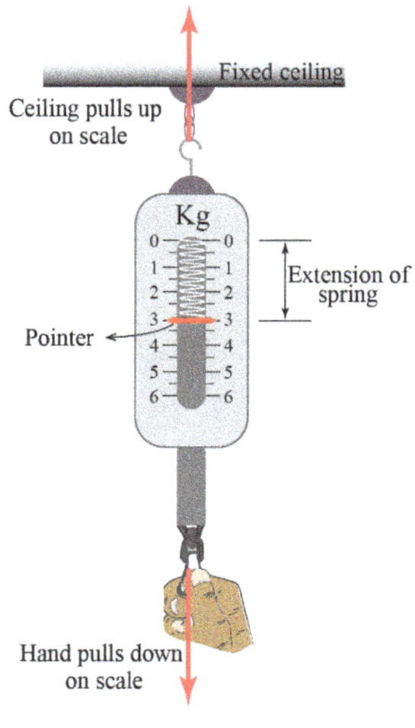

Figure 9.3: Spring scale: when the spring scale is pulled downward, the spring stretches, which can be used to measure the force applied

etc. Spring scales are simple devices that use the deformation of a spring to measure the force applied to it. They are commonly used in everyday applications, such as weighing fruits and vegetables.

Figure 9.3 illustrates a basic spring scale, where the downward pull of the pan causes the spring to stretch. The greater the force applied, the more the spring extends, which causes an attached pointer to move. By calibrating the scale to the amount of extension, one can precisely measure the applied force. Since the extension of many springs is approximately proportional to the applied force, calibrating the scale is a simple process.

Scales are generally calibrated to measure forces in kilograms (kg). In the SI system, the unit of force is the newton (N). To convert kilograms to newtons, use the approximate conversion factors

$$1 \text{ kg} = 9.81 \text{ N} \qquad \text{or} \qquad 1 \text{ N} = 0.10197 \text{ kg} \tag{9.1}$$

9.3 Resultant of Forces

Force is a vector quantity, which implies that it possesses both magnitude and direction. The magnitude of the force corresponds to the strength of the push or pull exerted, while the direction indicates the line along which the force acts. Figure 9.4 (a) and (b) illustrate the pulling and pushing force $\vec{F}$ of magnitude F directed at an angle θ above the horizontal, respectively.

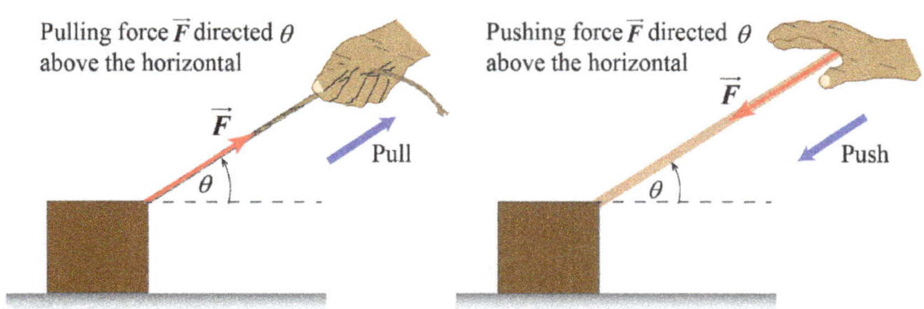

(a) (b)

Figure 9.4: Force is a vector quantity and can be represented by a vector arrow (a) the pulling force of magnitude F directed at an angle θ above the horizontal (b) and the pushing force F directed at an angle θ above the horizontal

When two or more forces = $\vec{F}_1 + \vec{F}_2 + \vec{F}_3$,... act on an object, the resultant force (or the net force) is the single force denoted by $\vec{R}$, which has the same effect on the object's motion as all the individual forces acting together. In other words, it is the vector sum of all the forces acting on an object (refer chapter 6, section 6.5). Thus,

$$\vec{R} = \vec{F}_1 + \vec{F}_2 + \vec{F}_3, \dots = \sum \vec{F} \tag{9.2}$$

Here, the Greek letter $\sum$ (capital sigma) is used as a shorthand notation for a sum.

The magnitude and direction of the resultant force can be calculated using the Pythagorean theorem and trigonometry, respectively.

Let us consider two or more coplanar forces $\vec{F}_1, \vec{F}_2, \vec{F}_3,\dots$ acting on an object as shown in figure 9.5, the resultant of these forces could be calculated using two perpendicular axes components, typically the x axis and y axis. Suppose $F_{1x}, F_{2x}, F_{3x}\cdots$ and $F_{1y}, F_{2y}, F_{3y}\cdots$ are the x axis and y axis components of the coplanar forces $\vec{F}_1, \vec{F}_2, \vec{F}_3,\dots$, respectively, then the x axis and y axis components of the resultant forces R_x and R_y could be written as (refer chapter 6, section 6.6)

$$\begin{aligned} R_x &= F_{1x}, F_{2x}, F_{3x}\dots = \sum F_x \\ R_y &= F_{1y}, F_{2y}, F_{3y}\dots = \sum F_y \end{aligned} \tag{9.3}$$

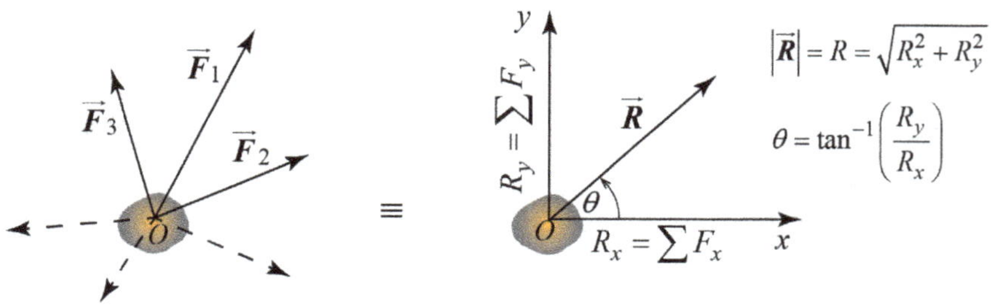

Figure 9.5: Two or more coplanar forces $\vec{F}_1, \vec{F}_2, \vec{F}_3,\dots$ acting on an object and the components of the vector sum to find the resultant force $\vec{R}$

Here F_x is the sum of the x components and $\sum F_y$ is the sum of the y components. Once we get the components R_x and R_y, we can find the magnitude and direction of the resultant force vector $\vec{\boldsymbol{R}}$ (figure 9.5).

The magnitude is $\left|\vec{\boldsymbol{R}}\right| = R = \sqrt{R_x^2 + R_y^2}$, and the direction, which is represented by the angle between $\vec{\boldsymbol{R}}$ and the x axis could be written as $\theta = \tan^{-1}\left(\dfrac{R_y}{R_x}\right)$, θ is counterclockwise from the x axis.

It should be noted that the orientation of the x and y coordinate axes is not always fixed, with the x axis necessarily being horizontal and the y axis vertical. As an example, let us consider the scenario of a heavy box being pulled up along an inclined plane by a force, $\vec{\boldsymbol{F}}$ as shown in figure 9.6.

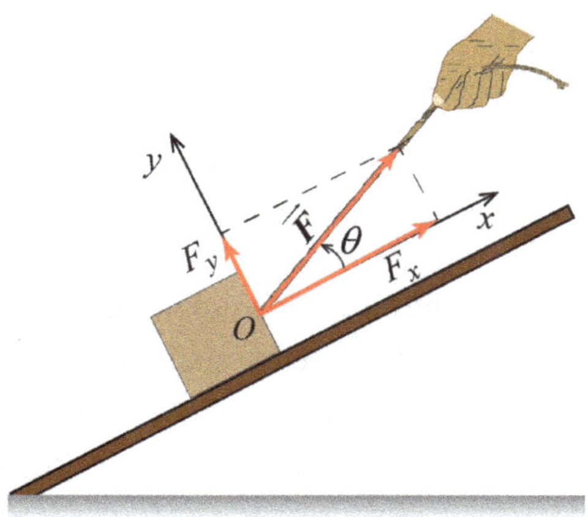

Figure 9.6: A heavy box being pulled up a ramp by force $\underset{\sim}{F}$. The components F_x and F_y are parallel and perpendicular to the sloping surface of the inclined plane.

In this case, it is most advantageous to select one axis that is parallel to the ramp (such as the x axis) and another that is perpendicular to the ramp (such as the y axis). In figure 9.6, both the x component (F_x) and component (F_y) of the force are positive. However, in different scenarios, depending on the orientation of the force and the chosen axes, either F_x or F_y could be zero or negative.

Activity 9.1

Consider a situation where two coplanar forces $\vec{F}_1$ and $\vec{F}_2$ act simultaneously on an object at angles 60° and 30° above the horizontal, as shown in the figure below. The magnitudes of these forces are $200\ \text{N}$ and $150\ \text{N}$, respectively.

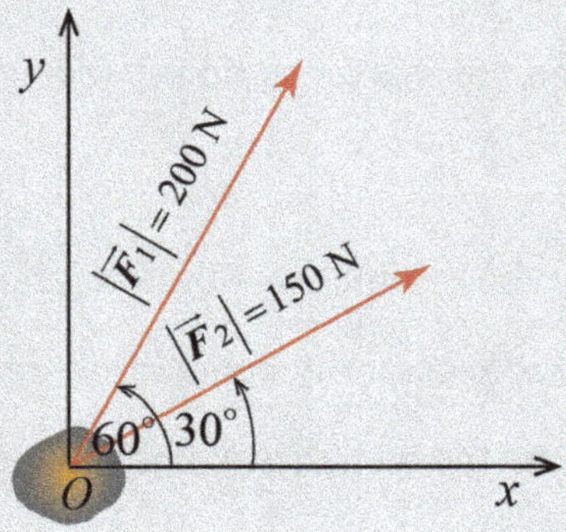

Determine

(i) the net horizontal and vertical forces along x, y axes, respectively.

(ii) the magnitude and direction of the resultant force.

We can also use the principle of the parallelogram of forces to determine the resultant force of two coplanar forces $\vec{F}_1$ and $\vec{F}_2$ acting simultaneously on an object, as shown in figure 9.7. θ is the angle between the forces $\vec{F}_1$ and $\vec{F}_2$ and α is the angle of inclination of the resultat force $\vec{R}$ above the forces $\vec{F}_2$ on the anticlockwise direction.

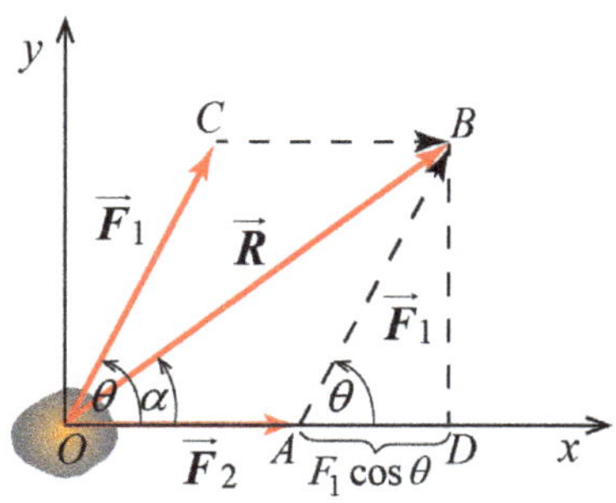

Figure 9.7: Vector addition using parallelogram law

Using the parallelogram law for adding vectors (refer chapter 6, section 6.5) we get the resultant force $\vec{R}$ as

$$\vec{R} = \vec{F}_1 + \vec{F}_2 \tag{9.4}$$

The magnitude and direction of the resultant force $\vec{R}$ could be determined as follows.

In figure 9.7, if we consider the right angled triangle ODB and use the Pythagorean theorm, we have

$$OB^2 = OD^2 + DB^2$$

$$\left|\vec{R}\right|^2 = \left(\left|\vec{F}_2\right| + \left|\vec{F}_1\right|\cos\theta\right)^2 + \left(\left|\vec{F}_1\right|\sin\theta\right)^2$$

$$R^2 = (F_2 + F_1\cos\theta)^2 + (F_1\sin\theta)^2$$

$$R^2 = F_2^2 + 2F_1F_2\cos\theta + F_1^2\left(\cos^2\theta + \sin^2\theta\right)$$

$$R = \sqrt{F_1^2 + F_2^2 + 2F_1F_2\cos\theta}$$

and .

$$\tan\alpha = \frac{F_1\sin\theta}{F_2 + F_1\cos\theta} \qquad \Rightarrow \qquad \alpha = \tan^{-1}\left(\frac{F_1\sin\theta}{F_2 + F_1\cos\theta}\right)$$

Activity 9.2

Determine the resultant force of two coplanar forces 20.0 N and 30.0 N which are inclined to each other with an angle 60.0°.

In addition to the previously mentioned methods, a geometric construction technique can be used to add force vectors in an organised way. To apply this technique, a polygon is drawn with the magnitude and direction of the force vectors to scale, and the polygon is completed (refer to Chapter 6, Section 6.6). The magnitude and direction of the resultant force can be determined from the final side of the polygon. For instance, if we have four force vectors $\vec{F}_1, \vec{F}_2, \vec{F}_3, \vec{F}_4$ the resultant vector $\vec{R}$ of these vectors is the vector sum $\vec{F}_1 + \vec{F}_2 + \vec{F}_3 + \vec{F}_4$ that completes the polygon, as depicted in figure 9.8.

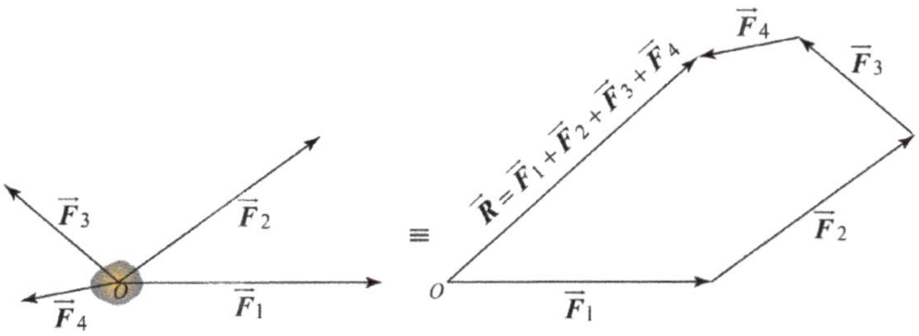

Figure 9.8

9.4 Newton's Laws of Motion

In this section, we examine the principles of dynamics using the concepts of force and mass, which form the basis of Newton's laws of motion. These laws were first introduced by Sir Isaac Newton (1642-1727) and are widely considered one of the most significant scientific achievements in history. Although some of the ideas in the laws were previously explored by scientists such as Galileo Galilei, it was Newton who developed them into a coherent theory. His laws revolutionised the way people understand and study the movement of objects, providing a fundamental framework for understanding their behaviour in motion. Newton's laws have been instrumental in explaining a wide range of physical phenomena, from the motion of celestial bodies to the behaviour of microscopic particles. While refinements and extensions to the laws have been made over time, they continue to be a critical tool for scientists and engineers in various fields of study. The following are the three laws of motion:

First Law of Motion – Law of Inertia:
"An object at rest tends to stay at rest, and an object in motion tends to stay in motion with the same speed and in the same direction unless acted upon by an external force."

Second Law of Motion – Law of Acceleration:
"The acceleration of an object is directly proportional to the net force acting on it and is inversely proportional to its mass. The direction of the acceleration is in the direction of the net force acting on the object."

Third Law of Motion – Law of Action-Reaction:
"If body A exerts a force on body B *(an action),* then body B exerts a force on body A *(a reaction).* These two forces have the same magnitude, but are opposite in direction and act on different bodies along the same line of action."

The laws of motion put forth by Newton are considered fundamental principles and cannot be derived or demonstrated from any other principles. These laws serve as the basis for classical mechanics and are often referred to as Newtonian mechanics. By utilising these laws, we can gain a better understanding of many common types of motion. Let us consider the laws of motion in detail.

9.5 First Law of Motion – Law of Inertia

Newton generalised the view of Galileo Galilei of motion and formulated Newton's first law of motion. It states that

> *an object at rest tends to stay at rest, and an object in motion tends to stay in motion with the same speed and in the same direction unless acted upon by an external force.*

Before Newton's first law of motion was proposed, the meaning of force was not well defined. However, the first law allowed us to precisely define force as that which changes the state of motion of an object that is at rest or moving uniformly in a straight line. This law can be explained through the example of a book resting on a table. The book will remain at rest until an external force is applied to it, such as when someone picks it up, or a strong gust of wind blows it off the table. Once a force is applied, the book will move in the direction of the force until another external force acts on it, such as gravity pulling it back down to the table.

(i) Inertia and Inertial Mass

Newton's first law of motion is also known as the law of inertia. Let us take a closer look at Newton's first law of motion. It states that an external force is required to alter the existing state of a body, whether it is at rest or moving uniformly in a straight line. Why is this necessary? Typically, force is needed when there is resistance to an attempted action. Upon examining the phrasing of Newton's first law, we realise that it requires an external force to change the state of any object. This means that all matter has an inherent tendency to resist any change in its present state, a characteristic known as inertia. That is why, Newton's first law is commonly referred to as the law of inertia. Inertia is a universal attribute of matter, and it is measured by the mass of an object. The greater the mass, the greater its inertia. In this context, we refer to an object's mass as its inertial mass.

To explain this further, imagine a scenario in which a mosquito is flying towards you at high speed, but you can easily stop it by intercepting it with your hand without causing any harm to yourself. However, if a car approaches you at the same speed, trying to stop it with your hand would have severe consequences. This demonstrates that velocity alone does not determine motion, and there must be another property of the body that must also be involved. This property is the mass of the body and is referred to as the inertial mass. Mass is a characteristic that all objects possess that determines how they respond to a force applied to them. The larger the mass, the more force is required to produce a given change in motion.

The first law has several real-world applications. For example, it helps explain why seat belts are important in a car. When a car suddenly stops, the passengers inside continue moving forward due to their inertia. Without a seat belt, they would keep moving forward and could be seriously injured. However, a seat belt applies an external force that prevents the passengers from continuing to move forward, thus reducing their risk of injury.

(ii) Inertial Frame of Reference

Newton's laws are the laws that determine how matter moves through space as a function of time. Therefore, in order to give these laws a precise meaning, we have to specify how we measure the position of some material object (say a particle) and the time at which it is at that position. We do this by introducing the concept called a frame of reference. The frame of reference (or reference frame) is a frame consisting of an abstract coordinate system and a set of physical reference points that uniquely locate and orient the coordinate system and regulate measurements.

Frames of reference can be classified as either inertial or non-inertial frames. An inertial reference frame is a reference frame in which Newton's law of inertia (Newton's first law) is valid. Accordingly, an inertial reference frame is one in which the acceleration of the frame is zero. Also, any frame of reference, which moves with a constant velocity relative to the inertial frame of reference, is also inertial. For example, the Earth is approximately an inertial frame of reference. Therefore, any reference frame moving at a uniform velocity relative to the Earth may also be considered as an inertial frame of reference. A reference frame in which Newton's first law is not valid can be considered as a non-inertial reference frame. Accordingly, the reference frame with an acceleration is a non-inertial frame.

(iii) Gravitational Mass

In subsequent sections, we will examine the phenomenon of gravitation. According to the law of gravitation, every pair of particles of matter exerts an attractive force on each other that is proportional to their masses. Therefore, by measuring the gravitational force acting on a body, its mass can be determined. For example, the weight of a body with mass m, which is the gravitational force exerted on it by the Earth, is given by $m\vec{g}$, where $\vec{g}$ is the acceleration due to gravity (note that weight is a vector quantity with magnitude and direction, while mass is a scalar). The mass of a body determined in this way is termed its gravitational mass. In particular, the numerical value and units of inertial and gravitational mass are equivalent.

Activity 9.3

Consider a situation where a worker shovels sand on the construction side, as shown in the figure below.

During the shovelling process, the worker fights against the inertia of the sand. Initially, the sand remains at rest on the ground. Explain how the worker could make the inertia of the sand easier to shovel.

9.6 Second Law of Motion – Law of Acceleration

According to Newton's first law, if no net force acts on an object at rest, the object remains at rest, or if it is moving, it continues to move with constant speed in the same direction. However, if a net force is applied to an object, its velocity will change. The direction and magnitude of the change in velocity depend on the direction and magnitude of the net force. If the net force is in the same direction as the motion, it increases the object's velocity. If the net force is in the opposite direction, it reduces the object's velocity. If the net force acts sideways, the direction and magnitude of the object's velocity change. Since the change in velocity per unit of time is acceleration, we can say that net force causes acceleration.

To explain the relationship between force and acceleration, consider the situation in which you push a trolley with a constant force. If you push it twice the force, the acceleration will double, and if you triple the force, the acceleration will triple. Therefore, the acceleration of an object is directly proportional to the net force applied to it. However, acceleration also depends on the mass of the object. If you push an empty trolley and it is filled with goods with the same force, you will notice that the filled trolley accelerates more slowly. The greater the mass, the less acceleration for the same net force. Thus, Newton argued that acceleration is inversely proportional to an object's mass. In general, these relationships are observed to be true and can be summarised as follows.

The acceleration of an object is directly proportional to the net force acting on it and is inversely proportional to its mass. The direction of the acceleration is in the direction of the net force acting on the object.

Mathematically this could be written as

$$\vec{a} = \frac{\sum \vec{F}}{m} = \frac{\vec{F}_{net}}{m} \tag{9.6}$$

where, $\vec{a}$ is the acceleration which is a vector quantity, m is the mass of the object and $\sum \vec{F}$ is the net force acting on the object and could be defined as $\vec{F}_{net}$. Here, $\sum \vec{F}$ means the vector sum of all the forces acting on the object.

If we rearrange equation 9.6, we get the standard form of the equation for Newton's second law as

$$\vec{F}_{net} = \sum \vec{F} = m\vec{a} \tag{9.7}$$

Equation 9.7 is expressed in vector form, which means that we can break it down into components along the Cartesian coordinate system. This involves writing a separate equation for each component of a force and its corresponding component of acceleration. Accordingly, for an object that moves in a two-dimensional plane, the individual components of the total force are equal to the product of mass and the corresponding component of acceleration.

$$\sum F_x = ma_x \quad \text{and} \quad \sum F_y = ma_y \tag{9.8}$$

Equations 9.7 and 9.8 are valid only when the mass m is constant. Additionally, like the first law, Newton's second law is valid only in inertial frames of reference.

The SI unit of force, Newton, is abbreviated as N. It is defined so that a net force of 1 N gives a 1kg mass an acceleration of 1 m s^{-2}. That is,

$$1 \text{ N} = (1 \text{ kg})(1 \text{ m s}^{-2}) = 1 \text{ kg m s}^{-2} \tag{9.9}$$

By defining the unit of force in this manner, it becomes feasible to express equations 9.7 and 9.8 without requiring a proportionality constant to convert between the force unit and kg m s^{-2}.

9.7 Object in Equilibrium

When an object modelled as a particle is acted on by no forces or by several forces whose vector sum (i.e. resultant) is zero, we say that the object is in equilibrium. This is commonly known as translational equilibrium. Thus, for an object in equilibrium, i.e. when equilibrium under zero resultant force,

$$\vec{R} = \sum \vec{F} = 0 \tag{9.10}$$

This means each component must be zero, therefore

$$\sum F_x = 0, \quad \sum F_y = 0 \quad \text{and} \quad \sum F_z = 0, \tag{9.11}$$

In other words, from equation 9.10, the object is in equilibrium, meaning it does not have acceleration. It also says that when the object is in equilibrium, it is either at rest or moving at constant velocity. In such cases, any forces acting on the object balance each other. This does not mean that the forces cease to exist. The forces act on the object but cannot change its velocity.

9.8 Free-Body Diagrams

A Free Body Diagram (FBD) is a graphical representation used to analyse the forces acting on an object in a given situation. It is a simple diagram that shows the object as a point or a simple shape, with all the forces acting on it represented by arrows. The direction and magnitude of each arrow represent the direction and magnitude of the force, respectively.

In an FBD, the object is typically represented as a point or a shape, and all external forces acting on the object are drawn as arrows pointing away from the object. These forces may include weight, friction, tension, normal force, and any other forces that are present in the situation. The arrows are labelled with the magnitude of the force and its direction.

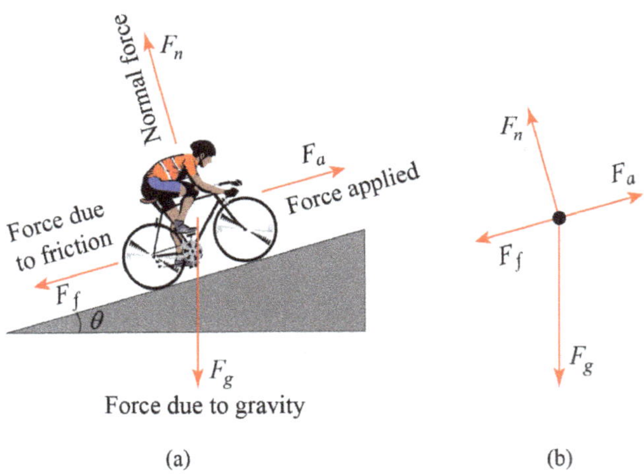

Figure 9.13: (a) Forces acting on a cyclist when climbing on a slope and (b) the FBD

The FBD is useful for understanding how different forces interact with each other and affect the motion of the object. By analysing the FBD, we can determine the net force acting on the object and the resulting acceleration. This information is then used to solve problems related to motion and force. When with more than one object, we draw separate free-body diagrams for each object. Figure 9.19 (a) shows the forces that act on a cyclist when climbing a slope, and figure 9.19 (b) shows its FBD.

Activity 9.4

The figure shows three scenarios in which forces $\vec{F}_1, \vec{F}_2$ and $\vec{F}_3$ act on a smooth flat disc of mass $160\ \text{g}$ that moves over a frictionless surface along an x axis in one-dimensional motion. If the forces $\vec{F}_1$ and $\vec{F}_2$ have magnitudes of $4.0\ \text{N}$ and $2.4\ \text{N}$, respectively, and act on the disc horizontally, as shown, the force

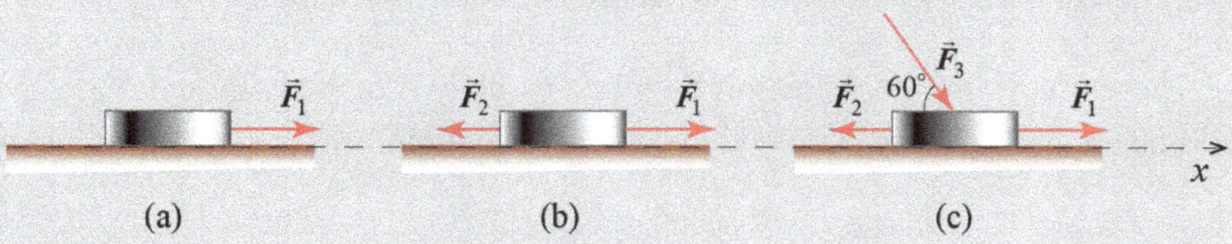

$\vec{F}_3$ is directed at an angle 60° with a magnitude of 1.6 N, as shown. In each situation, determine the acceleration of the disc.

9.9 Third Law of Motion – Law of Action-Reaction

Newton's third law of motion expresses that for every action, there is an equal and opposite reaction. For example, consider a situation in which a football player kicks a football, as shown in figure 9.10.

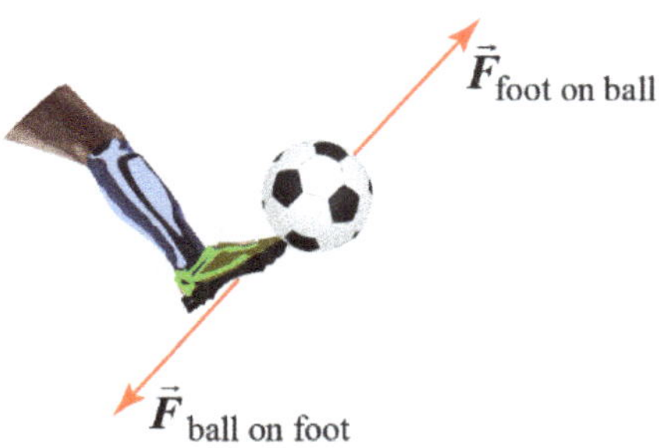

Figure 9.10: The forces between the ball and the foot act on each other when the ball is being kicked by the foot

When the player kicks the ball, his foot exerts a force on the ball $(\vec{F}_{\text{foot on ball}})$ in the forward direction. This is the action force. According to Newton's third law, the ball exerts an equal and opposite force back on the player's foot $(\vec{F}_{\text{ball on foot}})$ in the backward direction. This is the reaction force. This is what allows the player to feel the impact of their foot hitting the ball. It also explains that forces always come in pairs and that the magnitude and direction of each force in the pair are equal and opposite. Further, these two forces act on different bodies and act along the same line of action.

This means that when an object exerts a force on another object, the second object exerts an equal and opposite force on the first object. In other words, these relationships are observed to be true and can be summarised as follows.

If body A exerts a force on body B (an action), then body B exerts a force on body A (a reaction). These two forces have the same magnitude, but are opposite in direction and act on different bodies along the same line of action.

Therefore, if $\vec{F}_{A \text{ on } B}$ is the force applied by body A on body B, and $\vec{F}_{B \text{ on } A}$ is the force applied by body B on body A, then in equation form, we could express as

$$\vec{F}_{A \text{ on } B} = -\vec{F}_{B \text{ on } A} \tag{9.12}$$

We sometimes refer $\vec{F}_{A \text{ on } B}$ and $\vec{F}_{B \text{ on } A}$ which are the action–reaction pair acting on different bodies.

Newton's third law applies to all objects, regardless of their size, mass, or motion and is sometimes referred to as the law of action–reaction. This law has practical applications in a wide range of fields, including engineering, physics, and even sports. For example, a rocket accelerates by expelling hot gases out of its engines in one direction, which causes an equal and opposite reaction force that propels the rocket in the opposite direction.

Activity 9.5

Consider a plant that is grown in a container and suspended by a cord from a hook on the fixed ceiling. Draw FBD for each of the following with appropriate labelling:

(i) The container consists of a plant, soil, and pot.

(ii) The cord.

(iii) The hook.

(iv) The system consists of a container (with plant, soil, pot), cord, and hook.

9.10 Gravitational, Normal, Friction, Tension and Other Forces

As we have discussed earlier, a force is a push or pull acting upon an object as a result of its interaction with another object. There are various types of forces in the physical world, including normal force, friction force, tension force, gravitational force, electromagnetic force, and others. These forces play an essential role in the behaviour and motion of objects in the physical world. Understanding the properties and characteristics of each type of force is crucial in many areas of science and engineering, including mechanics, dynamics, and materials science. In this context, we will focus on the properties and characteristics of gravitational, normal, friction, and tension forces.

Gravitational force

As discussed in Section 7.5 of Chapter 7, Galileo asserted that if air resistance can be disregarded, all objects dropped close to the surface of the Earth would fall with the same acceleration $\vec{g}$. The force that leads to this acceleration is known as the gravitational force or the force of gravity.

Let us apply Newton's second law to an object of mass m falling under the influence of gravity. For acceleration, we use downward acceleration due to gravity $\vec{g}$. If we neglect the air resistance, the gravitational force on the object denoted by $\vec{F}_g$ can be written as

$$\vec{F}_g = m\vec{g} \tag{9.13}$$

The direction of this force is down towards the centre of the Earth. The magnitude of the gravitational force on the object is commonly called the object's weight and is denoted by W. Thus,

$$W = mg \tag{9.14}$$

This same gravitational force, with the same magnitude, still acts on the object even when an object is not in free fall and is at rest on a surface or moving across it.

We are usually concerned with the weight of objects on Earth. However, it is important to note that the weight of the same object would differ on the Moon, other planets, or in space. For instance, on the Moon, the acceleration due to gravity is approximately one-sixth of what it is on Earth, which means that the weight of an object on the Moon would also be one-sixth of its weight on Earth.

Activity 9.6

If the mass of a person is $70\ \text{kg}$ determine his weight on:

(i) Earth, where $g = 9.8\ \text{m s}^{-2}$

(ii) Earth's Moon, where $g = 1.6\ \text{m s}^{-2}$.

(iii) Venus, where $g = 8.9\ \text{m s}^{-2}$ and

(iv) Mars, where $g = 3.7\ \text{m s}^{-2}$.

Normal force

Normal force is a type of force that arises when two objects come into contact with each other. Specifically, the normal force is the force that a surface exerts on an object that is pressing against it. The normal force is called "normal" because it is perpendicular to the surface that the object is pressing against.

For example, when you place a book on a table, as shown in figure 9.11, the book exerts a downward force on the table due to the force of gravity $(\vec{F}_g)$. The table, in turn, exerts an equal and opposite force back on the book, called the normal force $(\vec{F}_n)$. This normal force prevents the book from falling through the table and provides the support needed to keep it in place.

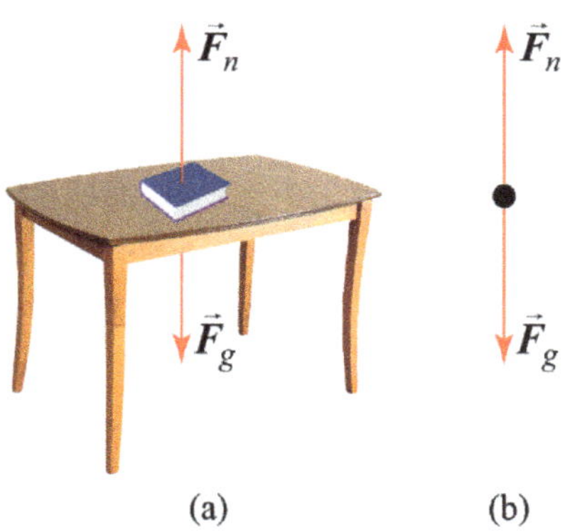

Figure 9.11: (a) When a book is at rest on a table, the forces acting on the book are the normal force $\vec{F}_n$ and the gravitational force $\vec{F_g}$ (b) A free-body diagram shows the book as a black dot with the forces acting on it

Because the book is at rest and has zero acceleration, Newton's second law applied to the book gives us the following.

$$\sum \vec{F} = \vec{F}_g + \vec{F}_n = 0$$

$$\vec{F}_n = -\vec{F}_g \qquad (9.15)$$

This means that the normal force balances the gravitational force on the book, so the net force on the monitor is zero. Figure 9.11 (b) is the free-body diagram showing the forces that act on the book. Remember that Newton's third-law action and reaction forces act on different objects. In figure 9.11 (b), the forces $\vec{F}_n$ and $\vec{F}_g$ are equal in magnitude and opposite in direction, but they do not represent an action–reaction pair because both forces act on the same object, the book.

The magnitude of the normal force depends on various factors, such as the weight of the object and the angle at which it is placed on the surface. If an object is at rest, the normal force is equal and opposite to the force of gravity, which is pulling the object downward. In other words, the normal force balances the weight of the object. However, if an object is placed on an inclined surface, the normal force will be less than the force of gravity acting on the object since a component of the force of gravity acts parallel to the surface.

External forces that act on the object also affect the normal force. For example, if an object is being pushed into the surface, the normal force will increase to prevent the object from sinking further into the surface. On the other hand, if an object is pulled away from the surface, the normal force will decrease.

Activity 9.7

(i) (ii) (iii)

(i) A container of mass 10.0 kg rest on a frictionless table, as shown in figure (i), determines the weight of the container and the normal force exerted on it by the table.

(ii) Suppose that you push the container with your hand perpendicularly in the downward direction with the force of $50.0\ \mathrm{N}$, as shown in figure (ii), determine the normal force exerted on the container by the table.

(iii) If you pull the container with the help of a massless cord perpendicularly in the upward direction, as shown in figure (iii), determine the normal force exerted on the container by the table. Assume $g = 10\ \mathrm{m\ s^{-2}}$.

Friction force

When an object is in motion either on a surface or in a viscous medium such as air or water, it encounters resistance to its motion because the object interacts with its surroundings. We call such resistance a friction force. Friction is an important aspect of our daily lives. In machinery, every rotating shaft will come to a halt if left to itself. About 20% of the engine's power in a motor vehicle is used to overcome friction forces. Due to friction, most machines undergo considerable wear and tear on their moving parts. However, friction is also a necessary factor in everyday life. Without it, we could not hold a pencil in our hand, walk or move the wheels of the vehicle.

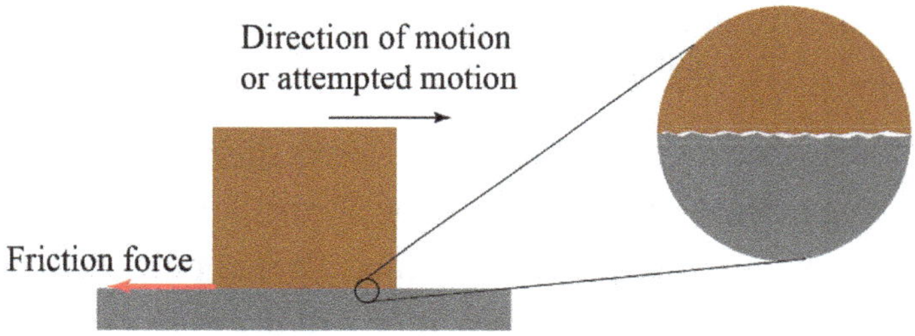

Figure 9.12 shows the pictorial representation of how friction occurs at the interface between two objects. A close-up inspection of these surfaces shows them to be rough

Figure 9.12: Friction forces, such as f always oppose the motion or attempted motion between surfaces in contact. Friction arises in part due to the roughness of the surfaces in contact, as seen in the expanded view.

Suppose that we push a block of mass m along a horizontal table and leave it. It will eventually come to rest. This means that, in the course of its motion, the block is experiencing an average deceleration $\underset{\sim}{a}$ in the direction of its motion or an acceleration in the opposite direction. We have already learnt from Newton's laws of motion that whenever a body experiences an acceleration, it must be acted upon by a force in the direction of the acceleration. Thus, the table exerts a force of friction on the moving block opposite to its direction of motion and equal to $m\underset{\sim}{a}$. Whenever the surface of any body slides over the surface of another, each body exerts a frictional force on the other, parallel to the surfaces. Frictional forces always oppose the motion of one body relative to the other.

If two surfaces are in contact and moving relative to one another, then the friction between them is called kinetic friction. But when objects are stationary, static friction can act between them; the static friction is usually greater than the kinetic friction between the surfaces.

Consider a block of mass m at rest on a horizontal table. When we apply a small force $\vec{F}$ to the block, it will not move, as shown in figure 9.13 (a). Obviously, according to Newton's laws of motion, there cannot be a resultant force acting on the block and $\vec{F}$ must be balanced by an equal and opposite force $\underset{\sim}{f_s}$, as illustrated in figure 19.13 (a), (b). This force $\underset{\sim}{f_s}$ is the force of friction exerted on the block by the table at their surface of contact. As we increase $\vec{F}$, there comes a stage when the block just begins to move.

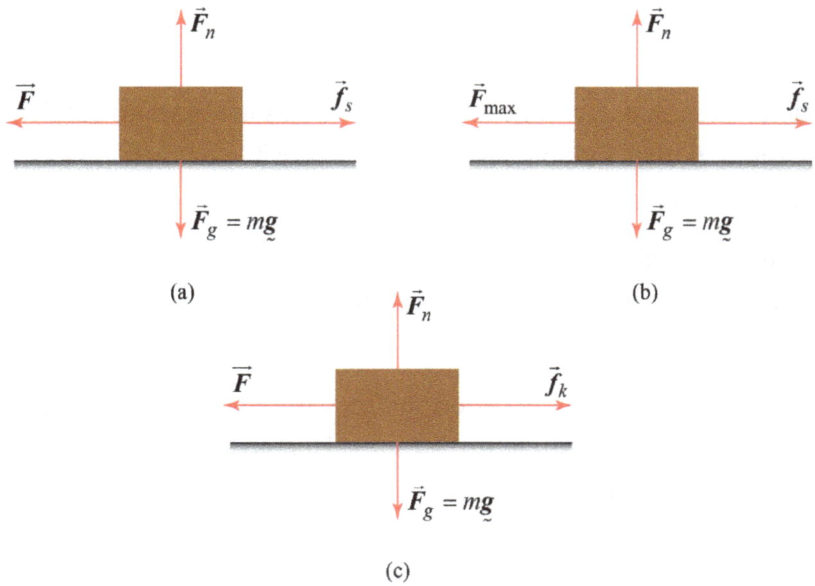

Figure 9.13: (a) When applying a small force $\vec{F}$ to the block and not moving (b) maximum force $\vec{F}_{max}$; just initiate its motion, but it will keep the block at rest (c) the frictional force acting between surfaces in relative motion.

The frictional force that acts on the block as long as it is at rest is called a force of static friction. The maximum force $\vec{F}_{max}$, which would still keep the block at rest, is also the minimum force that would just initiate its motion. $\vec{F}_n$ is the normal force exerted on the block by the table and it is the reaction to the weight $\left(\vec{F}_g = m\underset{\sim}{g}\right)$ of the block. Since there is no vertical motion, $\vec{F}_n$ always remains constant.

Now, the maximum force of friction $\vec{F}_{max}$, i.e. the maximum force that would keep the block at rest, obeys two empirical laws:

(i) It is approximately independent of the area of the surfaces in contact, and

(ii) It is proportional to the normal force $\vec{F}_n$.

The ratio of the magnitude of the maximum force of static friction $(f_s)_{max}$ to the magnitude of the normal force F_n is called the coefficient of static friction μ_s for the surfaces considered.

$$\mu_s = \frac{(f_s)_{max}}{F_n} \tag{9.16}$$

Thus, we can write

$$f_s \le (f_s)_{max} = \mu_s F_n \tag{9.17}$$

Once motion has started, the same force $\vec{F}_{max}$ causes the block to accelerate. If we reduce the force $\vec{F}$ once motion has begun, it is possible to keep the block moving in uniform motion without acceleration.

When the surfaces are in relative motion, it can be shown that the frictional forces between them tend to decrease. Thus, to maintain uniform motion, a smaller force $\vec{F}$ would be sufficient.

The friction force acting between surfaces in relative motion is the force of kinetic friction $\underset{\sim}{f_k}$, as shown in figure 9.13 (c). The force of kinetic friction follows the same laws as the force of static friction.

(i) It is approximately independent of the area of contact of the two surfaces, and

(ii) It is proportional to the normal force $\vec{F}_n$.

The force of kinetic friction is also generally independent of the relative speed of the two surfaces. The coefficient of kinetic friction μ_k is defined as the ratio of the magnitude of the force of kinetic friction f_k to the magnitude of the normal force F_n. Thus,

$$f_k = \mu_k F_n \tag{9.18}$$

Usually for two given surfaces, $\mu_s > \mu_k$. Their actual values depend on the nature of the surfaces in contact. Table 9.1 lists some typical values of μ_k and μ_s.

Table 9.1: Coefficients of static and kinetic friction

Materials	**Coefficient of Static Friction, μ_s**	**Coefficient of Kinetic Friction, μ_k**
Rubber on concrete	1.0	0.8
Steel on steel	0.74	0.57
Rubber on wet concrete	0.7	0.5
Wood on wood	0.25–0.5	0.2
Copper on steel	0.53	0.36
Ice on ice	0.1	0.03
Synovial joints in humans	0.01	0.003

Figure 19.14 shows that the magnitude of the force of friction as a function of the magnitude of the external force applied to an object initially at rest.

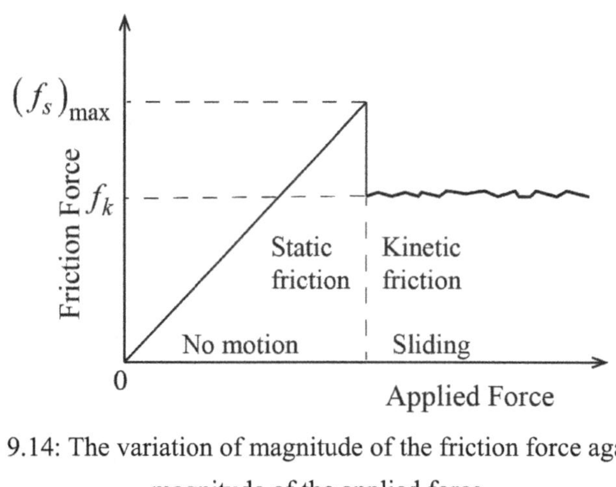

Figure 9.14: The variation of magnitude of the friction force against the magnitude of the applied force

When the applied force is increased in magnitude, the force of static friction increases linearly to the maximum value, called $(f_s)_{\max}$, until it just matches the applied force equals to $\mu_s F_n$. If we continue to increase the applied force further, the object will begin to move, and the friction force drops to a roughly constant value characteristic of kinetic friction.

Activity 9.8

A wooden block of weight W rests on a rough inclined plane, which makes an angle θ with the horizontal. Show that the difference between the minimum force necessary to start the block moving down the plane with that necessary to start the block moving up the plane is $2W\sin\theta$.

Activity 9.9

A wooden block is placed on a plank and one end is gradually raised. When the angle of inclination with horizontal reaches 30°, the block starts to slip and slide $5.0\ \text{m}$ down the plank in $2.0\ \text{s}$. Determine the coefficients of static and kinetic friction between the wooden block and the plank.

Tension force

Tension is a force transmitted along a flexible medium such as a rope, cable or cord. Any flexible connector (that can be easily bent or flexed without requiring much force), such as a string, rope, cord, chain, wire, or cable, can exert pulls only parallel to its length. The force carried by a flexible connector is called a tension force, where its direction is parallel to its connector. It is important to understand that tension is a pull on the connector and, on the contrary, it cannot be pushed.

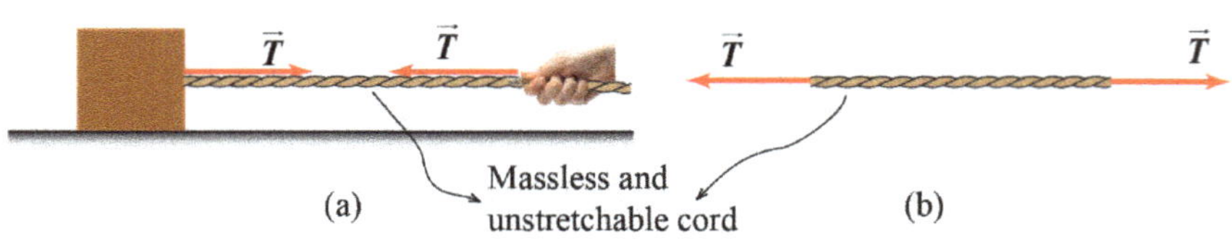

Figure 9.15: (a) The cord pulled the object. If the mass of the cord is negligible compared to the mass of the object, the cord pulls on the object and the hand with tension force $\vec{T}$ (b) The tension force $\vec{T}$ acting on the two ends of the rope

Figure 9.15 (a) shows the cord pulled the object which is sitting on a flat surface. If the mass of the cord is negligible compared to the mass of the object that is being pulled, the force exerted by the cord (pull) on the object and the hand is equal and is denoted by the tension force $\vec{T}$. The tension force pulls outward along the two ends of a cord, as shown in figure 9.15 (b).

A flexible connector (rope, cable, cord, etc.) is often said to be massless, which means that its mass is negligible compared to the mass of the objects being pulled and it is unstretchable. The flexible connector then exists only as a connection between two objects. It pulls on both objects with the same force magnitude T, even if the objects and the flexible connector are accelerating and even if the flexible connector runs around a massless, frictionless pulley. Figure 9.16 shows two situations where a cord runs through a massless, frictionless pulley. This pulley has negligible mass compared to the objects that are being pulled and negligible friction on its axle opposing its rotation.

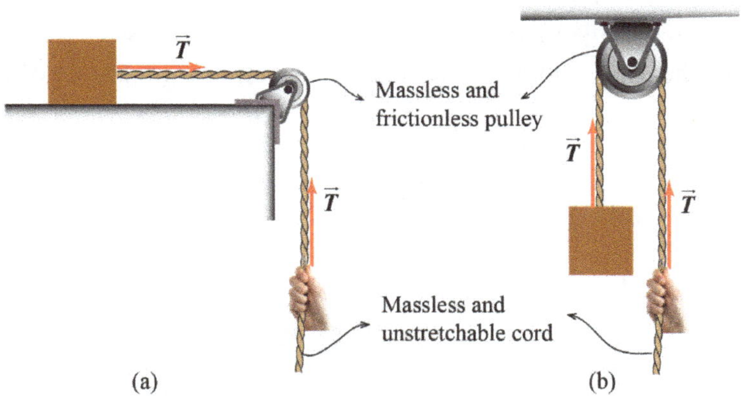

Figure 9.16: (a) The cord pulled the object, where the cord runs around a massless, frictionless pulley. (b) The cord wraps halfway around a massless, frictionless pulley and the object is pulled up by pulling the other end of the cord

Figure 9.16 (a) shows the cord pulled the object, where the cord runs through a massless, frictionless pulley. If the mass of the cord is negligible compared to the mass of the object that is being pulled, the force exerted by the cord (pull) on the object and the hand is equal and is denoted by the tension force $\vec{T}$. If the cord wraps halfway around a pulley, as shown in figure 9.16 (b), the net force exerted on the pulley by the cord has the magnitude $2T$.

Activity 9.10

Airbags are inflatable cushions built into a vehicle that protect the occupants during a crash. They are normally hidden, but inflate instantly when a crash begins. Consider a case where a $70\ \text{kg}$ passenger sits in the front seat of a car and experiences a crash. According to safety standards, the bags produce a maximum acceleration of $60g$, and act in a very short time, where $g = 10\ \text{m s}^{-2}$ is the acceleration due to Earth's gravity.

(i) Draw the free-body diagram for the passenger.

(ii) What force does the airbag exert on the passenger?

Activity 9.11

As shown in the figure at a construction site, a bucket filled with mixed cement having $15\ \text{kg}$ is hung from one end of a massless rope that passes over a small, frictionless and massless pulley. A $25\ \text{kg}$ counterweight is suspended from the other end of the rope, and the system is set in motion from a state of rest.

(i) Draw the FBDs for the bucket filled with mixed cement and the counterweight.

(ii) Determine the magnitude of the upward acceleration of the bucket filled with mixed cement.

(iii) Determine the tension on the rope while the load is moving.

(iv) How does tension compare with the weight of the bucket filled with mixed cement?

(v) How does tension compare with the weight of the counterweight?

Spring Force

The force exerted by a spring when it is stretched or compressed is commonly known as the spring force. It follows Hooke's law, which establishes a linear relationship between the applied force and the resulting displacement.

Let us consider the situation where a coil spring with one end attached to a fixed point and the other end attached to an object that can slide without friction along the x axis, as shown in figure 9.17 (a), (b) and (c).

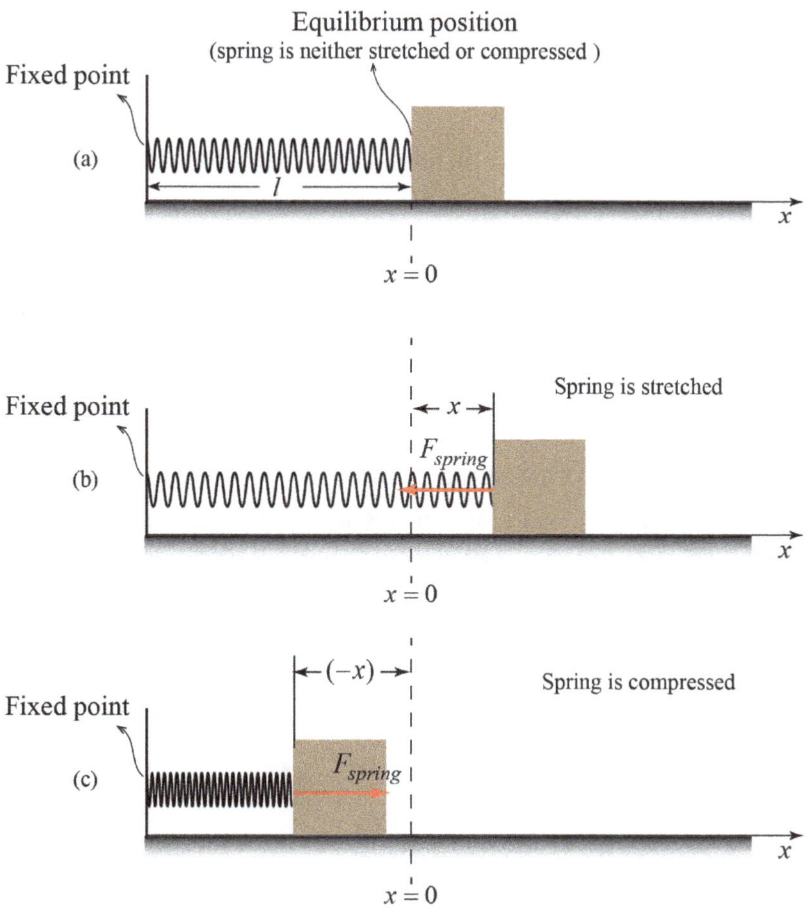

Figure 9.17: The force exerted by a spring on an object : (a) when $x = 0$ the spring is neither stretched nor compressed, it will have its natural length (b) the spring is stretched by an amount x (c) the spring is compressed by an amount x

Figure 9.17 (a) shows that when $x = 0$ the spring remains at its natural length l, without being stretched or compressed. As shown in figure 9.17 (b), when the object is displaced in the positive x direction by a distance x, the spring stretches by an amount x, causing the spring force F_{spring} to act in the leftward direction. On the contrary, as shown in figure 9.17 (c), when the object is displaced in a negative x direction by a distance x, the spring compresses by an amount x, causing the spring force F_{spring} to act in the rightward direction. In both scenarios, the spring exerts a force of magnitude F_{spring} on the object while applying an 'equal and opposite' force on the fixed point. This spring force F_{spring} can be mathematically modelled as

$$F_{spring} = -kx \tag{9.19}$$

and is known as Hooke's law. Accordingly, the spring force (F_{spring}) is directly proportional to the displacement (x) from the spring's equilibrium position and k is the proportionality constant known as the spring constant (a measure of the stiffness of the spring). The negative sign in the equation indicates that the spring force opposes the direction of displacement, acting as a restoring force. The restoring force will attempt to restore the spring to its original equilibrium position.

The magnitude of the spring force depends on the spring constant k and the amount of displacement x. If force is measured in newtons (N) and x in metres (m), the units of k are N m^{-1}. A large k value means a stiffer spring requiring greater force to stretch or compress. Similarly, larger displacement x results in larger spring forces.

Springs are commonly used in suspension systems, mechanical devices, and even in items such as mattress coils, where their ability to store and release energy through spring force proves beneficial.

Activity 9.12

As shown in the figure, a spring balance is used to weigh vegetables. The spring balance is built with a spring that stretches. When a 15.0 N weight is placed in the pan, the spring stretches 1.25 cm. Take $g = 10 \text{ m s}^{-2}$.

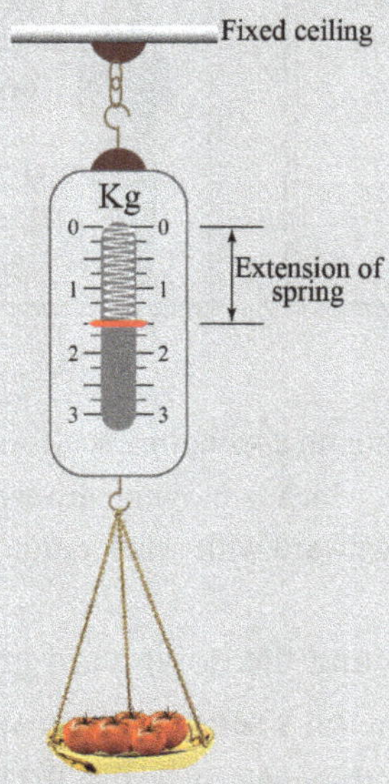

(i) Determine the spring constant k of the spring.

(ii) What distance does the spring stretch if a 2.50 kg vegetable is placed in the pan?

9.11 Apparent Weight

The actual weight of an object is the gravitational force that the Earth exerts on it and is measured when the object is not accelerating vertically relative to the ground. For example, you can measure your weight using a scale. However, if you measure your weight on the same scale in an accelerating elevator, the reading differs from your actual weight due to acceleration. Such a measure of weight is called an apparent weight.

Consider a situation where a student is standing on a weighing scale. The weighing scale measures the student's apparent weight, which is equal to the student's actual weight only if the student and the scale have zero acceleration (figure 9.18(a)).

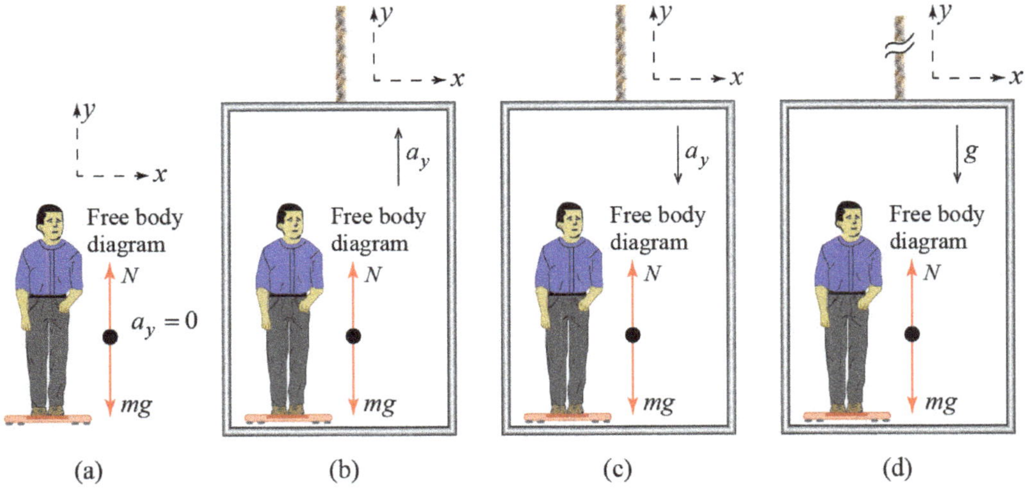

Figure 9.18: A student is standing on a weighing scale (a) the weighing scale has zero acceleration (b) inside the elevator, which is moving upward with acceleration a_y (c) inside the elevator, which is moving downward with acceleration (d) the elevator cable is broken

If the student's mass is m and the downward gravitational acceleration is g, we have the following from Newton's second law if we apply it in the $+y$ direction. If there is no acceleration, as in the case of figure 9.18(a), we have

$$\sum F_y = N - mg = ma_y = 0 \tag{9.20}$$

where N is the normal force of the weighing scale pushing up.

The apparent weight W' is the reading of the weighing scale, which is the magnitude of the normal force N.

i.e. $$W' = N = mg$$

Consider a situation where the student is standing on a weighing scale inside the elevator, which is moving upward with acceleration a_y, as shown in figure 9.18(b). In this case, the normal force N must be greater than the actual weight mg of the student for the net force to be in the upward direction. If we apply Newton's second law in the $+y$ direction, we get

$$\sum F_y = N - mg = ma_y \tag{9.21}$$

i.e. $$N = m(g + a\ )$$

Therefore, the apparent weight W' will be

$$W' = N = m(g + a_y) \tag{9.22}$$

Since the elevator's acceleration is upward, and $a_y > 0$, the apparent weight W' is greater than the actual weight of the student. Therefore, the student will experience the sensation of heaviness. Additionally, students will feel that the surface of the weighing scale is pushing more on the feet than when the elevator is stationary or moving with constant velocity.

Now, consider the student standing on a weighing scale inside the elevator, which is moving downward with acceleration a_y, as shown in figure 9.18 (c). Since the acceleration is downward, the net force must also be downward. Therefore, the normal force N must be less than the actual weight mg of the student for the net force to be in the downward direction. If we apply Newton's second law in the $+y$ direction, we get

$$\sum F_y = N - mg = -ma_y \tag{9.23}$$

i.e. $$N = m(g - a_y)$$

Therefore, the apparent weight W' will be

$$W' = N = m(g - a_y) \tag{9.24}$$

The apparent weight W' is less than the actual weight of the student. Therefore, the student will experience a feeling of weightlessness. Additionally, students will feel that the surface of the weighing scale is pushing less on the feet than when the elevator is stationary or moving with constant velocity. For example, if the elevator accelerates downward with the value of $\frac{1}{2}g$, then the apparent weight W' will be $\frac{1}{2}mg$. This means that the weighing scale reads half the actual weight of the student.

Suppose the elevator cable breaks, as shown in figure 9.18 (d), then the elevator, along with the student and the weighing scale, plunges straight down in free fall. In free fall, the elevator acceleration will be the same as the downward acceleration of gravity g, and from equation 9.24, the normal force $N = 0$. The students will feel that the surface of the weighing scale does not exert any force on the feet of the student. This means that the weighing scale will read zero and the apparent weight of the student will be zero. This phenomenon is called weightlessness. In fact, gravity is still pulling down the student, that is why the student is falling. But the student will not feel the sensation of weight as everything floats around him in free fall.

Activity 9.13

A $60.0\ \text{kg}$ student stands on a weighing scale in an $860.0\ \text{kg}$ (including the student and the weighing scale) elevator that is supported by a cable. As the elevator starts to move, the scale reads $500\ \text{N}$.

(i) Find the acceleration of the elevator (magnitude and direction).

(ii) Determine the acceleration of the elevator (magnitude and direction) if the scale reads $700\ \text{N}$.

(iii) If the weighing scale reads zero, should the student worry? Explain.

(iv) Determine the tension in the cable in parts (i), (ii) and (iii).

Another example of apparent weight can be observed when objects are submerged in a fluid, such as water. The apparent weight of an object in water is less than its actual weight due to the upward buoyant force exerted by the fluid. This phenomenon explains why objects appear lighter when immersed in water compared to when in air.

Roller coasters and amusement park rides often subject riders to rapid changes in apparent weight. The thrill and sensation of weightlessness experienced during these rides are the result of carefully designed acceleration patterns.

Pilots and astronauts experience changes in apparent weight during take-off, landing, and when manoeuvring aircraft and spacecraft. Knowledge of apparent weight is crucial when designing safe and comfortable transportation systems.

Questions

NOTE: Although the solutions are given on the last page, it is in your own interest to attempt the questions on your own as an indication of how far you have understood the material of the section.

1. Based on Newton's law, explain the scientific reason for the regulation that mandates the driver and passengers to wear seat belts while driving a car.

2. When a dog gets wet, it shakes its body all over to get dry. What principle of physics aids in the drying process? Explain.

3. Can an object in free fall be in equilibrium? Explain.

4. Can an object exert a force on itself? Explain.

5. As shown in the figure, a fishing trawler is stranded in the middle of the sea and is being towed by two small boats, *A* and *B*. The trawler engine does not provide any force. If the tension in each cable between *A* and *B* and the trawler is 4500 N, determine the direction and magnitude of the resultant force acting on the trawler. Each cable attached to the boats makes an angle 30° with the horizontal.

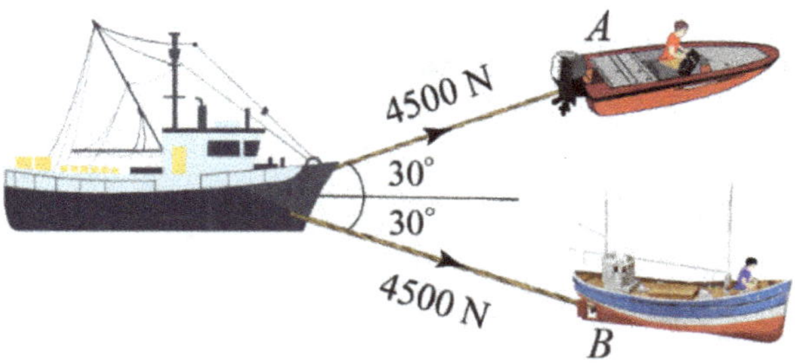

6. Three competitors, *A, B,* and *C,* participate in a two-dimensional tug-of-war competition, where they pull horizontally on an automobile tyre at the angles shown in the figure.

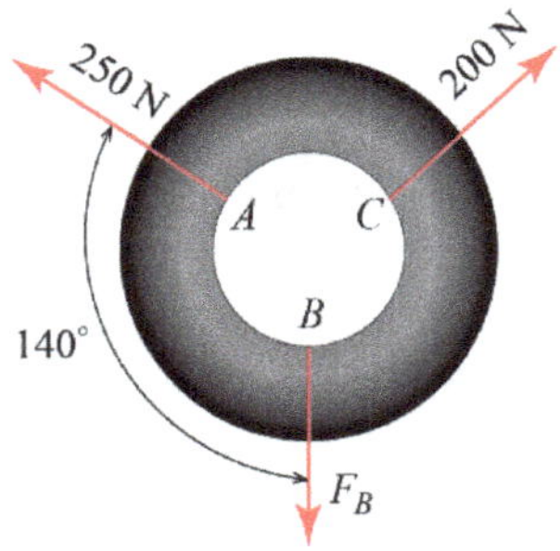

Despite the three pulls, the tyre remains stationary. Competitor *A* pulls with a force of $250\ \mathrm{N}$, and competitor *C* pulls with a force of $200\ \mathrm{N}$. Competitor *B* pulls with force F_B at an angle of 140° with the force of competitor *A*. Determine the magnitude of the force F_B exerted by competitor *B*.

7. The figure shows a stone suspended by two strings, *A* and *B*, at rest. String *A* is at an angle of 60° above the horizontal and has a tension of $1.20\ \mathrm{N}$. String *B* is at an angle of 30° above the horizontal, and its tension is T.

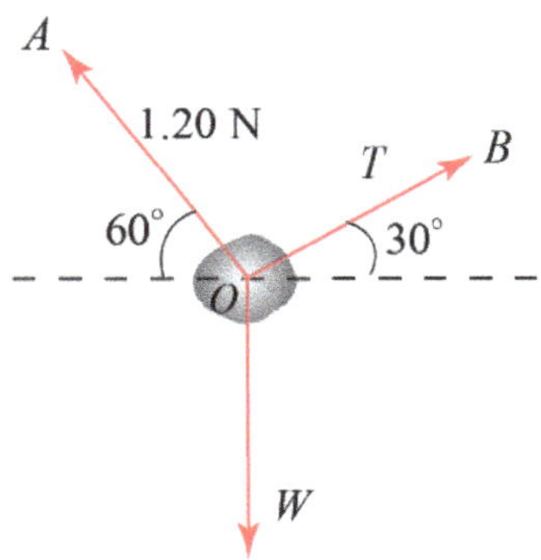

Determine the following:

(i) The tension T in string *B*.

(ii) The weight *W* of the stone.

8. A factory worker pulls a 50 kg wooden crate across a wooden floor using a rope that is fixed to the crate, as shown in the figure.

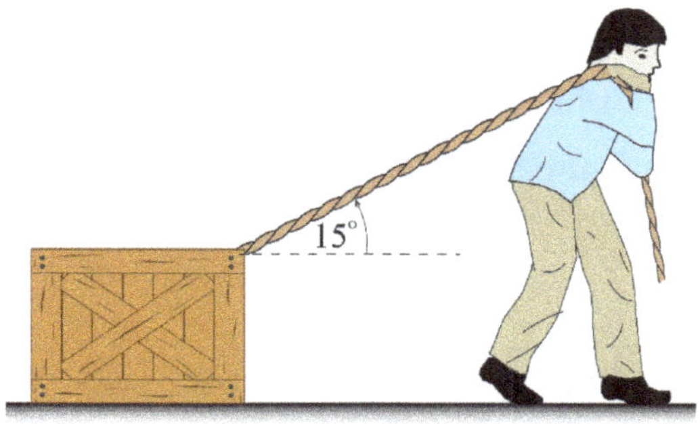

When pulling the rope with a force of 120 N, it makes an angle of 15° with the horizontal. Take the kinetic friction between the wooden floor and the wooden crate as 0.20 and the crate is under a uniform motion along the horizontal floor.

(i) Draw a free-body diagram for the crate.

(ii) Determine the normal force acting on the crate.

(iii) Determine the acceleration of the crate while pulling it.

(iv) Determine the time that elapses between when the crate just starts to move and when he pulls it with a velocity of 1.5 m s^{-1}.

9. Two blocks *A* and *B*, having masses m_1 and m_2 respectively $(m_1 > m_2)$, lie on a horizontal frictionless table and are in contact as shown in the figure below.

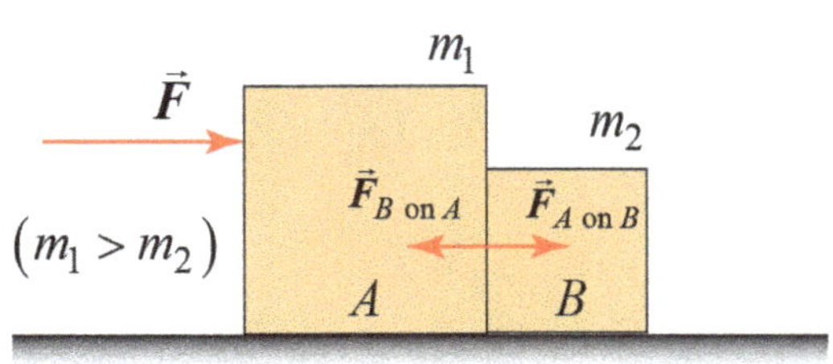

(i) A horizontal force $\vec{F}$ is applied to block *A* as shown in the figure and the blocks move together in the direction of the applied force $\vec{F}$.

(a) Draw a free-body diagram for block *A* and block ***B***.

(b) Find the magnitude of the force of contact between the two blocks A and B ($F_{A \text{ on } B}$ and $F_{B \text{ on } A}$) in terms of m_1, m_2 and the magnitude of the force F.

(iii) Now consider the situation where the same force $\vec{F}$ is applied to block B as shown in the figure below and the blocks move together in the direction of the applied force $\vec{F}$.

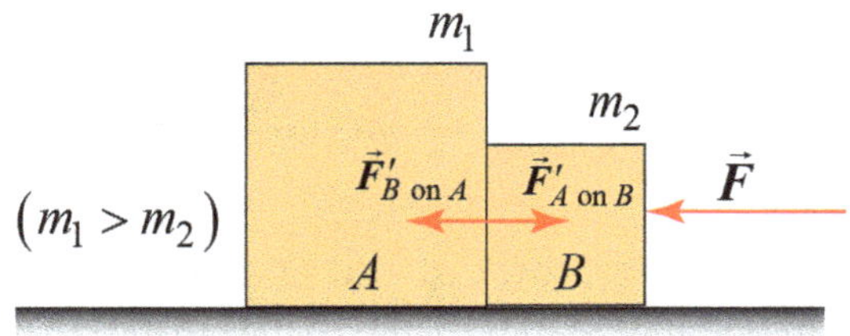

(a) Show that the magnitude of the acceleration of this system is the same as the magnitude of the acceleration of the system considered in part (i).

(b) Is the force of contact between blocks A and B ($F'_{A \text{ on } B}$ and $F'_{B \text{ on } A}$) the same, greater, or less than the contact force obtained in part (i)? Explain.

10. The figure shows that a 100 kg load of bricks is pulled upward at a constant speed at a construction site using the massless rope and two massless and frictionless pulleys. Take $g = 10 \text{ m s}^{-2}$.

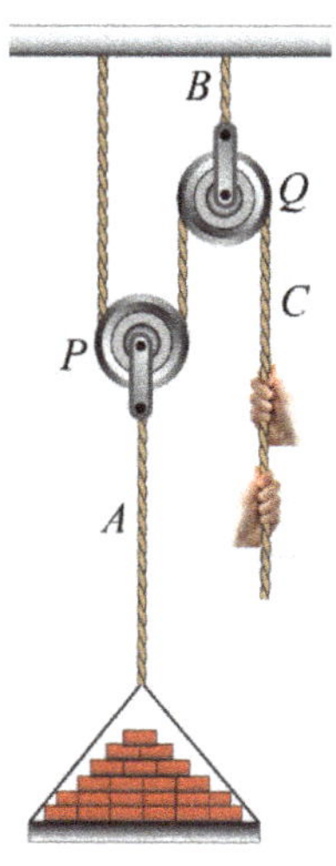

(i) Draw a free-body diagram for bricks, pulley *P* and pulley *Q*.

(ii) Determine the tensions in the three ropes labelled as *A, B* and *C*.

(iii) Explain why the workers have used two pulleys instead of one.

11. The figure shows a situation where a crow rests on a clothesline made of a massless cord, a device for hanging clothes on for the purpose of drying them. Each end of the cord makes an angle of θ below the horizontal where it connects to the pole. If the mass of the crow is m, show that the tension in the cord is $\dfrac{mg}{2\sin\theta}$. Here g is the acceleration due to Earth's gravity.

12. Consider two wooden boxes labelled A and B, weighing 5.0 kg and 8.0 kg respectively, connected by a light rope and placed on a frictionless floor, as shown in the figure.

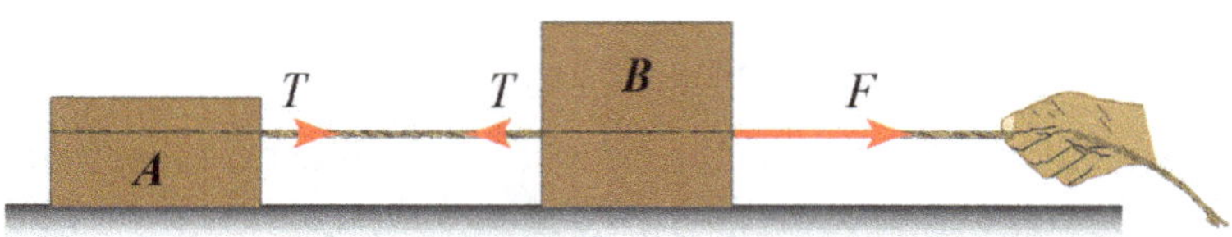

A person pulls box *B* horizontally with force *F*, and box *B* moves with acceleration 2.0 m s^{-2}. Assume that both boxes slide without toppling. Take $g = 10 \text{ m s}^{-2}$.

(i) What is the acceleration of the box A?

(ii) Draw FBD for box A.

(iii) Using the FBD in part (b) and Newton's second law, determine the tension T in the rope that connects the two boxes.

(iv) Draw FBD for box B.

(v) Which is larger in magnitude, the tension force T or the applied force F ? Explain your answer without calculation.

(vi) Using the FBD drawn in part (e) and Newton's second law, determine the magnitude of the force F.

13. Figure shows the front view of a person of mass m walking with constant velocity using crutches.

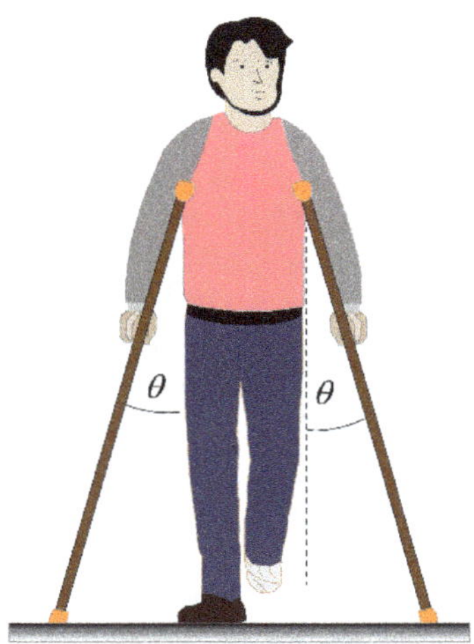

Whilst walking, each crutch forms an angle θ with vertical, and the half of the person's weight is supported by the crutches and while the other half is supported by the vertical forces of the ground on the person's feet. It is assumed that the force exerted by the ground on the crutches acts along the crutches.

(i) Draw a free-body diagram for the person.

(ii) Show that the magnitude of the compression force F is same in each crutch.

(iii) Show that $F = \dfrac{mg}{4\cos\theta}$.

(iv) Draw a free-body diagram for the crutch tip.

(v) Show that the minimum possible coefficient of friction μ_s between the crutches tip and the ground is $\mu_s = \tan\theta$.

14. Consider a spring having a natural length of 15.0 cm. When an object having a mass of 900 g is hanged on it, it stretches to a length of 18.0 cm. Take $g = 10 \text{ m s}^{-2}$.

   (i) Determine the spring constant of the spring.

   (ii) What total mass must be hung from the spring to stretch it to a total length of 22.0 cm?

15. As shown in the figure, two iron balls, each of mass m, are hung from the ceiling of an elevator. T_1 and T_2 are the tensions in the upper and lower strings, respectively, when the elevator moves with an upward acceleration a_y.

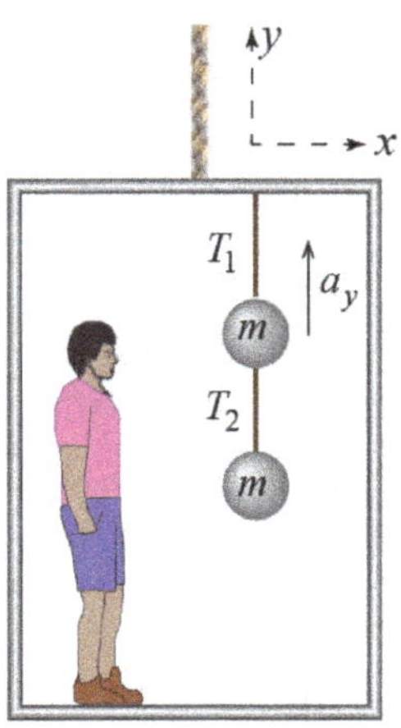

(i) Draw the free-body diagrams of the two iron balls.

(ii) If $m = 3.0 \text{ kg}$, and the elevator moves with an upward acceleration a_y of magnitude 1.50 m s^{-2}, determine the tensions T_1 and T_2.

(iii) If the strings can withstand a maximum tension of 80.0 N, what maximum acceleration can the elevator have before a string breaks?

Chapter 10

Centre of Gravity, Moment and Static Equilibrium

10.1 Introduction

This chapter will discuss the concepts of the centre of mass (CM) and equilibrium forces in a system. These important concepts in physics help us to understand the motion and stability of objects and systems. The centre of mass refers to the point within a system or object where the mass is evenly distributed. Equilibrium forces, on the other hand, are the forces that maintain balance and stability in a system.

Initially, we shall focus on the concept of the centre of mass of an object or system, which can be thought of as the average position of all the mass in the system. The centre of mass is located at the geometric centre for objects with uniform density and symmetrical shapes. However, the centre of mass may be located at a different point for irregularly shaped objects or systems with non-uniform density. Furthermore, the centre of mass is the point at which the mass is evenly distributed, regardless of the presence or absence of a gravitational field. However, the centre of gravity (CG) is the point at which the entire weight of an object or system can be considered to act vertically downward. It is the point where the gravitational force on an object is balanced.

Next, the chapter will focus on equilibrium forces that act on an object or system in such a way that it remains in a state of equilibrium. Equilibrium is characterised when there is an absence of acceleration or rotation. In other words, the net force and torque acting on the system are zero. Understanding equilibrium forces allows us to analyse and predict the behaviour of objects and systems, ranging from simple stationary objects to complex structures or systems in motion. It is fundamental in fields such as mechanics, engineering, and physics, where the stability and motion of objects are of utmost importance.

10.2 Centre of Mass

The centre of mass (CM) is the average mass position within an object or system, and it is the point at which the mass is evenly distributed in all directions. Then we can treat the mass of the system as if it were all concentrated at the centre of mass, like a point particle. By identifying the centre of mass, we can treat a complex system as a simplified point particle, greatly simplifying its analysis.

For solid homogeneous objects such as a cube, a ball or a can, we often model them as having a continuous distribution of matter and having its centre of mass at their respective geometric centre, shown in figure 10.1 (a). However, the centre of mass of an object is not necessarily located within the object. For example, the centre of mass of objects, such as for a boomerang, is located outside the boomerang; for a circular ring, it is located in the middle of the hole, as shown in figure 10.1 (b).

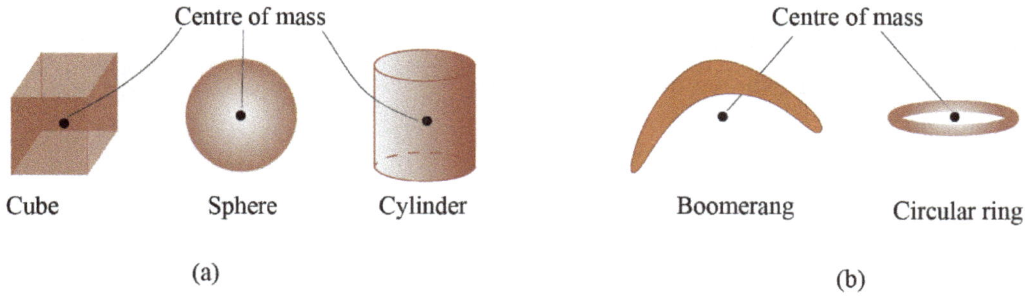

Figure 10.1: (a) Homogeneous objects (a cube, a ball, or a can) have their centre of mass located at their respective geometric centre. (b) The centre of mass of a boomerang is located outside the boomerang, and that of a circular ring is located in the middle of the hole

Consider a situation where an object (say, a cricket bat) is thrown vertically upward with pure transitional motion without any rotation or vibration, as shown in figure 10.2 (a). In this case, every particle in the object travels in the same way along a parabolic trajectory. When the same object (say, a cricket bat) is thrown vertically upward as before, but this time it also undergoes a rotation, as shown in figure 10.2 (b). In this case, only the centre of mass follows a parabolic trajectory, and every particle in the object rotates about its centre of mass. In both cases, one point in the object (shown as C), called the centre of mass, performs a purely translational motion along the parabolic trajectory.

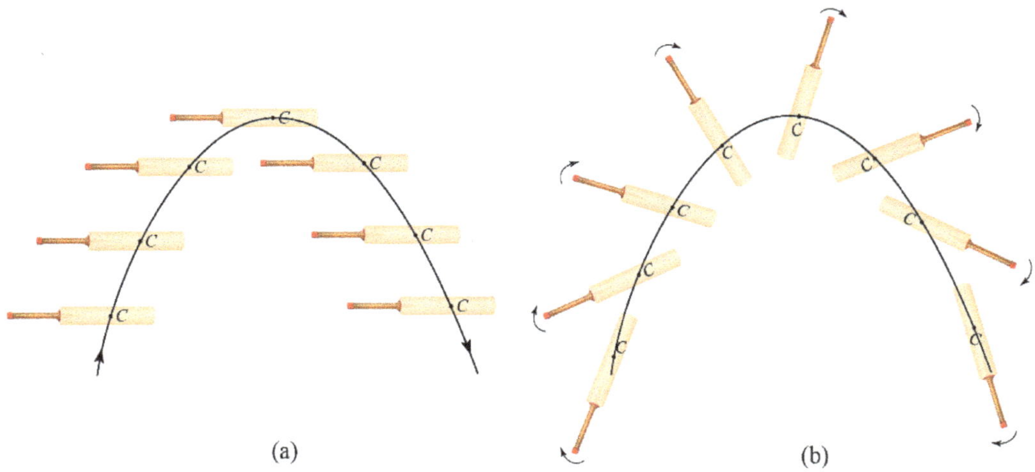

Figure 10.2: (a) The object (say a cricket bat) is thrown vertically upwards with pure transitional motion without any rotation or vibration (b) The object (say a cricket bat) is thrown vertically upwards as same as (a), but this time it also undergoes a rotation

Centre of mass in a system of particles

Consider two particles of mass m_1, and m_2 at distances x_1 and x_2 respectively, from an origin O as shown in figure 10.3 (a).

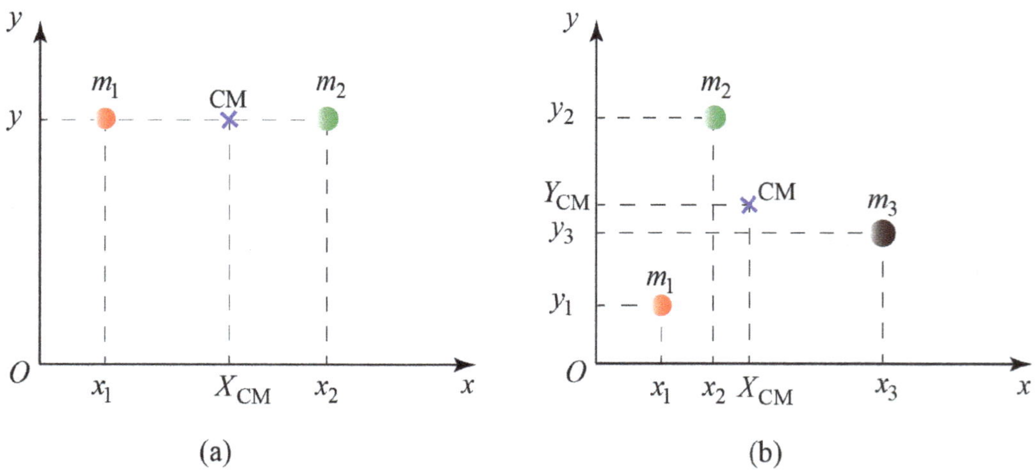

Figure 10.3: (a) (a) Two particles of mass m_1 and m_2 are placed in a straight line (b) three particles of mass m_1, m_2 and m_3 in a plane, but not in a straight line

The centre of mass of the two particles is located on the x axis and lies somewhere between the particles. Its x coordinate of the centre of mass X_{CM} is given by

$$X_{CM} = \frac{m_1 x_1 + m_2 x_2}{m_1 + m_2} \tag{10.1}$$

If, then $X_{CM} = \dfrac{x_1 + x_2}{2}$. This means that the centre of mass lies midway between the two particles. Suppose $m_2 = 2m_1$, and $x_1 = 0$, x_2 d, the $X_{CM} = \dfrac{2}{3}d$. This means that the centre of mass of the two particles lies closer to the heavier particle.

If there are n number of particles of mass $m_1, m_2, m_3 ..., m_n$ placed along a straight line at distances $x_1, x_2, x_3 ..., x_n$ respectively from the origin *O*, then its x coordinate of the centre of mass X_{CM} is given by

$$X_{CM} = \frac{m_1 x_1 + m_2 x_2 + m_3 x_3 + \cdots, + m_n x_n}{m_1 + m_2 + m_3 + \cdots, m_n} = \frac{\sum m_i x_i}{\sum m_i} \tag{10.2}$$

Hence, we could write

$$X_{CM} = \frac{\sum m_i x_i}{\sum m_i} \tag{10.3}$$

The sum $\sum m_i = M$ is the total mass of the system. Thus we have

$$X_{CM} = \frac{\sum m_i x_i}{M} \tag{10.4}$$

Now consider three particles of mass m_1, m_2 and m_3 in a plane but not in a straight line, as shown in figure 10.3 (b). Let their coordinates be $(x_1, y_1), (x_2, y_2)$ and (x_3, y_3) respectively. Its x and y coordinate of the centre of mass X_{CM} and Y_{CM} is given by

$$X_{CM} = \frac{m_1 x_1 + m_2 x_2 + m_3 x_3}{m_1 + m_2 + m_3} \quad \text{and} \quad Y_{CM} = \frac{m_1 y_1 + m_2 y_2 + m_3 y_3}{m_1 + m_2 + m_3} \tag{10.5}$$

If we consider a system of many particles with masses m_i (the mass of the ith particle) in three dimensions having its coordinate (x_i, y_i, z_i), the coordinates of their centre of mass X_{CM}, Y_{CM} and Z_{CM} are given by

$$X_{\mathrm{CM}} = \frac{\sum m_i x_i}{M}, \quad Y_{\mathrm{CM}} = \frac{\sum m_i y_i}{M} \quad \text{and} \quad Z_{\mathrm{CM}} = \frac{\sum m_i z_i}{M} \tag{10.6}$$

Activity 10.1

Three particles of mass $m_1 = 1.0\text{ kg}$, $m_2 = 2.0\text{ kg}$ and $m_3 = 3.0\text{ kg}$ are placed at the corners of an equilateral triangle whose side is 1.0 m in length. Determine the position of their centre of mass.

10.3 Centre of Gravity

In an extended object, the mass is distributed throughout its volume and each small mass element is acted upon by the gravitational force. The net effect of all these forces is equivalent to the effect of a single force acting through a special point called the Centre of Gravity (CG), where all the weight of the object could be considered to act. If the variation of the acceleration due to gravity (g) with altitude can be ignored, then the object's centre of gravity is identical to its centre of mass. If an extended object is pivoted at its centre of gravity, it balances in any orientation.

The centre of gravity of an irregularly shaped object could be determined by suspending it first from one point and then from another.

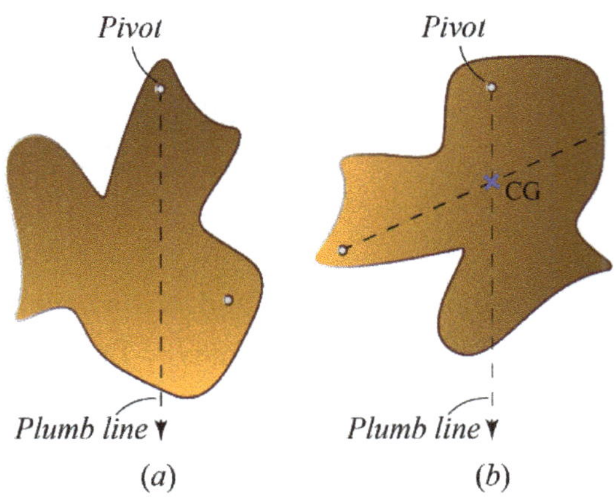

Figure 10.4: The experimental setup to find the centre of gravity of a flat thin object (a) The object is hung from a pivot and identified the plumb line (b)The same object is hung from another pivot and identified the plumb line. The centre of gravity is located at the intersection of these plumb lines.

Figure 10.4 shows the experimental setup to find the centre of gravity of a flat thin object. As a first step, firmly hang the object from a pivot using a pin so that the object can swing freely, as shown in figure 10.4 (i). Take a length of string and attach one end to the pin and the other to a heavy mass (called a plumb bob). The object will stop swinging once its centre of gravity is vertically below the pivot. Draw a vertical line downward from the pivot on the object along the vertical string, called a plumb line. As a second step, the process is repeated with the same object hanging from a different pivot, as shown in figure 10.4 (ii). The object's centre of gravity (CG) will be at the point of intersection of the plumb line drawn on the object.

Centre of gravity of regular shaped objects

Table 10.1 illustrates the centre of gravity of commonly known regular shaped objects.

Table10.1: The centre of gravity of commonly known regular shaped objects

Object	Centre of Gravity (CG)	Object	Centre of Gravity (CG)
Uniform rod	CG lies at mid point 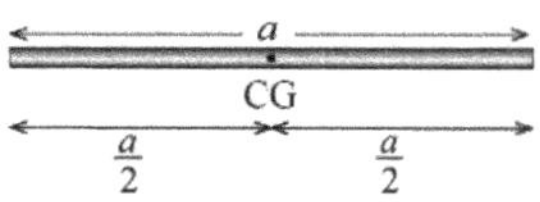	Uniform solid hemisphere	CG lies at the central radius at a distance $\frac{3}{8}r$ above the base 
Uniform cirular ring	CG lies at the centre of the ring 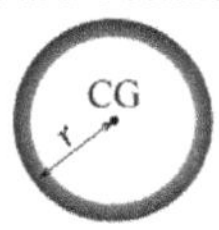	Uniform solid/hollow cylinder	CG lies at the mid point along the axis 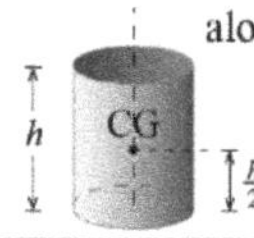
Uniform cirular disc	CG lies at the centre of the disc 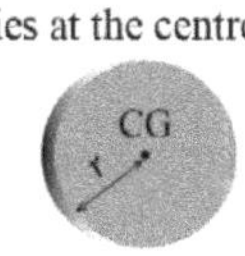	Uniform triangular lamina	CG lies at the point of intersection of the medians 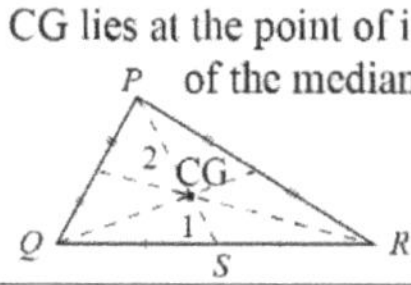
Uniform solid/hollow sphere	CG lies at the centre of the sphere 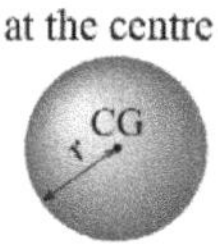	Uniform hollow cone	CG lies at the height $h/3$ above the base along the axis 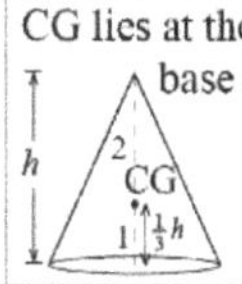
Uniform rectangular lamina	CG lies at the point of intersection of its diagonals 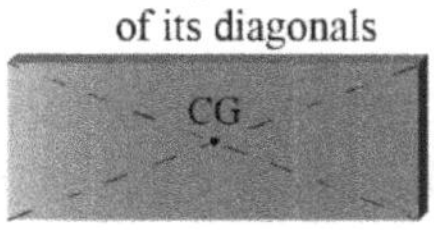	Uniform solid cone	CG lies at the height $h/4$ above the base along the axis 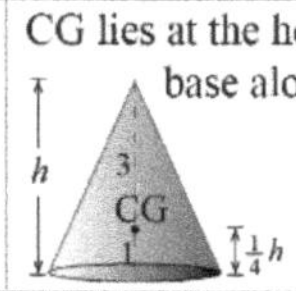

10.4 Moment

When we think of force, we usually think of a push or pull that causes an object to move or stop moving. But what if the force acts on an object in such a way that it causes it to rotate around a fixed point? This is where the concept of the moment of a force comes into play. The moment of a force is simply referred to as a moment and is also known as a torque with the symbol τ . It describes the ability of a force to rotate an object around a specific axis or pivot point. Physicists and engineers rely heavily on the moment of a force to analyse and design complex mechanical systems, as it helps them comprehend the behaviour of objects when subjected to external forces. In this context, it is essential to understand the basic principles of the moment and how it is calculated. This section will explore the basic principles of moments and how they can be applied in various situations, ranging from simple levers to complex machinery.

Figure 10.5 shows the typical effect caused by the tendency of the force to rotate the object around a certain point. This point is often called the pivot or fulcrum.

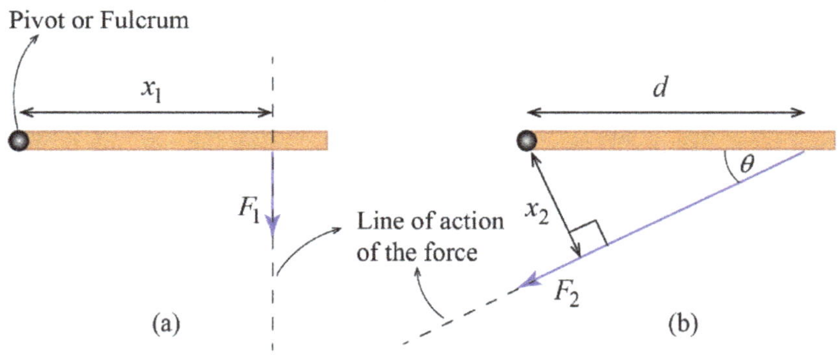

Figure 10.5: Calculating moment of a force: the involved quantities

The moment of force, commonly known as the moment, depends on two quantities:

(i) The magnitude of the force (the bigger the force, the greater its moment) and

(ii) The perpendicular distance from the pivot point to the line of action of the force (the further the force acts from the pivot, the greater its moment).

Thus, the magnitude of the moment is equal to the product of the force and the perpendicular distance from the pivot point to the line of action of the force. i.e.

Moment = Force × perpendicular distance from the pivot to the line of action.

In figure 10.5 (a), the magnitude of the force F_1 is pushing down on the object (the lever) at a perpendicular distance x_1 from the pivot point to the line of action of the force. Hence, the moment of the force F_1 about the pivot is then given by

$$\text{moment } (\tau) = F_1 \times x_1$$

Figure 10.5 (b) shows a situation where the magnitude of the force F_2 is pushing at an angle θ to the object (the lever) rather than perpendicular to it. This makes it have less turning effect, i.e. less moment. There are two ways in which we could calculate its moment.

(i) By drawing a perpendicular line from the pivot to the line of action of the force, we have, $\text{moment } (\tau) = F_1 \times x_2$

From the right-angled triangle, we get $x_2 = d\sin\theta$. Thus,

$$\text{moment } (\tau) = F_2 \times (d\sin\theta) = F_2 d\sin\theta$$

(ii) By taking the component of F_2, i.e. 90° to the object (the lever)

perpendicular to the lever, the component of F_2 is $F_2 \sin\theta$. Thus,

$$\text{moment } (\tau) = (F_2 \sin\theta) \times d = F_2 d\sin\theta$$

$F_2 \cos\theta$ is the other component of F_2, and it passes through the lever. Therefore, this component will not have a turning effect because the perpendicular distance from the pivot to the line of action of the force is zero.

Although the moment of a force could be calculated about any point, it is usually easiest to calculate the moment of a force around the pivot point because there may be an unknown force acting at that point, such as a contact force on the object.

The unit of the moment is newton metre (N m) and does not have a special name.

The moment of force is a vector quantity. It has both magnitude and direction and can be represented graphically as an arrow. The direction of the moment vector is perpendicular to the plane of the force vector, and the line connecting the pivot point to the point of the force application. Its direction can be determined using the right-hand screw rule, as shown in figure 10.6 (a). Accordingly, if an object rotates due to a moment and the gripped fingers of the right hand indicate the direction of rotation, then the direction of the moment can be determined by the pointing direction of the thumb.

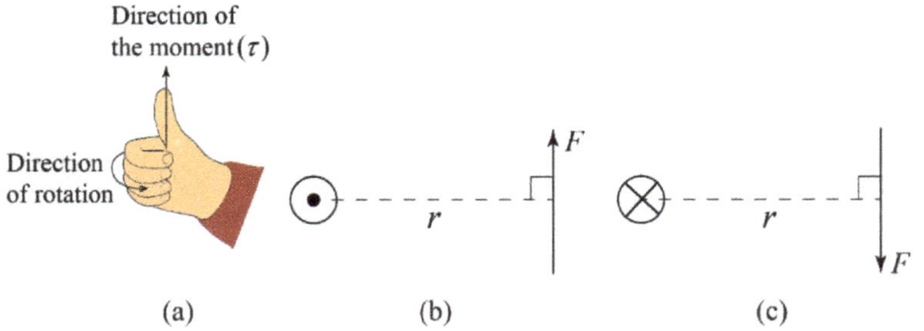

Figure 10.6: (a) Right-hand screw rule (b) clockwise moment (c) anticlockwise moment

The moment can be clockwise or counterclockwise, depending on the direction of rotation. If the direction of rotation is counterclockwise, then the direction of the moment will be pointing inward towards the paper, as shown in figure 10.6 (b). Conversely, if the direction of rotation is clockwise, then the direction of the moment will be pointing outward from the paper, as shown in figure 10.6 (c).

Couple Forces

Couple forces are a pair of equal and opposite forces that are parallel to each other and act on different points of an object. These forces do not produce a net force on the object, but they do produce a turning effect or moment. The magnitude of the moment produced by a couple force is equal to the product of one of the forces and the perpendicular distance between the two forces. Consider a case where a couple of forces F are acting across a pivot at a distance d, as shown in figure 10.7.

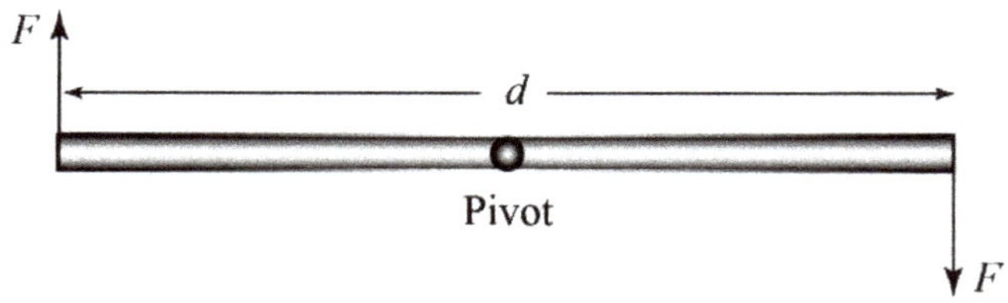

Figure 10.7: A couple acting across a pivot.

The moment (τ) produced by the couple forces can be calculated as

$$\tau = F \times d \tag{10.7}$$

Couple forces produce a pure rotational motion without any translational motion. The direction of the moment or torque is perpendicular to the plane of the forces and follows the right-hand rule. This principle is widely used in physics and engineering to analyse and design mechanical systems such as bridges, buildings, and machines. They are also important in understanding the behaviour of rigid bodies and their equilibrium.

Activity 10.2

Calculate the torque of the couple force when two equal forces of $20\ \text{N}$ are applied in opposite directions to turn the steering wheel of a car with a diameter of $0.38\ \text{m}$ as illustrated in the figure.

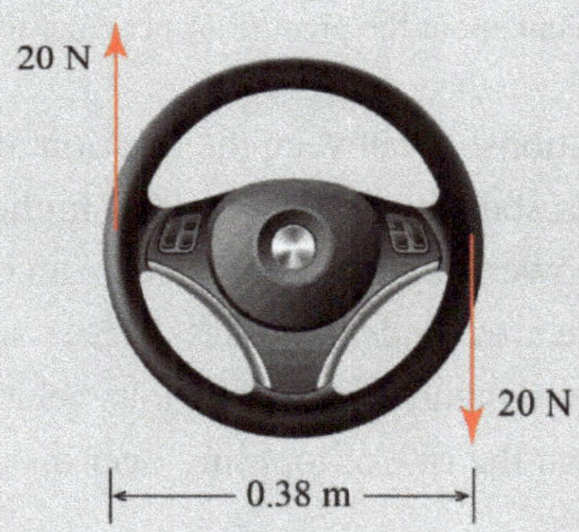

Centre of gravity and moments

In an object, if the line of action of its weight lies outside its base area, then due to the presence of a resultant moment that creates a turning force, the object will experience toppling. Consider the motion of a bus having mass m , as shown in figure 10.8. If there is no resultant moment, as shown in figure 10. 8(a), the bus will be stable. When a net anticlockwise moment about the pivot starts to appear, as shown in figure 10.8(b), the bus will rotate to the left and right. When a net clockwise moment about the pivot starts to appear, as shown in figure 10.8(c), where the line of action of the bus weight (mg) lies outside its base area, and it will rotate to the right and fall over.

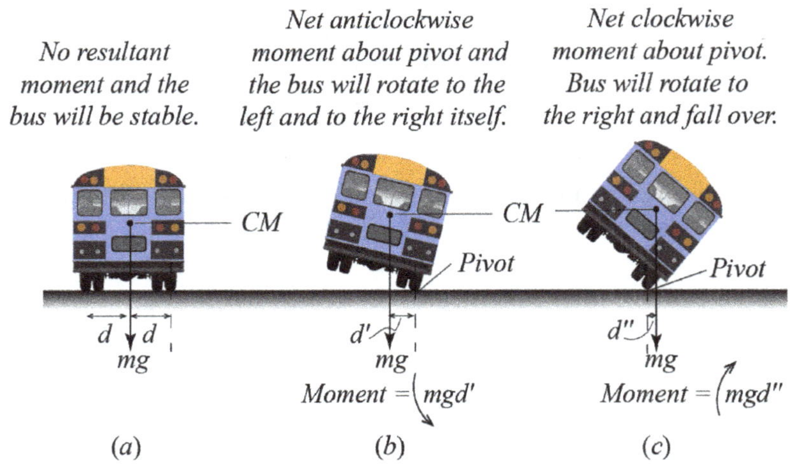

Figure 10.8: Moments acting on a bus at different angle of tilt. (a) No resultant moment (b) A net anticlockwise moment about the pivot (c) A net clockwise moment about the pivot

For an object, the relationship between the location of the centre of gravity and the base area influences its stability. Objects with a higher centre of gravity and a smaller base area exhibit lower stability. Conversely, an object with a lower centre of gravity and a wider base area tends to possess higher stability. This fundamental concept is extensively utilised in the design of various vehicles, including racing cars, buses, and trucks, to prevent the risk of toppling over during turns.

Activity 10.3

A uniform rectangular block having dimensions of width a and height $2a$ is placed on top of a plank and gradually raised through an angle θ, as shown in the figure.

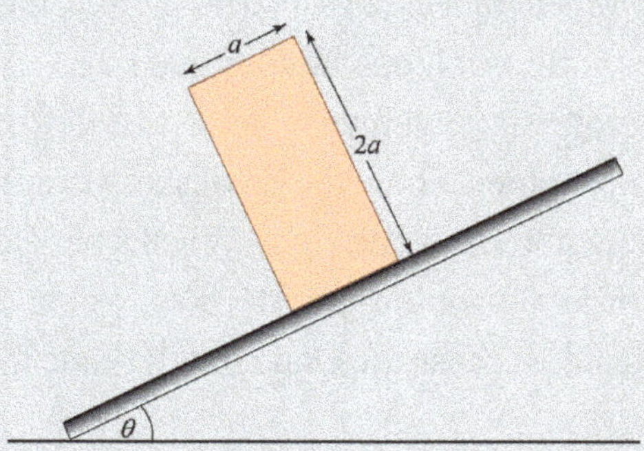

If the block is prevented from sliding while raising through the angle , at what value of angle θ will the block topple over?

10.5 Static equilibrium

In the previous chapter, we learnt that an object modelled as a particle is in translational equilibrium when the net force acting on it is zero. In real situations, objects often cannot be modelled as particles. If we consider a rigid object, the sum of the external forces, like the gravitational force that acts at every point in the object, may be treated as acting at the centre of mass. However, it matters where the force is applied on the object because, even though the sum of the forces on the object may be zero, a non-zero sum of torques on the object may produce rotation. Therefore, a second condition must be satisfied for a rigid object to be in equilibrium. This second condition involves the rotational motion of the rigid object. This is commonly known as the condition for rotational equilibrium.

The two sufficient and necessary conditions for a rigid object to be in equilibrium (usually known as static equilibrium) are the following:

(i) The net external force on the object must equal zero:

$$\sum \vec{F}_{ext} = 0 \tag{10.8}$$

(ii) The net external torque on the object about any axis must be zero:

$$\sum \vec{\tau}_{ext} = 0 \tag{10.9}$$

It should be noted that any axis can be used as a reference line because every point in the rigid object is at rest in an inertial frame. As said in the second condition, the sum of the torques about any axis must be zero.

Therefore, the rigid object does not have to be pivoted about an axis through the chosen point. If a rigid object is in equilibrium, it cannot have any tendency to begin to rotate about any axis. Thus, the sum of the torques must be zero, no matter what axis is chosen.

When solving a given problem, although the choice of the axis is arbitrary, once we choose it, we must use the same axis to calculate all of the torques to simplify the calculations as much as possible. This is to simplify the calculations as much as possible.

Principle of Moments

The principle of moments, also known as the law of moments, states that when an object is balanced (in equilibrium), the sum of the clockwise moments about any point is equal to the sum of the counterclockwise moments about that same point (figure 10.9). In other words, if an object is not moving and is balanced, then the total torque (or turning effect) acting on the object must be zero.

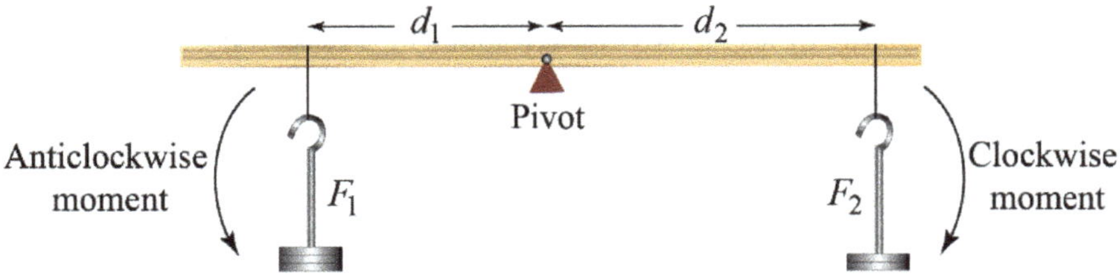

Figure 9.11: Counterclockwise and anticounterclockwise moments

If we take the moment about the pivot

$$F_1 \times d_1 = F_2 \times d_2$$

This principle of moments is used in many practical applications, such as designing structures, machines, and vehicles, as well as in everyday situations such as balancing a seesaw or a ladder.

Activity 10.4

Two children weighing 300 N and 200 N are seated on a seesaw as shown in the figure below. To balance the seesaw, the father of the children applied a force F on the edge of the beam. Determine the magnitude and the direction of the applied force F.

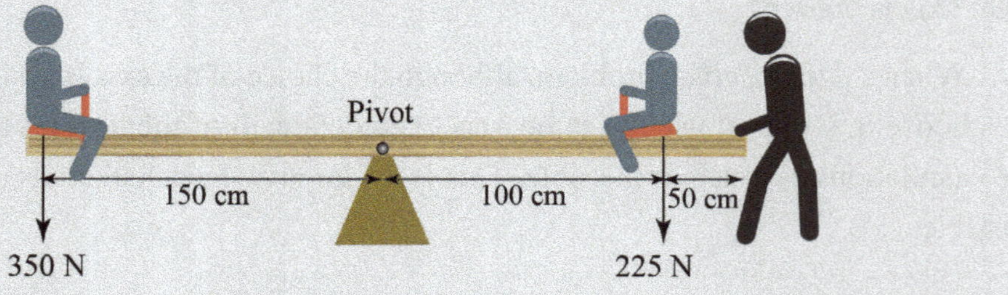

Questions

NOTE: Although the solutions are given on the last page, it is in your own interest to attempt the questions on your own as an indication of how far you have understood the material of the section.

1. State what is meant by (a) a couple and (b) torque.

2. Explain why it is easier to open a swinging door with a force near the edge, away from the hinges, rather than in the middle of the door.

3. Show that the centre of mass of two particles is on the line joining them at a point whose distance from each particle is inversely proportional to the mass of that particle.

4. Three particles of masses $m_1 = 2.0$ kg , $m_2 = 4.0$ kg and $m_3 = 6.0$ kg are located at the vertices of an equilateral triangle of side length 0.5 m as shown in the figure below.

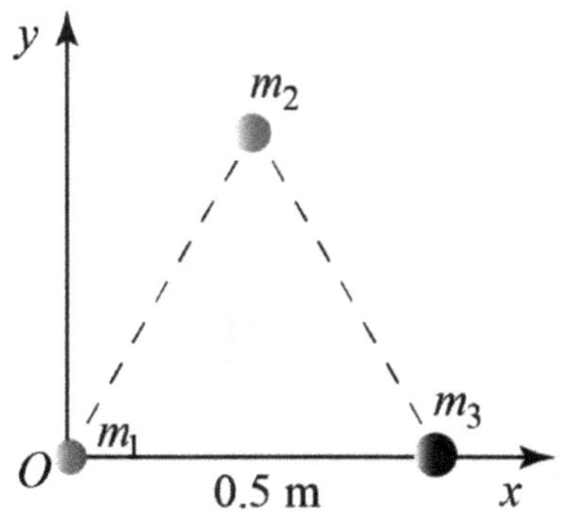

Determine the centre of mass of these three masses.

5. The figure shows a composite object that has a uniform square-shaped lamina labelled *ABCD* and a uniform triangular-shaped lamina labelled *BCE*.

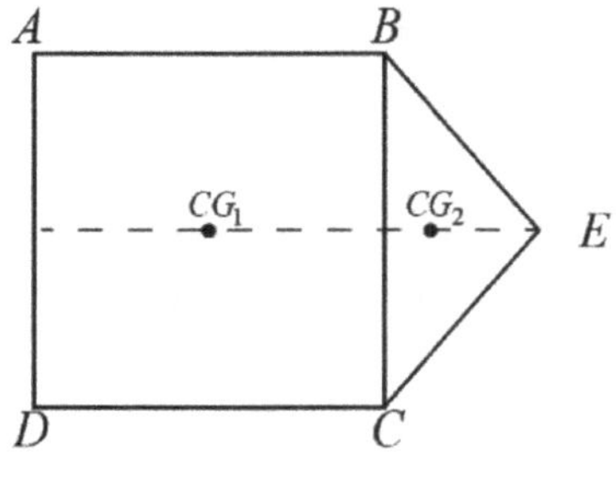

(i) If CG_1 is the centre of gravity of $ABCD$ and CG_2 is the centre of gravity of BCE, draw the diagram of the composite object on paper and mark its possible centre of gravity.

(ii) If $AB = 32.0$ cm, $AD = 32.0$ cm, $BE = 20.0$ cm and $CE = 20.0$ cm, determine the position of the centre of gravity of the composite object.

6. A force of 200 N is applied at a distance of 76 cm from the hinge O and an angle of 30° to the door, as shown in the figure.

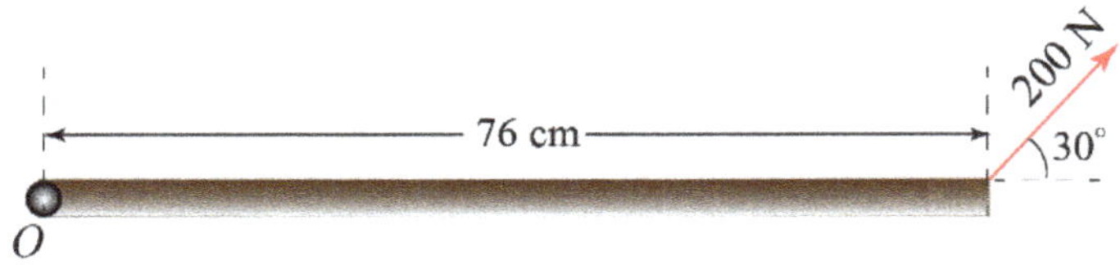

Determine the moment of the force about the hinge O.

7. A cyclist is taking a sharp right turn by exerting equal and opposite forces of 30 N on the ends of the handlebars, as shown in the figure. Determine the moment applied to the handle if the length of the handlebar is 70 cm.

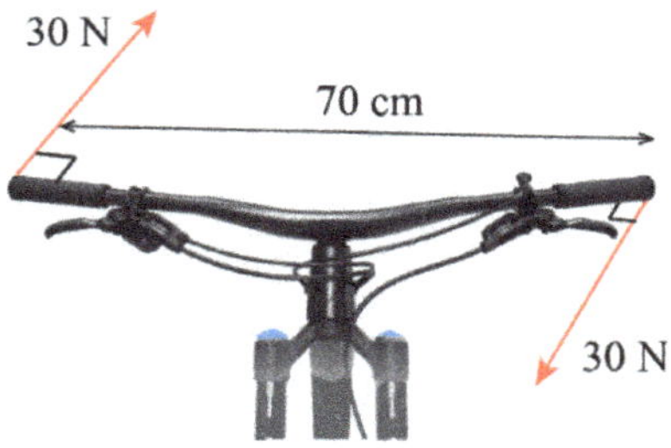

8. The engine of a car generates a torque of 250 N m on the axle of the wheel in contact with the road, allowing the car to move at a constant velocity in the forward direction as shown in the figure. Answer the following:

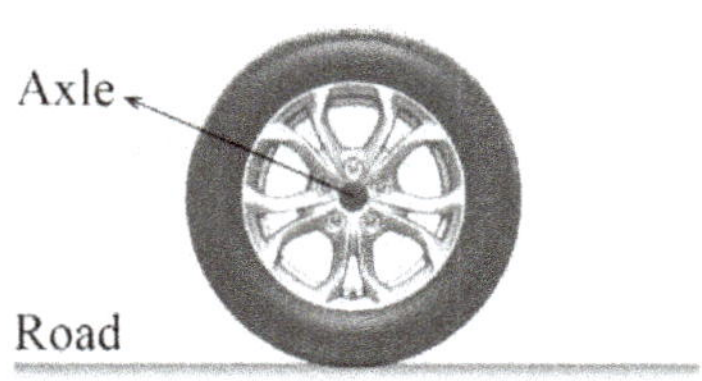

(i) Draw a diagram of the wheel and indicate the direction of rotation of the wheel and the horizontal component of the force exerted by the road on the wheel.

(ii) What is the resultant torque acting on the wheel? Provide an explanation for your answer.

(iii) Given that the diameter of the car's wheel is 68.9 cm, what is the magnitude of the horizontal component of the force exerted by the road on the wheel?

9. The figure shows the internal structure of a person's arm while holding a rectangular solid block of mass of 2.5 kg (denoted as m_1). The person's upper arm is positioned vertically, while the forearm remains horizontal. The biceps muscle, attached to one of the forearm bones, exerts a force to support the combined weight of the forearm and the block. The triceps muscle, responsible for extending the limb, is assumed to be relaxed and not contributing to the force.

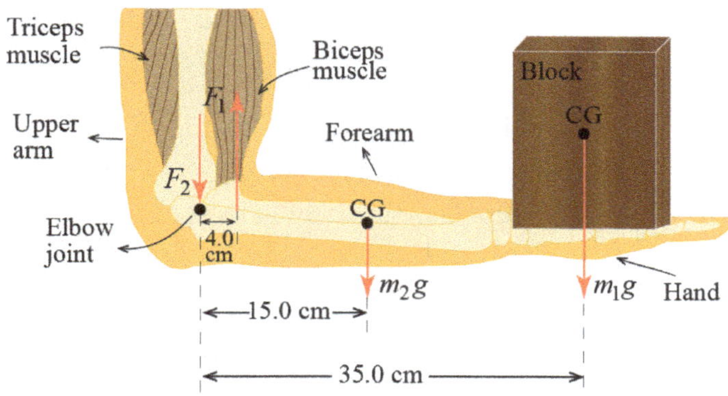

As shown in the figure, the magnitude of the force exerted by the biceps in the upward direction is labelled as F_1, while the force exerted at the elbow joint in the downward direction is labelled as F_2. The horizontal distance between the elbow joint and the centre of gravity of the block is 35.0 cm. The biceps is attached to the forearm at a position 4.0 cm from the elbow joint. The forearm and the empty hand together have a mass of 1.5 kg (denoted as m_2). The centre of gravity of the forearm is situated at a distance of 15.0 cm from the elbow joint. Take $g = 10 \text{ m s}^{-2}$.

(i) Determine the force F_1 that the biceps muscle must exert to support the weight of the forearm and the block.

(ii) Determine the force F_2 between the bony structures at the elbow joint.

(iii) Compare the forces F_1 and F_2 with the combined weight of the forearm and the block. What can you infer from this comparison?

10. In a factory, two workers carry a uniform wooden beam of length 2.40 m and a mass of 49.0 kg in a straight line as shown in the figure.

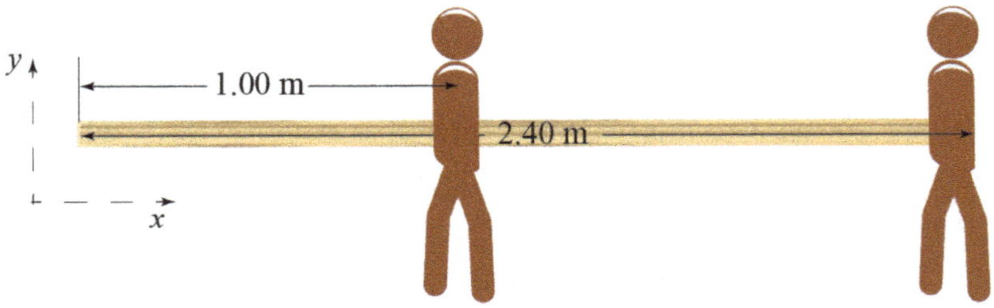

One of the workers, possessing slightly greater strength, agrees to bear the beam at a distance of 1.00 m from one end while the other worker supports it from the opposite end. Determine the force exerted on the beam by each worker. Take $g = 10 \text{ m s}^{-2}$.

11. The figure shows that a 100.0 cm long uniform rod *AB* of weight 12 N is pivoted at point *A*. An irregular solid object of weight W is suspended 30.0 cm from the end *B*. The end *B* is supported by a spring balance that reads 20 N. Determine the mass of the irregular solid object. Take $g = 10 \text{ m s}^{-2}$.

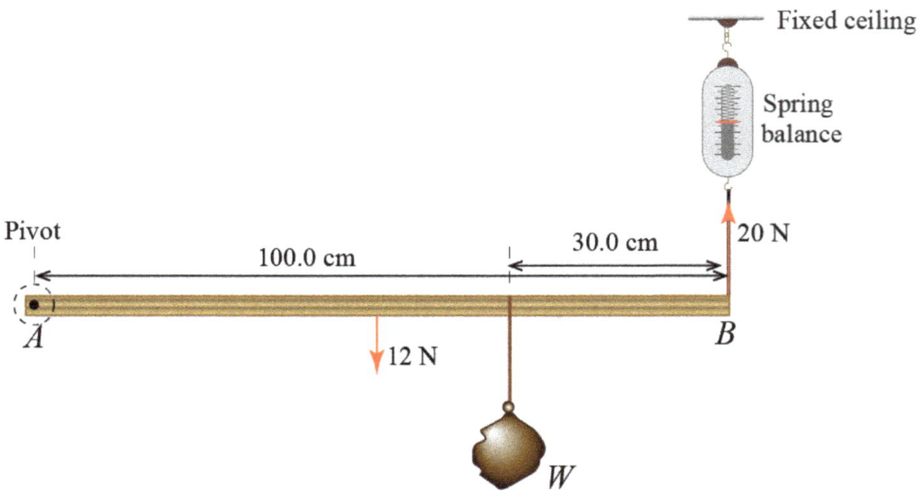

12. A painter weighing 60.0 kg mass painting a billboard standing on a wooden frame in the form of a trestle, consisting of a uniform plank of mass 20.0 kg and length 6.00 m balanced on two saw horses as shown in the figure. Each saw horse is placed 1.50 m from the end of the plank. The 4.0 kg mass paint bucket, having a base diameter of 0.30 m, has been placed at the right-hand

edge of the plank. Assume that the entire base of the paint bucket is in contact with the plank.

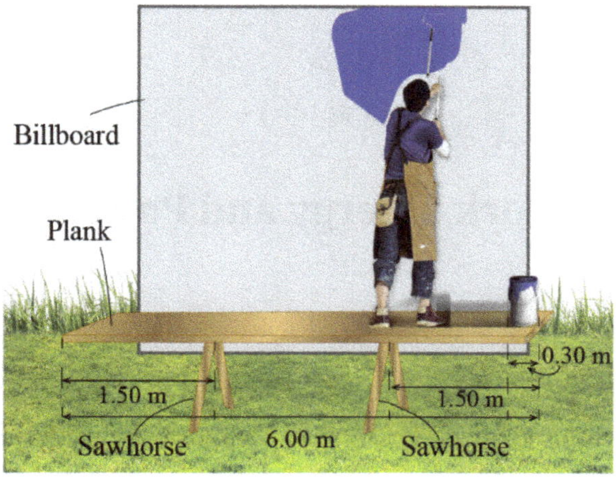

(i) How close could the painter walk towards the right-hand edge of the plank before the plank gets toppled and spills the paint?

(ii) How close could the painter walk towards the left-hand edge before causing the plank to topple?

Chapter 11

Work, Energy and Power

11.1 Introduction

This chapter will discuss the three fundamental concepts in physics, called work, energy, and power, which are closely related to each other. Work is the transfer of energy from one object to another by the application of force. Energy is the ability to do work, and it can exist in many different forms, such as kinetic energy, potential energy, thermal energy, and electrical energy. Power is the rate at which work is done and measured in watts.

Initially, you will learn how to compute the work done by a force or a system of forces. Then you will be focussing on the energy which is possessed by a system by virtue of its motion, called the kinetic energy, which will be discussed in detail. The detail of the work-energy theorem, which relates the work done to the change in kinetic energy of a moving object, will also be discussed. Finally, the concept of power will be discussed in detail.

Another purpose of this chapter is mainly to draw your attention to one of the most fundamental conservation laws in physics, the law of conservation of energy. The concepts of conservative and nonconservative forces and the meaning of a conservative system will also be discussed.

11.2 Work done by a Constant Force

The concept of work is fundamental in the field of physics. However, it is different from how it is used in everyday conversation. In everyday conversation, work is often used to describe any kind of physical or mental effort. For example, we would say that getting good grades in college takes a lot of hard "work". However, in physics, work is defined as the measure of energy transfer that occurs when an object is moved over a distance by an external force. This means that work is done only if the object moves in the direction of the force.

Take, for instance, the scenario of a person holding a heavy load on his hand without movement for a while. In colloquial terms, the person might claim to be exerting considerable effort or engaged in work. In fact, we can observe that he is exhausted. This may be due to physiological senses as a result of muscular effort. But in physics, we do not consider him to be doing any work. This is because the object is not moving and, therefore, no energy is being transferred. In physics, work is a very specific concept with a precise definition involving a force acting on an object while the object undergoes a displacement. It allows us to quantify the amount of energy that is transferred when a force acts on an object. A force acting on an object is said to be doing work only if the object undergoes some kind of displacement in the direction of the force.

Let us consider a very simple example. Suppose a constant force $\vec{F}$ acts on an object, which consequently undergoes a displacement $\vec{d}$ in the direction of the force, as shown in figure 11.1(a).

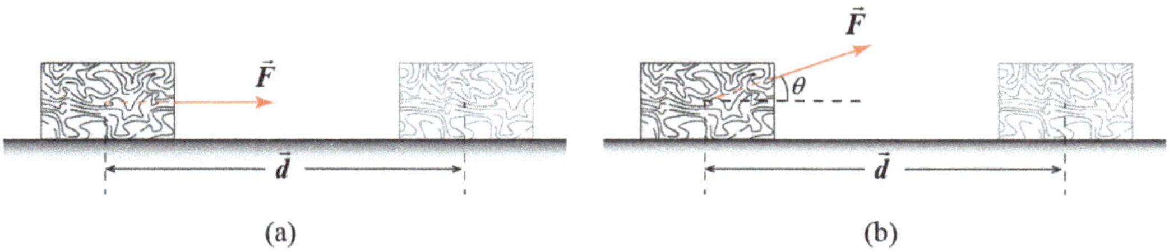

Figure 11.1

We say that the magnitude of the work done W by the force is the product of the magnitude of the force, i.e. F, and the magnitude of the displacement, i.e. d.

Thus, $$W = Fd \tag{11.1}$$

Often the constant force $\vec{F}$ may not act in the direction of the object's displacement. Now consider the situation where constant force $\vec{F}$ is acting on the object, making an angle θ with the horizontal as shown in figure 11.1(b). Then we define the work done W as the product of the displacement of the object and the component of the force in the direction of the object's displacement. If W is the work done by the force $\vec{F}$ during the displacement $\vec{d}$ of the object, then,

$$W = (F\cos\theta)d \tag{11.2}$$

where, $F\cos\theta$ is the component of the force in the $\vec{F}$ direction of the object's displacement.

If $\theta = 0$, the work done by the force is, $W = Fd$, in agreement with the simple case considered earlier in figure 11.1(a) and equation 11.1. When $\theta = 90°$, the force $\vec{F}$ has no component in the direction of motion of the object, and therefore, the force does not do any work. For example, when you walk with constant velocity on a horizontal level floor while carrying a heavy load on your hand, you do not work on the load because the (vertical) supporting force you exert on the load has no component in the direction of the (horizontal) motion. Furthermore, no horizontal force is required to move the load horizontally with constant velocity.

Equation 11.2 can be written either as $W = (F\cos\theta)d$ or $W = F(d\cos\theta)$. Therefore, the work done can be defined as (a) the product of the magnitude of the displacement and the component of the force in the direction of the displacement, i.e. $W = (F\cos\theta)d$ or (b) the product of the magnitude of the force and the component of the displacement in the direction of the force, i.e. $W = F(d\cos\theta)$. It is important to understand that even though $\vec{F}$ and $\vec{d}$ both are vectors, the work W is a scalar quantity. Additionally, the work can be positive or negative. For instance, the work W is positive when the force has a component in the same direction as the displacement. The work W is negative when the force has a component opposite to the displacement.

Furthermore, when discussing the work done by a force on a particular object, always be sure to specify exactly what force is doing the work. When a force is used to lift an object vertically upward, the work done on the object by the lifting force is positive. However, in this situation, the object is lifted against the gravitational force, and since the gravitational force is in the downward direction, it is opposite to the displacement, which is in the upward direction. Therefore, the work done by the gravitational force on the object that is being lifted is negative.

When two or more forces act on an object, the net work done on the object is the sum of the work done by the individual forces. This means that we can find the work done by each force and then sum those works. Alternatively, we can find the net force acting on the object and the angle between the direction of the net force and the object's displacement, and we can calculate the net work done on the object.

The unit of work is defined as the work done by a unit force when an object moves a unit distance in the direction of the force. Hence, in the S.I unit system, the unit of work is 1 newton-metre, which is called 1 joule (J). In the c.g.s. unit system, the unit of work is 1 dyne-centimetre, which is called 1 erg. From the relations between the newton and the dyne, the metre, and the centimetre, it follows that 1 joule $= 10^7$ ergs.

Activity 11.1

As shown in the figure, a 15.0 kg mass crate is raised through a vertical height of 3.00 m by sliding it along a distance of 5.00 m up a given inclination θ with the horizontal floor. This is done by means of a force $\vec{F}$ parallel to the incline, which pushes the crate at a constant speed up the incline. Assume that the surface in contact is frictionless and that the acceleration due to gravity g is 10 m s^{-2}.

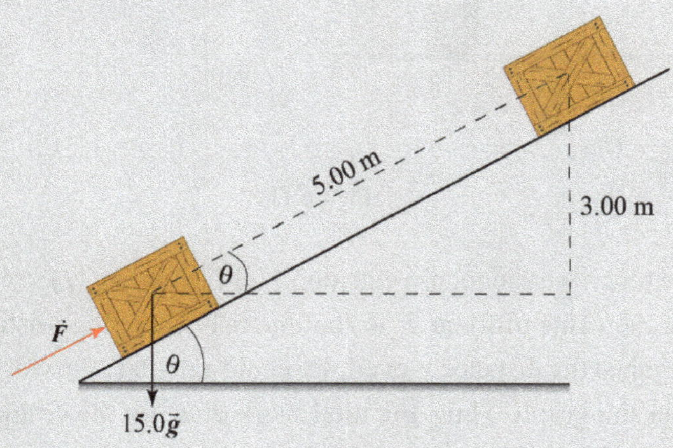

(i) Determine the work done by the force $\vec{F}$.

(ii) If the crate is raised vertically up with the constant speed through the same height of 3.50 m, determine the work done by the force $\vec{F}$.

11.3 Work Done by a Varying Force

So far, we have considered only constant forces acting on an object and discussed how to determine the work done by that force on the object. But in many practical situations, we experience that the force varies in magnitude during a process. For example, the force exerted by a spring increases with the amount of stretch (see Section 9.10 in Chapter 9).

Consider an object moving from the initial position x_i to the final position x_f in response to a varying force in the +ve x direction, as shown in figure 11.2(i). Figure 11.2 (ii) represents a graph of the x component of force F_x as a function of the object's position coordinate x.

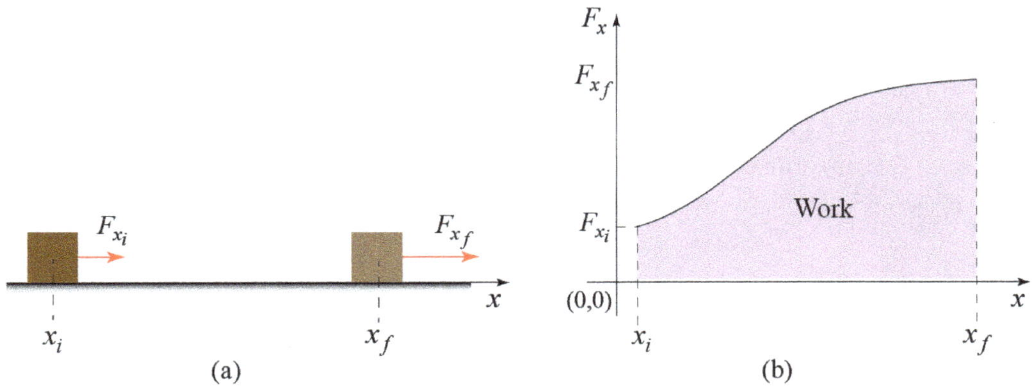

Figure 11.2

For a variable force, we cannot use equation 11.1 (i.e. $W = Fd$) to calculate the work done because the value of force F is continuously changing. Instead, by plotting a force graph against the distance moved, we could calculate the work done by finding the area under the graph. Thus, the total work done by the component F_x of the varying force as the object moves from the initial position x_i to the final position x_f is exactly equal to the area under the curve.

For example, consider the situation where a coil spring with one end attached to a fixed point and the other end attached to an object that can slide without friction along the x axis, as shown in figure 9.21(a) and (b). If the spring is either stretched or compressed by a small distance x from its equilibrium, it exerts a force F_{spring} on the object that can be mathematically modelled as

$$F_{spring} = -kx \tag{11.3}$$

where k is the spring constant.

The negative sign in this equation signifies that the force exerted by the spring on the object is always directed opposite the displacement from equilibrium.

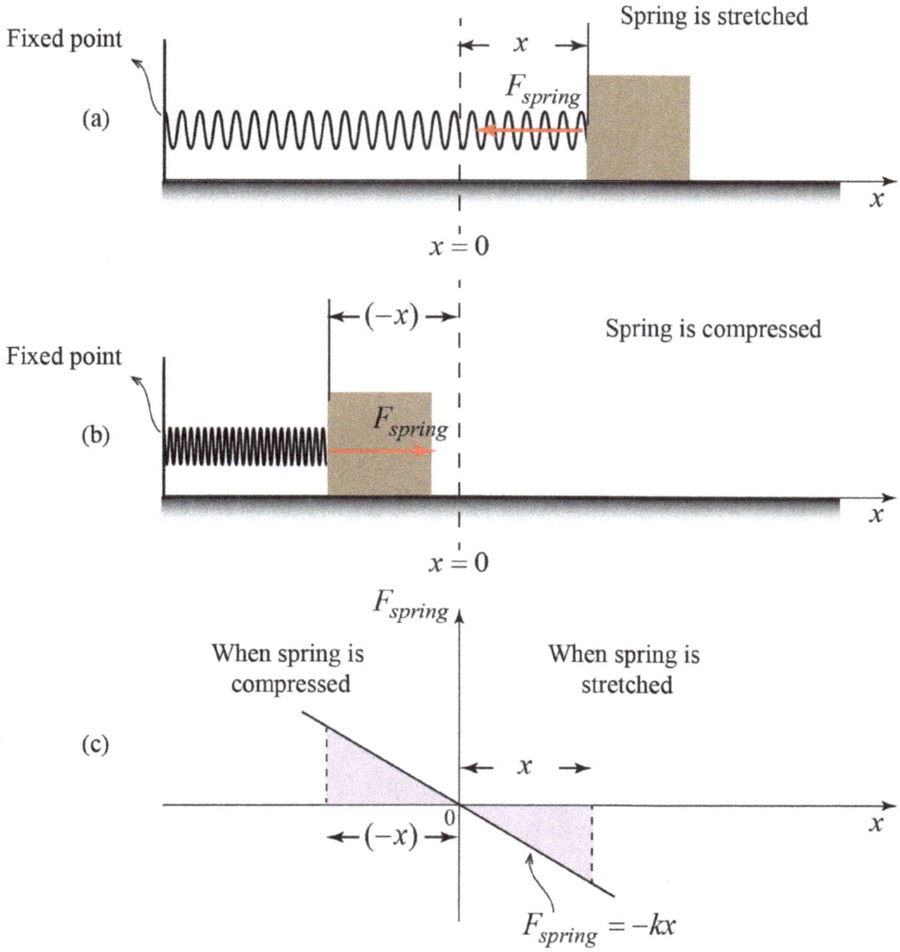

Figure 11.3: The force exerted by a spring on an object: (a) the spring is stretched by an amount x (b) the spring is compressed by an amount x (c) graph of the varying force F_{spring} *versus* the displacement x; work done on the object is the area of the shaded triangle, when the spring is stretched (right-hand side) and when the spring is compressed (left-hand side).

The total work done by the spring on the object when it is stretched from zero to a value x as shown in figure 11.3 (b), is given by the area under the graph (the shaded triangle in figure 11.3 (d) – right-hand side). Since the force exerted on the object F_{spring} is opposite to the displacement x, the work done on the object W is negative and is given by the area under the graph. The area is equal to half the product of the base (x) and the height (kx). Thus, $W = -\frac{1}{2}kx^2$. On the other hand, the total work done by the spring on the object when it is compressed from zero to a value x as shown in figure 11.3 (c), is given by the area under the graph (the shaded triangle in figure 11.3 (d) – left-hand side). Since the force exerted on the object F_{spring} is in the same direction as the displacement x, the work done on the object W is positive and is given by the area under the graph. The area is equal to half the product of the base (x) and the height (kx). Thus, $W = \frac{1}{2}kx^2$.

Activity 11.2

An ideal spring has a spring constant $k = 21.0 \text{ N m}^{-1}$. Determine the amount of work that must be done to stretch the spring 0.50 m from its relaxed length.

11.4 Kinetic Energy and the Work – Kinetic Energy Theorem

So far, we have considered examples of unaccelerated objects (moving with constant velocity) in which the net force acting on the object is zero. Now consider a situation where an object of mass m moves along the x axis under the action of a constant net force $\vec{F}_{net}$ that points in the positive x-direction and produces a constant acceleration $\vec{a}$ in the same direction as shown in figure 11.4. Since the motion is in the positive x-direction, let the magnitude of the constant net force $\vec{F}_{net}$ be F_{net_x} , and the magnitude of the constant acceleration $\vec{a}$ be a_x.

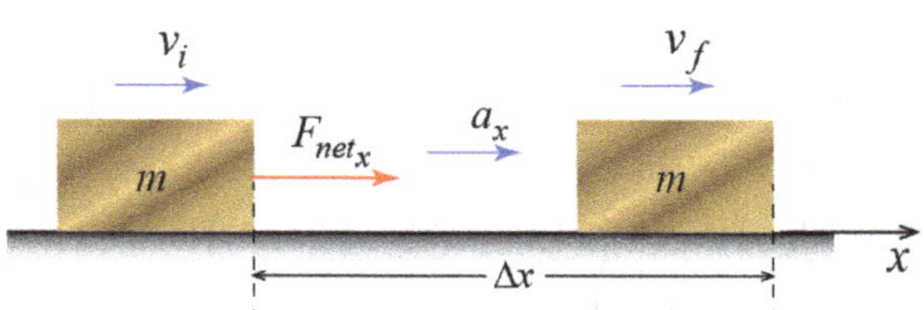

Figure 11.4: Work on a moving object by a constant net force $\vec{F}_{net}$.

The object moves through the distance Δx , and its initial and final speed in the positive x-direction is v_i and v_f , respectively. If we use the kinematic equation $v^2 = u^2 + 2as$ learnt from Section 7.3, we have

$$a_x = \frac{v_f^2 - v_i^2}{2\Delta x} \tag{11.4}$$

From Newton's second law, we have

$$F_{net_x} = ma_x \tag{11.5}$$

The net work done on the object by the constant net force F_{net_x} will be

$$W = F_{net_x}\Delta x \tag{11.6}$$

Using Equations 11.4, 11.5 and 11.6, we will get an expression for the net work done on the object by the constant net force F_{net_x} as,

$$W = \frac{1}{2}mv_f^2 - \frac{1}{2}mv_i^2 \tag{11.7}$$

This equation 11.7 can be interpreted as the work done by the net force F_{net_x} on the object of mass m and is equal to the difference between the initial and final values of a quantity $\frac{1}{2}mv^2$. The quantity $\frac{1}{2}mv^2$ is called the kinetic energy of the object and has been abbreviated as K (or KE). Furthermore, the kinetic energy represents the energy associated with the motion of the object and is sometimes referred to as the translational kinetic energy. The kinetic energy is a scalar quantity and is always positive if the object is moving or zero if it is at rest. It has the same unit as work.

Equation 11.17 is an important result known as the work–kinetic energy theorem and can be stated as follows.

The work done by the net force acting on an object is equal to the change in the object's kinetic energy.

i.e.
$$W = \frac{1}{2}mv_f^2 - \frac{1}{2}mv_i^2 = K_f - K_i = \Delta K \tag{11.8}$$

where, K_i and K_f are the initial and final kinetic energies, respectively and ΔK is the change in kinetic energy.

Although we proved this result for a constant force, it can also be shown that it is true for a variable force. Equation 11.8 leads to the conclusion that the kinetic energy of an object decreases by an amount equal to the work done by the object. The kinetic energy cannot be a negative quantity. However, a change in kinetic energy can be negative. When the total work done is negative, the object's speed decreases.

Activity 11.3

If the work required to be done by an external force on an object on a frictionless horizontal surface to accelerate it from a speed zero to v is W, calculate the work required to be done to accelerate it from speed v to $3v$ in terms of W.

Activity 11.4

A driver of a vehicle of mass 1750 kg (including the driver and passengers) travelling at 25.0 m s^{-1} on a horizontal dry road suddenly applied the brake and locked the wheels to stop the vehicle. The coefficient of kinetic friction between the vehicle's tyre and the road is 0.720.

(i) Calculate the initial kinetic energy of the vehicle before the brake is applied.

(ii) How far will the car travel before stopping?

(iii) How far would the car travel if it were travelling twice as fast?

(iv) What happened to the car's initial kinetic energy?

11.5 Potential Energy

In the above section, we learnt how an object is said to have energy by virtue of its motion, and we called it kinetic energy. It is also possible to have another kind of energy called potential energy which is associated with forces that depend on the position or configuration of an object (or objects) relative to the surrounding, generally abbreviated as U (or PE). We can define various types of potential energy, each associated with a particular force. The most common type is gravitational potential energy, which is associated with the object's position relative to the Earth and the gravitational force due to Earth's gravity acting on the object (its weight).

Consider an object of mass m that moves along the vertical direction (y-axis), as shown in figure 11.5. The forces acting on the object are its weight, with magnitude mg , where g is the Earth's gravitational acceleration, and possibly some other forces, which we call the resultant of all the other forces, say $\vec{F}_{other}$.

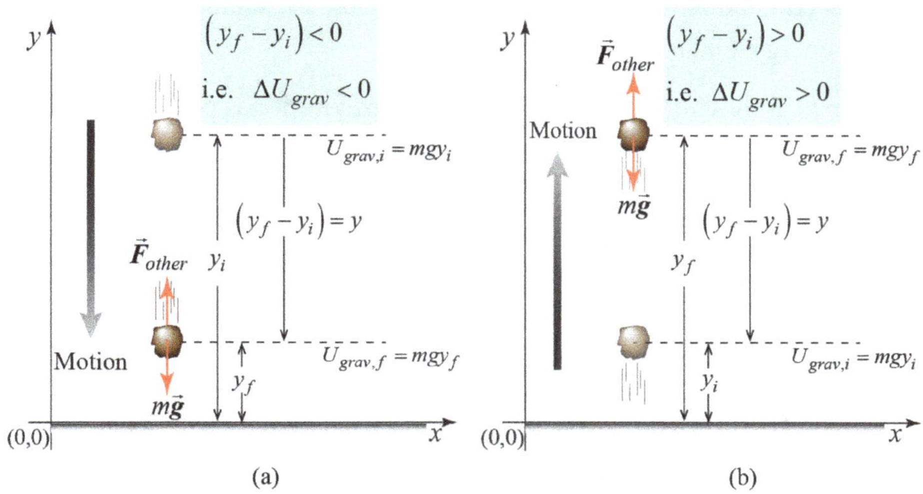

Figure 11.5: (a) When an object moves vertically up, the gravitational potential energy increases. (b) When an object moves vertically down, the gravitational potential energy decreases.

Figure 11.5 (a) shows a situation in which an object of mass m is dropped from a height y_i above the horizontal plane (representing the x axis) to a smaller height y_f . The weight of the object mg and displacement $(y_f - y_i)$ are in the same direction. Therefore, the work W_{grav} done on the object by its weight is positive and is given by

$$W_{grav} = mg(y_i - y_f) = mgy_i - mgy_f \tag{11.9}$$

In the above expression, $mgy_i > mgy_f$.

Figure 11.5 (b) shows a situation in which an object of mass m moves vertically upward from a height y_i above the horizontal plane (representing the x axis) to a smaller height y_f . The weight of the object mg and displacement $(y_f - y_i)$ are in the opposite direction. Therefore, the work W_{grav} done on the object by its weight is negative and in equation 11.9 $mgy_i < mgy_f$.

In equation 11.9, the W_{grav} is expressed in terms of the values of the quantity mgy at the beginning and at the end of the displacement and is called the gravitational potential energy. Therefore, we can define the gravitational potential energy as follows.

When an object with mass m is at a vertical distance y above some reference level in a uniform gravitational field g, the gravitational potential energy U_{grav} is defined as

$$U_{grav} = mgy \tag{11.10}$$

The units of gravitational potential energy are joules, the same as the units of work and kinetic energy, and a scalar quantity.

Figure 11.5 indicates the initial and final gravitational potential energies as $U_{grav,i} = mgy_i$ and $U_{grav,f} = mgy_f$ respectively.

Therefore, the work W_{grav} done by the gravitational force during the displacement from y_i to y_f can be expressed as

$$W_{grav} = U_{grav,i} - U_{grav,f} = -\Delta U_{grav} \tag{11.11}$$

where ΔU_{grav} is the change in gravitational potential energy, i.e. the final value minus the initial value of the gravitational potential energy. The negative sign in front of ΔU_{grav} in equation 11.11 is important. When the object moves downward, as shown in figure 11.5 (a), y decreases (i.e. $y_f < y_i$), the gravitational force does positive work, and the potential energy decreases. When the object moves upward, as shown in figure 11.5 (a), y increases (i.e. $y_f > y_i$), the gravitational force does negative work, and the potential energy increases.

The other important potential energy that we commonly use in mechanics is the elastic potential energy. In the previous session, we saw that when a spring is stretched or compressed by a small distance x from its equilibrium, it exerts a force F_{spring} on the object, and this force can be expressed as $F_{spring} = -kx$, where k is the spring constant.

Furthermore, in session 11.3, we learn that the total work done by the spring on the object when it is stretched from zero to a value x is $W = -\frac{1}{2}kx^2$. The work done on the object is negative because the force exerted on the object F_{spring} is opposite to the displacement x. On the other hand, we learnt that the total work done by the spring on the object when it is compressed from zero to a value x is $W = \frac{1}{2}kx^2$. The work done on the object is positive since the force exerted on the object F_{spring}

is in the same direction as the displacement x. In both cases, the work done by the spring is expressed in terms of a quantity $\frac{1}{2}kx^2$, and we define this quantity as the elastic potential energy $(U_{elastic})$, i.e. $U_{elastic} = \frac{1}{2}kx^2$.

In each of the above examples that we have considered (an object raised vertically up to a height y above the reference level or the stretched/compressed spring), the energy can be stored for later use in the form of potential energy.

Activity 11.5

A student with mass 50 kg climbs a staircase of length 5.00 m that rises 60° above the horizontal at a constant speed.

(i) Calculate the increase in gravitational potential energy by climbing stairs.

(ii) How much work has gravity done on the student as the student climbed the stairs?

11.6 Conservative and Nonconservative Forces

Consider an object of mass m projected vertically upwards from the ground with an initial velocity v. Its kinetic energy at the instant of projection would be $\frac{1}{2}mv^2$. In rising above the ground, the object has to do work against its weight mg, where g is the acceleration due to gravity. From the work-energy theorem, the work done $W = \Delta K$, where ΔK is the change in its kinetic energy. Thus, the energy expended as the object rises must be at the expense of its kinetic energy $\frac{1}{2}mv^2$. Suppose the object rises to a maximum height h, when its vertical velocity becomes momentarily zero. Thus, from the work-energy theorem, we have

$$mgh = \frac{1}{2}mv^2$$

The object begins to fall freely under gravity just after reaching the maximum height h. The force of gravity does work on the object. Therefore, its kinetic energy increases. Suppose that its velocity after falling through the height h. i.e., just as it reaches the ground, is v'. Using the third equation of motion, we get $v'^2 = 2gh$. Therefore, its kinetic energy just before reaching the ground would be $\frac{1}{2}mv'^2$.

i.e. $$\frac{1}{2}mv'^2 = \frac{1}{2}m \times 2gh = mgh = \frac{1}{2}mv^2$$

The above results reveal that the object has the kinetic energy it initially had when it left the ground. Therefore, if we assume air resistance to be negligible, the object has regained all the kinetic energy it lost in rising when it returns to its starting point. The force of gravity and other forces that act in this way are said to be conservative. i.e. the energy of the object is conserved during the round trip back to its starting point. The elastic force exerted by an ideal spring (i.e. one that obeys Hooke's law) is also a conservative force.

On the other hand, if an object on which one or more forces act returns to its starting point with more or less kinetic energy than it had initially, then at least one of the forces acting is said to be nonconservative.

Suppose that the air resistance is not negligible and acts on the object in the example that we considered earlier. This is obvious when the object's velocity through the atmosphere is large, as in the case of meteors entering the atmosphere or space vehicles re-entering the atmosphere. For the upward and downward journeys of the object, it has to do work against the force due to the air resistance. The energy expended in this process is dissipated in the form of heat. The object becomes heated and the surrounding air molecules acquire kinetic energy. Thus, the object's kinetic energy would be less than its initial value when it returns to its starting point. Forces, such as friction, are said to be nonconservative.

We could generally define conservative and nonconservative forces as follows.

> **A force is *conservative* if the work done by the force on an object that moves through any round trip is zero. A force is *nonconservative* if the work done by the force on an object that moves through any round trip is not zero.**

From the above definition of conservative and nonconservative forces, it is possible to distinguish between them in another way.

For example, consider an object that goes from A to B along path 1 and returns to A along path 2, as shown in figure 11.6 (i).

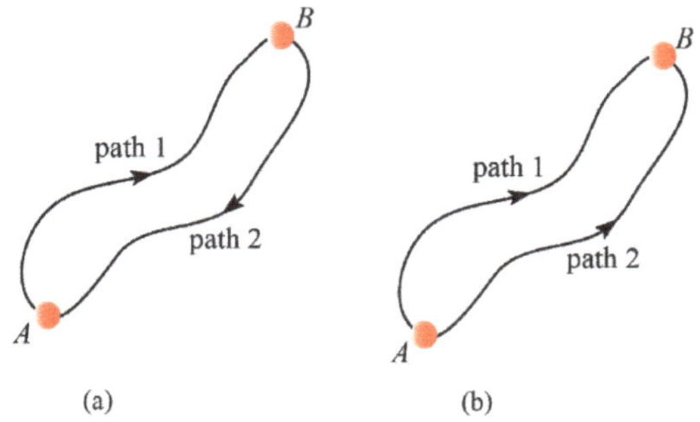

Figure 11.6: (a) Object goes from A to B along path 1 and returns to A along path 2 (b) object goes from A to B along path 1 and path 2

Several forces may act on the object, each being considered separately. If the particular force considered is conservative, the work done on the particle by that force from A to B and back again to A is zero.

i.e. $$W_{A\to B,\text{path 1}} + W_{B\to A,\text{path 2}} = 0 \quad \Rightarrow \quad W_{A\to B,\text{path 1}} = -\, W_{B\to A,\text{path 2}}$$

However, if the object goes from A to B along path 2, as in figure 3.6 b, then,

$$W_{A\to B,\text{path 1}} = W_{A\to B,\text{path 2}}$$

The above result shows that the work done on the object by a conservative force going from A to B is the same whether the movement is along path 1 or path 2.

Therefore, the work done by a conservative force on an object that moves between two points is independent of the path and depends only on the two points considered. If the work done by a force between two points is dependent on the path, such a force is nonconservative.

Activity 11.6

A 750 g wooden block slides on a horizontal wooden table. The kinetic friction force on the block has a magnitude of 2.2 N .

(i) Determine the kinetic friction between the wooden block and the wooden table.

(ii) Determine the work done on the block by friction force during displacement 2.0 m to the left.

(iii) The block now slides 2.0 m to the right, returning to its starting point. How much work is done on the block during this return trip by friction force?

(iv) What is the total work done on the block by friction force during the complete round trip?

(v) Based on your answer to parts (b), (c), and (d), is the friction force conservative or nonconservative? Explain.

11.7 Law of Conservation of Energy

The law of conservation of energy is one of the most fundamental in the field of physics. Energy can exist in several forms in addition to mechanical energy. Some other common forms of energy include sound, thermal, chemical, electrical, nuclear, electromagnetic, etc. We shall study the nature of these various forms of energy in later chapters. Any system capable of performing work has energy. Furthermore, energy can be converted from one form to another or transferred from one place to another. If we carefully account for all the energy transformations, the total energy remains the same.

A conservation law is a physical principle that identifies a quantity that does not change over time. The law of conservation of energy means that in every physical process, the total energy in the universe remains constant.

> A central feature of the energy approach is that energy can neither be created nor destroyed; it only changes form.

As we have already stated, energy can exist in many forms. While the total sum of energy remains constant, energy can still be transformed from one kind to another. Consider the example of hydroelectricity: Damming a river can produce a strong flow of water, which possesses mechanical energy, mainly kinetic energy. This mechanical energy is used to drive turbines, which in turn operate the dynamos that generate electricity. When distributed among consumers, this electricity can be used to produce light in our homes, operate electric kettles, irons, work fans, create sound in public address systems, etc. In other words, the mechanical energy of flowing water is transformed into electrical energy, which in turn is converted into light, heat, sound, mechanical energy, etc.

Conservation of Mechanical Energy

The conservation of mechanical energy states that the total mechanical energy of an isolated system remains constant and applies to any isolated system that is subject to only conservative forces. This means that the sum of the system's kinetic energy and potential energy does not change over time.

Mechanical energy is the energy of motion and position. It comprises the kinetic energy (K) and the potential energy (U). The work done by a conservative force depends only on the initial and final positions of the object and is independent of the path taken between these points. If only conservative forces act on a system, then the sum of the change in kinetic energy (ΔK) and the change in potential energy (ΔU) must be zero. Thus,

$$\Delta K + \Delta U = 0 \tag{11.12}$$

i.e

$$\left(K_f - K_i\right) + \left(U_f - U_i\right) = 0 \tag{11.13}$$

where K_i, U_i and K_f, U_f are the system's initial and final kinetic energy and potential energy, respectively. Thus, we have

$$K_i + U_i = K_f + U_f \tag{11.14}$$

If E is the total mechanical energy of the system expressed as the sum of the kinetic energy (K) and potential energy (U) at any moment, then,

$$E + U = \text{constant} \tag{11.15}$$

Equation 1.15 is a useful and profound result that emphasises that the total mechanical energy E of a system remains constant as long as only conservative forces act on it. This is the principle of conservation of mechanical energy for conservative forces. It should be noted that if nonconservative forces act in an isolated system, the total energy of the system is conserved, even though the mechanical energy is not.

The conservation of mechanical energy is a fundamental law of physics that has many applications in everyday life. For example, it is used to design roller coasters, which rely on the conversion of potential energy to kinetic energy to create a thrilling ride. It is also used to explain the motion of planets and stars that are subject to the force of gravity. The conservation of mechanical energy is a reminder that energy can neither be created nor destroyed, but can only be converted from one form to another.

Activity 11.7

A projectile is fired at an upward angle from the top of a 300.0 m cliff with a speed of 200.0 m s^{-1}. Using the principle of conservation of mechanical energy, determine its speed when striking the ground below.

11.8 Power

In physics, power is defined as the rate at which work is done or the rate of energy transfer in a system. When work is performed on an object, it results in a change in the object's energy. Power quantifies how quickly this work is being done or how fast energy is being transferred from one form to another. Thus, when a quantity of work ΔW is done during a time interval Δt, the average power P_{avg} or the work per unit time is defined as

$$P_{avg} = \frac{\Delta W}{\Delta t} \tag{11.16}$$

We can define the instantaneous power P as the limit of $\frac{\Delta W}{\Delta t}$ as Δt approaches zero. Thus, mathematically it can be represented as

$$P = \lim_{\Delta t \to 0} \frac{\Delta W}{\Delta t} = \frac{dW}{dt} \tag{11.17}$$

Furthermore, power can also be defined in terms of force and velocity. The quantity of work ΔW done by a force F on an object is the product of the magnitudes of the force, i.e. F, and the magnitude of the displacement, i.e. Δr, in the direction of the force applied. Thus,

$$P_{avg} = \frac{\Delta W}{\Delta t} = F\frac{\Delta r}{\Delta t} = Fv_{avg}$$

Therefore, the instantaneous power P as the limit of $F\frac{\Delta r}{\Delta t}$ as Δt approaches zero can be mathematically written as

$$P = F \lim_{\Delta t \to 0} \frac{\Delta r}{\Delta t} = F\frac{dr}{dt} = Fv \tag{11.18}$$

where $\frac{dr}{dt} = v$ the instantaneous velocity.

Like work and energy, power is a scalar quantity. The unit of power in the SI system is 1 joule/second = 1 watt (W). Remember that larger or smaller units can be stated in terms of watts using certain prefixes common to physics.

1 milliwatt (mW) = 10^{-3} W 1 kilowatt (kW) = 10^{3} W

1 microwatt (μW) = 10^{-6} W 1 megawatt (MW) = 10^{6} W

A unit of energy (or work) can now be defined in terms of a unit of power. Thus, we often use a unit known as the kilowatt-hour (kWh), which is the energy transferred in 1 hour at a constant rate of 1 kW ($=1\ 000\ \text{J s}^{-1}$). The amount of energy represented by 1 kW is

$$1\ \text{kW} = \left(10^{3}\ \text{W}\right)\left(3\ 600\ \text{s}\right) = 3.60\times10^{6}\ \text{J}$$

Please note that a kilowatt-hour is a unit of energy, not power. For example, when we pay our electric bill, we pay for the energy that we have used, i.e. the amount of energy transferred by electrical transmission into our home during the period represented by the electric bill. In Sri Lanka, it is usually expressed as the number of "units" of electricity that we use. These units are expressed in kilowatt-hours.

Traditionally, another popular unit is used for power known as horse power (hp).

$$1\ \text{hp} = 746\ \text{W}$$

Quantity power is also vital in understanding the performance of machines, engines, and mechanical devices. It determines the amount of work output a machine can produce in a given time, affecting the efficiency and effectiveness of these systems.

The efficiency of a system is a measure of how well it converts input energy or resources into useful output energy or work. It quantifies the ratio of useful output to the total input, indicating how effectively the system accomplishes its intended task. Mathematically, the efficiency (η) of a system is typically expressed as a percentage and can be calculated using the following formula:

$$\eta = \frac{\text{Useful Output}}{\text{Total Input}}\times100\% \tag{11.19}$$

where useful output refers to the desired or usable output obtained from the system, such as useful work, output energy, etc., and total input represents the total amount of energy, resources, or effort supplied to the system to achieve the desired output. The efficiency of a system is always less than 100% because some energy is invariably lost or dissipated in the form of heat, friction, etc., during the conversion process.

Activity 11.8

As shown in the figure, a vehicle of mass 1700 kg travels up a frictionless incline road at a constant velocity of 36 km h^{-1}. Determine the power output of its engine. Assume the incline has a slope of 1 in 20 and the acceleration due to gravity $g = 10\text{ m s}^{-2}$.

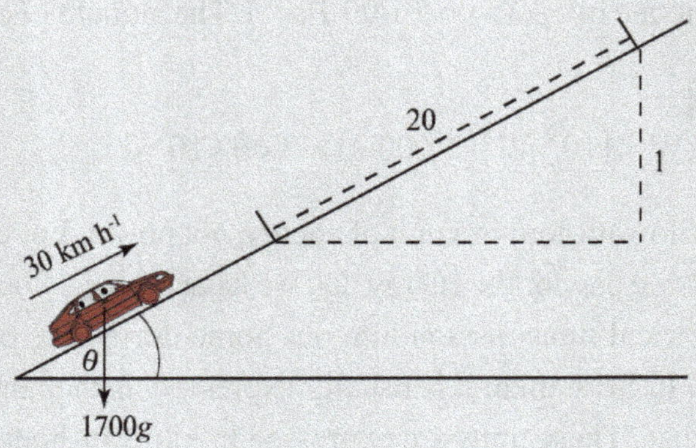

Questions

NOTE: Although the solutions are given on the last page, it is in your own interest to attempt the questions on your own as an indication of how far you have understood the material of the section.

1. Consider a situation where a student engaged in reading a book for more than an hour. Is this student doing any work? Explain.
2. In your garden, you see a mango fall to the ground. During the fall, does the Earth's gravitational field do positive or negative work on the mango? Explain.
3. Imagine a book on a conveyor belt that is carried up an incline. Static friction acts on the book, which prevents relative motion along the contact surface. Can this static friction do any work? Explain. What is the total work done on the book?

4. (i) Identify the difference between gravitational potential energy and elastic potential energy.

   (ii) Explain the energy transfers that occur when the rubber cord of a catapult is pulled back and a pellet is launched from the catapult.

5. A constant force $\vec{F}$ is applied to a solid block in four different ways, as shown in (a), (b), (c), and (d) below. In each case, the solid block slides on the horizontal frictionless floor from left to right through a distance x.

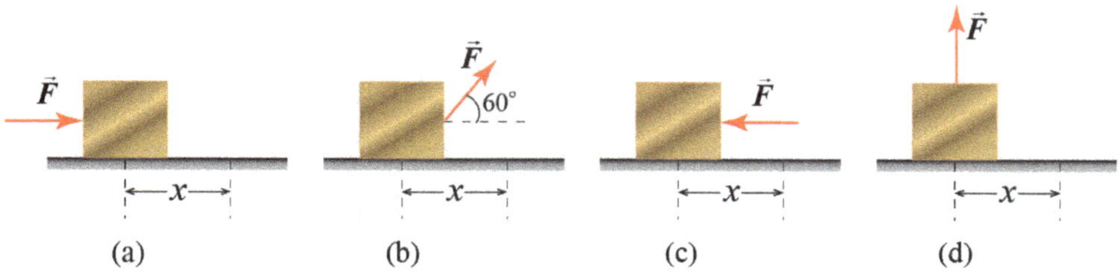

   (i) For each case, write down the work done on the solid block by the constant force $\vec{F}$.

   (ii) Rank the situations in the order of the work done by the force on the object, from most positive to most negative.

6. A block of mass 3.50 kg is drawn at a constant velocity through a distance 4.00 m along the surface of a horizontal table by a string which exerts a constant force of 7.50 N, making an angle 30° with the horizontal. Calculate

   (i) the total work done on the block.

   (ii) the work done by the string on the block.

   (iii) the work done by friction on the block.

   (iv) the coefficient of kinetic friction between the block and the table.

7. A nail is driven initially through 2.0 cm thick soft dry wall and then through 4.0 cm thick solid wood wall using a hammer. The force that has been exerted to drive the nail is shown in the figure.

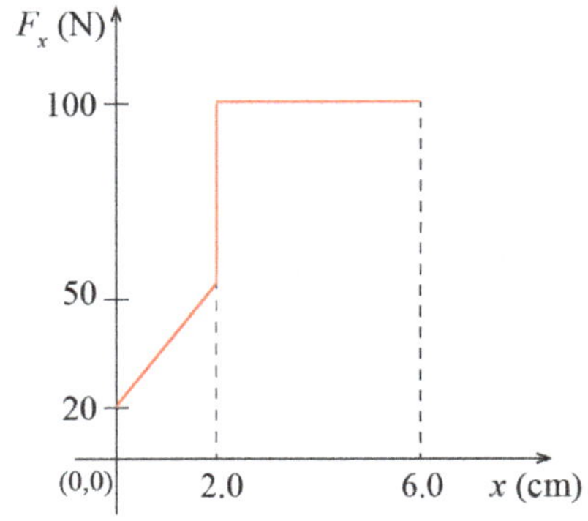

Determine the work that must be done with a hammer on the nail to drill a horizontal distance of 6.0 cm into the wall.

8. On a rainy day, a car travelling on a flat road skid off the road, met with an accident and came to an immediate stop. During the investigation, the traffic police found that the car had skidded 90.0 m before ceasing its motion. If the coefficient of kinetic friction between the car's tyre and the road is 0.400, calculate the car's speed when the driver forcefully applies the brakes.

9. (i) Draw a graph showing the variation of gravitational potential energy (P_E) and kinetic energy (K_E) when an object is projected vertically upward from a height (h) of zero to the maximum height h_{max} it reaches, where its velocity is zero, as shown in figure (a) below. Neglect the air resistance during the motion.

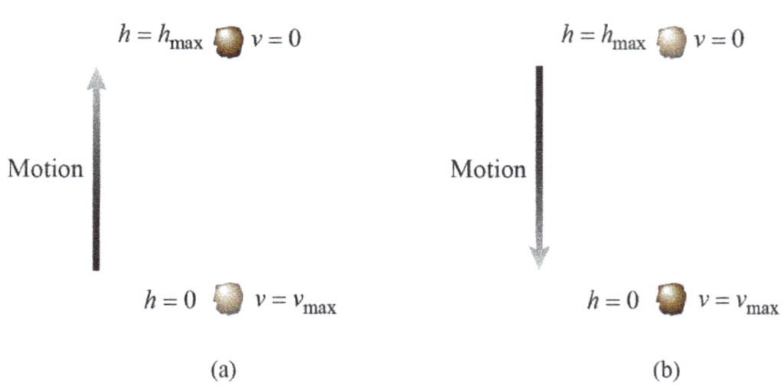

(ii) Draw a graph showing the variation of gravitational potential energy (P_E) and kinetic energy (K_E) when an object initially at rest falls vertically downward to a height (h) of zero where its velocity is maximum, as shown in figure (b) above. Neglect the air resistance during the motion.

10. A mechanic who started to repair a machine needed a spring that could store 900 J by compression. In his workshop, he found that the available spring is 8.70 cm long when no forces act on it, which will store 14.4 J energy when compressed by 1.20 cm from its equilibrium position.

    (i) Determine the spring's force constant, which is available in the workshop.

    (ii) Can this spring store 900 J by compression?

11. A falcon with a mass of 750 g, positioned at a certain height above the ground, dives vertically by 120 m to capture its prey.

    (i) Determine the amount of gravitational potential energy the falcon loses during its dive.

    (ii) While diving, 90% of the falcon's gravitational potential energy transforms into kinetic energy. If the falcon began from a rest position, determine its velocity at the end of the dive.

    (iii) Unfortunately, the falcon misses its prey and climbs back upwards, converting 75% of its kinetic energy into gravitational potential energy. How high will it climb before it flaps its wings?

12. An object of mass 3.0 kg starting from rest falls vertically through a distance of 45.0 cm before striking a vertical coiled spring whose one end is fixed to the floor, as shown in the figures. When the object strikes the spring, it compresses an amount of 15.0 cm, as illustrated in figure (c). Assume that the mass of the spring is negligible compared to the mass of the object, and all distances are measured from the point where the object first touches the upper end of the uncompressed spring.

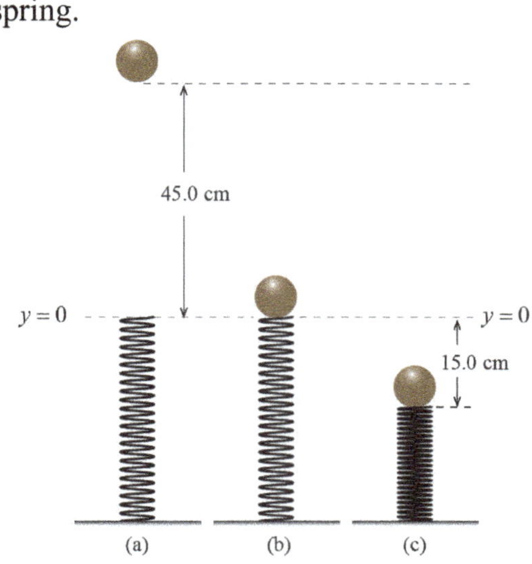

(i) Identify the forces acting on the objects.

(ii) Are the forces identified in part (a) conservative or nonconservative forces?

(iii) Determine the speed of the object when it touches the upper end of the vertical spring.

(iv) Determine the spring constant of the coiled spring.

13. A worker who repairs a roof climbs a roof that slants 30° above the horizontal. While climbing, he carried his 6.0 kg tool box. At 5.0 m from the roof's lower edge, he accidentally pushed his tool box with his hand, and it began to slide along the roof top starting from rest. If the 16.0 N kinetic friction force acts on the tool box while sliding, determine the speed of the tool box when it reaches the roof edge.

14. The specific end product of a factory is packaged in corrugated triple-walled boxes, each weighing 100 kg.

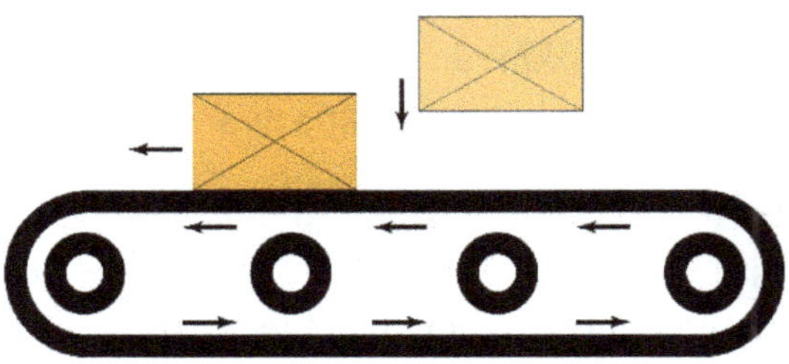

As shown in the figure, these packages are vertically dropped from a packing machine on a conveyor belt that moves with a constant velocity 1.40 m s^{-1} before being loaded for delivery. The coefficient of kinetic friction between the belt and each box is 0.400. After the box is dropped on the conveyor belt, the slip between the belt and the box stops quickly, and the box moves along with the belt. Calculate the following relative to the factory ground during the period in which the box is brought to rest relative to the belt.

(i) The kinetic energy supplied to the box.

(ii) The magnitude of the kinetic frictional force acting on the box.

(iii) The energy supplied by the motor.

Is the kinetic energy supplied to the box and the energy supplied by the motor the same? Explain.

15. A 70.0 kg bungee jumper starting from rest jumps from a bridge that is 60.0 m above the river below, as shown in the figure.

The bungee cord has an unstretched length of 30.0 m, and to be safe, the cord should stop the jumper's fall at least 2.5 m above the river. Calculate the following.

(i) The jumper's gravitational potential energy above the river just before he falls from the bridge.

(ii) The jumper's gravitational potential energy above the river after the jumper reaches the maximum stretched length of the bungee cord.

(iii) The maximum stretched length of the bungee cord.

(iv) The minimum spring constant of the bungee cord.

16. In a children's park, there are two playground slides with frictionless surfaces. Although these slides are shaped differently, they share the same length and height h, as shown in the figure. Two children, Nathan and Mary, start sliding from rest simultaneously; Nathan on slide 1 and Mary on slide 2, respectively.

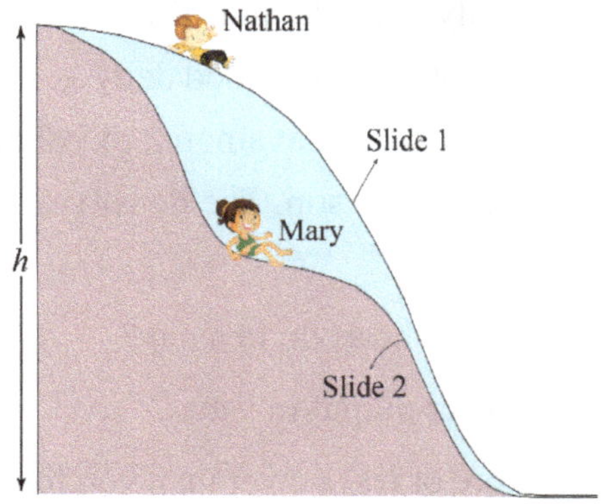

(i) Upon reaching the bottom of the slides, will Nathan travel faster than Mary, or will Mary travel faster than Nathan, or will both kids, Nathan and Mary, have the same speed? Explain your answer.

(ii) Between Nathan and Mary, who will reach the bottom first? Explain your answer.

17. An object of mass $3.0\ \text{kg}$ slides down a frictionless incline, starting from a height of $3.2\ \text{m}$ as shown in the figure.

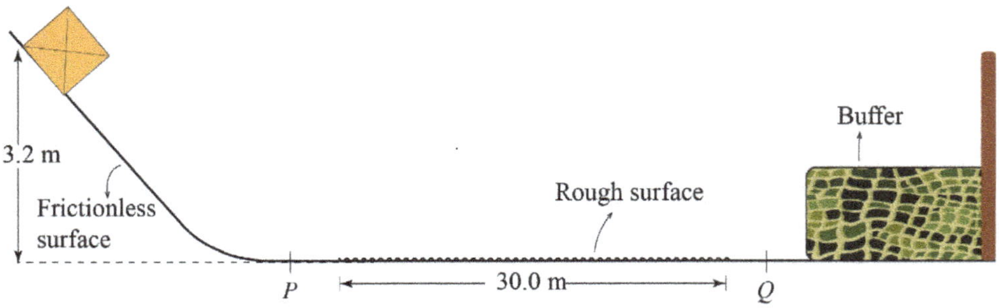

(i) Determine the velocity of the object as it passes through the point P.

(ii) If the object then slides $30.0\ \text{m}$ over a rough surface, determine the total energy dissipated by friction on the object when it gets to the point Q. The coefficient of kinetic friction of the rough surface is $\mu_k = 0.100$.

(iii) Determine the velocity of the object as it moves through the point Q.

(iv) The object then hits a crash buffer and comes to rest. If the work done by the buffer is $-150x^2$, where x is the distance the buffer is compressed, and at what value of x is the object brought to rest by the buffer?

18. The human heart is a highly reliable and powerful pump that takes in and discharges approximately 7500 litres of blood daily. Assume that the work done by the heart is comparable to lifting that amount of volume of blood to a height of about $1.60\ \text{m}$ by an average person. The density of blood is $1050\ \text{kg m}^{-3}$ and $g = 10\ \text{m s}^{-2}$.

(i) How much work does the heart do in a day?

(ii) What is the heart's power output in watts?

(iii) Is the heart put out more or less power than you found in part (b)? Explain.

19. As shown in the figure, an elevator with a mass of 1 800 kg carries passengers whose combined mass is 120 kg . The elevator's motion is being slowed down by a constant friction force of 3 500 N .

    (i) If the elevator and its passengers move upward at a constant speed, calculate the force exerted by the motor to raise the elevator.

    (ii) Calculate the power output of the motor required to lift the elevator and its passengers at a constant speed 2.5 m s^{-1} .

    (iii) Assume that the motor is designed to give the elevator an upward acceleration of 1.0 ms^{-2}. Determine the power that the motor delivers to lift the elevator and its passengers when it reaches an instantaneous speed 2.5 ms^{-1}.

    (iv) Is the power obtained in part (c) greater or less or equal to that in part (b)? Explain why.

20. Wind turbines are the modern version of a windmill that uses wind power to create electricity. Wind turbines use blades to collect the wind's kinetic energy. Consider one of such wind turbines located on a wind farm, as shown in the figure. At the location of the wind turbine, the average wind speed is 7.0 m s^{-1}, and the average air density is 1.2 kg m^{-3}.

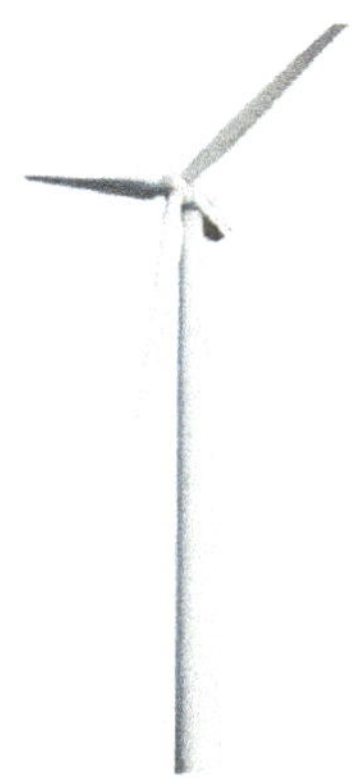

    (i) Calculate how much kinetic energy the wind contains per cubic metre at the wind turbine's location.

    (ii) Suppose that the turbine has blades with a radius of 40.0 m and can capture 35% of the available wind energy. What would the power output of this turbine be under above-average wind conditions?

Chapter 12

Momentum, Impulse and Collision

12.1 Introduction

This chapter will discuss the concepts of momentum and impulse, which play a crucial role in understanding how objects move and interact with each other. First, we will focus on the concept of momentum. Momentum can be thought of as the "quantity of motion" possessed by an object. It takes into account both the object's mass and its velocity. We will learn to calculate momentum and explore its relationship with Newton's laws of motion. You will be guided to develop the skills to quantify and analyse momentum for different objects and scenarios.

Next, we will focus on the concept of impulse, which is closely related to the concept of force. By studying impulse, we can understand how forces affect an object's motion and learn to calculate the resulting change in momentum. We also examine the impulse-momentum theorem and explore its implications through real-world examples.

Finally, we will focus on the principle of conservation of momentum, which is particularly useful when dealing with a system of two or more objects that interact with each other, such as in collisions.

12.2 Momentum

The momentum of an object or particle is defined as the product of its mass and velocity. The symbol used to represent the momentum is $\vec{p}$. The term momentum is commonly referred to as linear momentum to differentiate it from angular momentum, which you will study in Chapter 14.

Accordingly, if m represents the mass of the object and $\vec{v}$ represents its velocity, then its momentum $\vec{p}$ is defined as

$$\vec{p} = m\vec{v} \tag{12.1}$$

Since velocity is a vector quantity, momentum is also a vector quantity. Furthermore, the object's mass m is always a positive scalar quantity, therefore, the momentum $\vec{p}$ of the object and its velocity $\vec{v}$ have the same direction. The SI unit for momentum is kg m s^{-1}.

If we consider an object having a mass m that moves in an arbitrary direction with a velocity $\vec{v}$, then the three momentum components p_x, p_y, p_z of its momentum $\vec{p}$ along the x, y, z axes of a Cartesian coordinate are given by

$$p_x = mv_x \qquad p_y = mv_y \qquad p_z = mv_z \tag{12.2}$$

where v_x, v_y, v_z are the velocity components of the velocity $\vec{v}$ of the same x, y, z axes of the Cartesian coordinate.

From the definition of momentum, we can see that it provides a quantitative distinction between heavy and light objects moving at the same velocity. For example, the momentum of a heavy ball of mass $3\ \text{kg}$ is much greater than that of a tennis ball with a mass 0.057 kg moving at the same velocity. Thus, the product $m\vec{v}$, i.e. momentum, is sometimes referred to as the quantity of motion.

Total Momentum

The total momentum of two or more objects in a system is defined as the vector sum of the momenta of these objects. Therefore, if there are many number of objects (namely objects *A, B, C,…*) in the system, then the total momentum $\vec{p}$ of these objects is equal to the vector sum of the momenta of the individual objects.

i.e.
$$\vec{p} = \vec{p}_A + \vec{p}_B + \vec{p}_C + \dots\dots\dots \tag{12.3}$$

Generally, equation 12.3 is used in a component form when we solve problems. Therefore, components of the total momentum of the system could be written as

$$p_x = p_{A_x} + p_{B_x} + p_{C_x} + \dots\dots\dots \tag{12.4}$$

$$p_y = p_{A_y} + p_{B_y} + p_{C_y} + \dots\dots\dots$$

$$p_z = p_{A_z} + p_{B_z} + p_{C_z} + \dots\dots\dots$$

where $p_{A_x}, p_{A_y}, p_{A_z}$ are the components of object A, $p_{B_x}, p_{B_y}, p_{B_z}$ are the components of object B and $p_{C_x}, p_{C_y}, p_{C_z}$ are the components of object C along the x, y, z axes of a Cartesian coordinate.

Activity 12.1

A car having a mass of 1.2×10^3 kg travels northwards at a speed of 15.0 m s^{-1}. After turning through a sharp bend, the car travels eastward at 7.0 m s^{-1}.

Determine the following.

(i) The initial momentum of the car.

(ii) The car's momentum after turning through the sharp bend.

(iii) The magnitude and direction of the change in the momentum of the car.

12.3 Newton's Second Law and Momentum

In Chapter 9, we studied Newton's laws of motion. Newton's second law says that *"the acceleration of an object is directly proportional to the net force acting on it and is inversely proportional to its mass. The direction of the acceleration is in the direction of the net force acting on the object"*. Accordingly, we learnt that mathematically, the standard form of the equation for Newton's second law could be written as

$$\vec{F}_{net} = \sum \vec{F} = m\vec{a} \tag{12.5}$$

Where, $\sum \vec{F}$ is the net force acting on the object and could be denoted as $\vec{F}_{net}$, and $\vec{a}$ is the acceleration of the object of mass m, which is a vector quantity.

In Chapter 8, we studied that the instantaneous acceleration $\vec{a}$ of the object could be written as

$$\underset{\sim}{a} = \lim_{\Delta t \to 0} \frac{\Delta \underset{\sim}{v}}{\Delta t} \tag{12.6}$$

Therefore, Newton's second law can be written as

$$\vec{F}_{net} = \sum \vec{F} = m\vec{a} = m \lim_{\Delta t \to 0} \frac{\Delta \vec{v}}{\Delta t} \tag{12.7}$$

i.e.

$$\sum \vec{F} = m \lim_{\Delta t \to 0} \frac{\Delta \vec{v}}{\Delta t} \tag{12.8}$$

Now if m is constant, the right-hand side of equation 12.8 could be rewritten as

$$\sum \vec{F} = \lim_{\Delta t \to 0} \frac{m\Delta\vec{v}}{\Delta t} \tag{12.9}$$

If $\vec{v}_1$, $\vec{v}_2$ and $\vec{p}_1$, $\vec{p}_2$ are the initial and final velocity and momentum of the object, respectively, during the time interval Δt, the momentum change $\Delta\vec{p}$ of the object during this time interval Δt could be written as

$$\Delta\vec{p} = \vec{p}_1 - \vec{p}_2 = m\vec{v}_2 - m\vec{v}_1 = m\left(\vec{v}_2 - \vec{v}_1\right) = m\Delta\vec{v} \tag{12.10}$$

Therefore, equation 12.9 could be written as

$$\sum \vec{F} = \lim_{\Delta t \to 0} \frac{\Delta\vec{p}}{\Delta t} \tag{12.11}$$

This means, mathematically, net force $\sum \vec{F}$ or $\vec{F}_{net}$ acting on the object could be described as the limit of $\frac{\Delta\vec{p}}{\Delta t}$ as Δt approaches zero. Furthermore, $\lim_{\Delta t \to 0} \frac{\Delta\vec{p}}{\Delta t}$, mathematically expressed as the rate of change of the momentum of the object with respect to time, it could be denoted as $\frac{d\vec{p}}{dt}$.Thus, equation 12.11 could be written as

$$\vec{F}_{net} = \sum \vec{F} = \frac{d\vec{p}}{dt} \tag{12.12}$$

Therefore, Newton's second law in terms of momentum could be stated as follows.

> ***The net force acting on an object is equal to the rate of change in momentum of the object with respect to time. The direction of the object is in the direction of the net force acting on the object.***

It should be noted that this alternative form of stating Newton's second law in terms of momentum is more general than the form introduced in Chapter 9.

12.4 Principle of Conservation of Linear Momentum

The significance of momentum arises prominently when two or more objects engage in interactions. Let us now consider a definite collection of objects or particles that we generally call a system. In any system, the various objects exert forces on each other, which we call internal forces. Furthermore, forces may also be exerted on any part of the system by objects outside the system, which we call external forces. A system that is acted upon by no external forces is called an isolated system.

We studied in the previous section that Newton's second law in terms of momentum could be written as

$$\vec{F}_{net} = \sum \vec{F} = \frac{d\vec{p}}{dt} \tag{12.13}$$

Suppose that the vector sum of the external forces on the system is zero $\left(\vec{F}_{net} = \sum \vec{F} = 0\right)$, i.e. for an isolated system, equation 12.13 becomes

$$\frac{d\vec{p}}{dt} = 0 \tag{12.14}$$

From the definition of the time derivative of the momentum of the system (see Equation 12.11), we could say that the system's momentum change $\Delta\vec{p}$ for the time interval Δt during which the objects interact must be zero.

i.e. $$\Delta\vec{p} = 0 \tag{12.15}$$

For easy understanding, let us consider an isolated system of two objects with initial and final momentum $\vec{p}_{1i}$, $\vec{p}_{2i}$ and $\vec{p}_{1f}$, $\vec{p}_{2f}$ respectively, for the time interval Δt during which the objects interact.

Then, the total initial momentum of the isolated system would be

$$\vec{p}_i = \vec{p}_{1i} + \vec{p}_{2i}$$

and, the total final momentum of the isolated system would be

$$\vec{p}_f = \vec{p}_{1f} + \vec{p}_{2f}$$

We have already seen that when no external forces are acting on the system, i.e. for an isolated system,

$$\Delta\vec{p} = 0$$

i.e. $$\vec{p}_f - \vec{p}_i = 0 \quad \Rightarrow \quad \vec{p}_i = \vec{p}_f \tag{12.16}$$

Therefore, $$\vec{p}_{1i} + \vec{p}_{2i} = \vec{p}_{1f} + \vec{p}_{2f} \tag{12.17}$$

This could be extended to any number of objects in an isolated system, and if we have n number of objects in an isolated system, then we could write;

$$\vec{p}_{1i} + \vec{p}_{2i} + \vec{p}_{3i} + \ldots + \vec{p}_{ni} = \vec{p}_{1f} + \vec{p}_{2f} + \vec{p}_{3f} \ldots + \vec{p}_{nf} = \text{constant} \qquad (12.18)$$

Therefore, the principle of conservation of linear momentum could be stated as follows.

The total momentum of a system is constant whenever the vector sum of the external forces on the system is zero. In other words, the total momentum of an isolated system is constant.

It should be noted that since we used Newton's second law to derive the conservation of linear momentum, this principle should be used only in inertial frames of reference.

Equations 12.17 and 12.18 are vector equations, and these equations in components along the x, y, z axes of a Cartesian coordinate are all independently conserved. If we consider an isolated system of two objects, from equation 12.17, we have the following.

$$p_{1ix} + p_{2ix} = p_{1fx} + p_{2fx}$$

$$p_{1iy} + p_{2iy} = p_{1fy} + p_{2fy}$$

$$p_{1iz} + p_{2iz} = p_{1fz} + p_{2fz}$$

Activity 12.2

An archer stands at rest on a frictionless surface and fires a 300 g arrow horizontally at 45.0 m/s, as shown in the figure. The combined mass of the archer and bow is 70.0 kg.

(i) Determine the velocity of the archer after firing the arrow.

(ii) In which direction the archer will move after the arrow is fired? Explain.

12.5 Impulse and Collision

The change in momentum of an object depends on both the force exerted on it and the time interval during which the force is exerted. In our day-to-day experiences, we come across various interactions where significant forces are applied within a short time duration of impact. For example, a tennis ball is struck by a racket, a bat strikes a cricket ball, a hammer hitting a nail, etc. To analyse these specific effects, we define a quantity called impulse.

When a constant net force $\vec{F}_{net}$ acts on an object during a time interval Δt from t_i to t_f (where $\Delta t = t_f - t_i$), the impulse of the force, denoted by $\vec{J}$, is defined as the product of the net force $\vec{F}_{net}$ and the time interval Δt.

$$\vec{J} = \vec{F}_{net}\left(t_f - t_i\right) = \vec{F}_{net}\Delta t \tag{12.20}$$

The impulse is a vector quantity and its direction is the same as that of the net force $\vec{F}_{net}$. The SI unit of impulse is $\mathrm{N\,s}$.

The term collision is often used to describe an event in which two or more objects come into contact and exert strong forces on each other for a brief period of time. The forces exerted on the objects are strong enough to change their direction of motion, speed, or both. Furthermore, the interaction between objects during the collision is usually much stronger than any interaction with its environment. In such situations, we generally ignore the effects of any other forces acting on the objects during the very short time interval of the collision.

When a collision occurs between two objects, both objects get deformed due to the involvement of large forces on each other. The net force $\vec{F}_{net}$ exerted on the object usually varies from zero at the moment of contact to a large force within a very short time interval Δt and then rapidly returns to zero, as shown in figure 12.1(a). We often approximate such varying net force $F_{net}\left(t\right)$ as a time-averaged force $\left(F_{net}\right)_{avg}$ acting during a very short time interval Δt, as shown in figure 12.1(b).

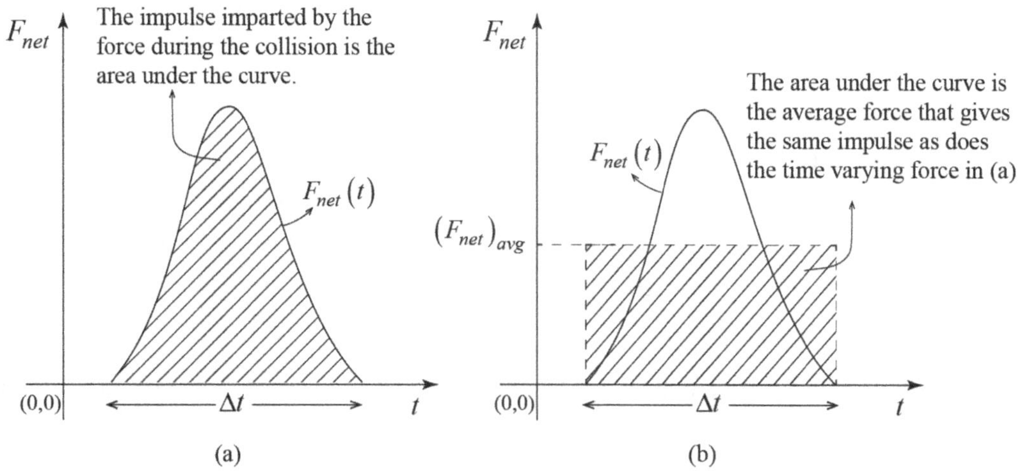

Figure 12.1: (a) Varying net force $F_{net}(t)$ acting on an object with time. (b) The average net force $(F_{net})_{avg}$ is chosen so that the area of the rectangle is the same as the area under the curve in (a)

From Newton's second law, the average net force on the object is equal to the rate of change of its momentum. Since the collision occurs during a very short time interval Δt, we have

$$\left(\vec{F}_{net}\right)_{avg} = \frac{\Delta\vec{p}}{\Delta t} \tag{12.21}$$

This equation applies to each of the two objects in the collision. This equation could be rewritten as

$$\left(\vec{F}_{net}\right)_{avg} \Delta t = \Delta\vec{p} \tag{12.22}$$

In the above equation, the left-hand side quantity $\left(\vec{F}_{net}\right)_{avg} \Delta t$ is called the impulse $\vec{J}$ (refer to equation 12.20). Thus

$$\vec{J} = \Delta\vec{p} \tag{12.23}$$

This could be expressed as follows;

The change in the momentum of an object is equal to the impulse of the net force acting on the object.

The above statement is known as the impulse-momentum theorem.

The time-averaged force, shown in figure 12.1(b), is the constant force that would give to the object in the time interval Δt the same impulse that the time-varying force gives over this same interval. The magnitude of the impulse is the area under the force-time curve shown in figure 12.1.

In a collision, if the interaction forces are much larger than any external forces, the colliding objects form an isolated system, neglecting the external forces entirely. The momentum of such a system must be conserved in any collision. However, the total kinetic energy of the system of objects may or may not be conserved. Collisions are categorised as either elastic or inelastic, depending on whether or not kinetic energy is conserved.

Nothing is called perfection in nature except our theoretical assumptions, which are often useful in obtaining approximate results for real situations. Only collisions between atomic, nuclear, and fundamental particles come very close to perfect collisions. All other collisions are inelastic to some extent. However, we often treat collisions between frictionless hard balls (e.g. billiard balls) as perfectly elastic to derive approximate results.

So far, we have considered the collisions that involve physical contact between two macroscopic objects. However, our picture of a collision is different on the microscopic and submicroscopic scales. When atoms collide, they do not "touch" each other. However, the collision model that we have discussed is still helpful for atoms and subatomic particles whenever there is an interaction in which the forces are strong over a short time interval. To understand this concept, consider a collision on an atomic scale, such as the collision of a proton with an alpha particle (the nucleus of a helium atom). Because both particles are positively charged, they repel each other due to the strong electrostatic force between them at close separations and never come into physical contact.

Elastic collision: In an elastic collision, there is no net loss of kinetic energy in the system due to the collision. This means that the total energy of the colliding objects before and after the collision is the same. Furthermore, in an elastic collision, the kinetic energy is converted from one form to another, but it is not lost. For example, when a tennis ball bounces off a wall, the kinetic energy of the ball is converted into potential energy as the ball compresses, and then back into kinetic energy as the ball rebounds (assume there is no loss of energy such as heat, sound, etc.).

Inelastic collision: In an inelastic collision, some of the kinetic energy of the system is lost during the collision. This energy loss can be due to heat, sound, or deformation of the objects involved. For example, when a car crashes into a wall, some of the kinetic energy of the car is converted into heat and sound.

When the colliding objects stick together and form a single object after the collision, such a collision is said to be perfectly inelastic. During these types of collision, some or all of the initial kinetic energy is converted into other forms of energy, such as heat, sound, or deformation.

Activity 12.3

A ball weighing 50 g travels horizontally at a speed of 2 m s^{-1}, collides perpendicularly with a rigid wall, and bounces off elastically with the same speed, as shown in the figure.

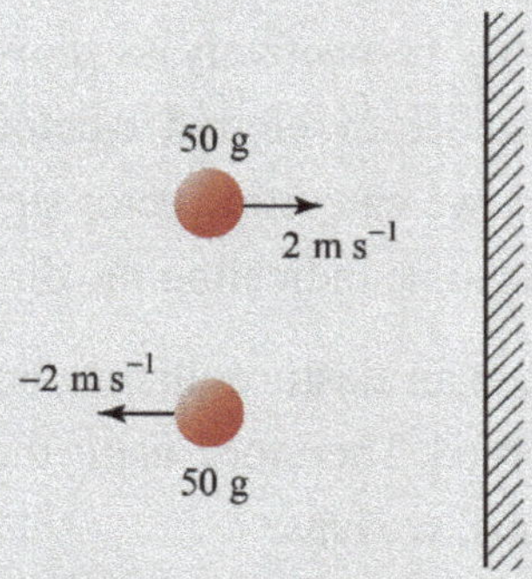

Given that the ball is in contact with the wall for 0.05 s, calculate the following.

(i) The impulse of the net force

(ii) The average force experienced during the collision.

12.6 Collisions in One Dimension

This section focuses on two objects in an isolated system and investigates the mathematical details of their collisions in one dimension. We intend to examine the two cases, (a) elastic collision and (b) perfectly inelastic collision.

Elastic collision

Let us consider an isolated system consisting of two objects (say smooth non-rotating spheres), A and B, of masses m_A and m_B moving with initial velocities $\vec{v}_{A_i}$ and $\vec{v}_{B_i}$ respectively along the same straight line as shown in figure 12.2 (a).

Elastic collision

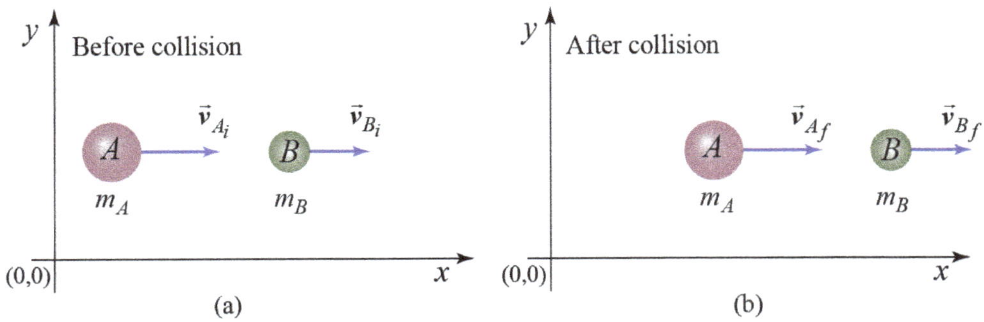

Figure 12.2: Schematic representation of an elastic collision of two objects of masses m_A and m_B in one dimension (a) before the collision (b) after the collision.

The two objects, A and B, collide head-on and then leave (without rotation) with different velocities $\vec{v}_{A_f}$ and $\vec{v}_{B_f}$ respectively, as shown in figure 11.2 (b). Here, the subscripts i and f indicated the velocities referred to before and after the collision. Since the motion is one-dimensional, we can drop vector signs (the overhead arrows) and use only components along the axis, indicating the direction with an algebraic sign.

We have seen earlier that, in an elastic collision, both the momentum and kinetic energy of the system are conserved. Therefore, using the law of conservation of linear momentum along the +ve x-axis, we have

$$p_i = p_f$$

i.e. $$m_A v_{A_i} + m_B v_{B_i} = m_A v_{A_f} + m_B v_{B_f} \tag{12.24}$$

Further, the total kinetic energy (KE) is conserved, and we can write that conservation as

$$KE_i = KE_f$$

i.e. $$\frac{1}{2} m_A \left(v_{A_i}\right)^2 + \frac{1}{2} m_B \left(v_{B_i}\right)^2 = \frac{1}{2} m_A \left(v_{A_f}\right)^2 + \frac{1}{2} m_B \left(v_{B_f}\right)^2 \tag{12.25}$$

Equation 12.24 could be rewritten as

$$m_A \left(v_{A_i} - v_{A_f}\right) = m_B \left(v_{B_f} - v_{B_i}\right) \tag{12.26}$$

Equation 12.25 could be rewritten as

$$m_A \left(v_{A_i} - v_{A_f}\right)\left(v_{A_i} + v_{A_f}\right) = m_B \left(v_{B_f} - v_{B_i}\right)\left(v_{B_f} + v_{B_i}\right) \tag{12.27}$$

If we divide equation 12.27 by equation 12.26 and rearrange it, we have

$$\left(v_{A_i} - v_{B_i}\right) = -\left(v_{A_f} - v_{B_f}\right) \tag{12.28}$$

Equation 12.28 delivers an interesting result. It says that 'in an elastic collision, the relative velocity of the two objects has the same magnitude before and after the collision, and have opposite directions'.

If we know the values of the masses m_A, m_B and their initial velocities v_{A_i}, v_{B_i} before the collision, using equations 12.24 and 12.25, we could find their final velocities v_{A_f}, v_{B_f} after the collision. Thus, we have

$$v_{A_f} = \left(\frac{m_A - m_B}{m_A + m_B}\right) v_{A_i} + \left(\frac{2m_B}{m_A + m_B}\right) v_{B_i} \tag{12.29}$$

and $$v_{B_f} = \left(\frac{2m_A}{m_A + m_B}\right) v_{A_i} + \left(\frac{m_B - m_A}{m_A + m_B}\right) v_{B_i} \tag{12.30}$$

It is important to note that in the above derivation, the positive direction is taken rightward, and when we use equations 12.29 and 12.30, appropriate signs should be used for v_{A_i} and v_{B_i}.

Some special cases:

(a) If $m_A = m_B$, equations 12.29 and 12.30 now become

$v_{A_f} = v_{B_i}$ and $v_{B_f} = v_{A_i}$

i.e. the objects simply exchange velocities during the collision.

(b) Suppose object B is initially at rest. i.e. $v_{B_i} = 0$, thus

$$v_{A_f} = \left(\frac{m_A - m_B}{m_A + m_B}\right) v_{A_i} \quad \text{and} \quad v_{B_f} = \left(\frac{2m_A}{m_A + m_B}\right) v_{A_i}$$

(c) If $m_A = m_B$, in addition to the object *B* initially at rest (i.e. $v_{B_i} = 0$).

Then, $v_{A_f} = 0$ and $v_{B_f} = v_{A_i}$

i.e. the first object *A* is "stopped", and the second object *B* moves with the original velocity of the first object *A*.

If $m_B \gg m_A$, in addition to the object *B* is initially at rest (i.e. $v_{B_i} = 0$).

Then, $v_{A_f} \approx -v_{A_i}$ and $v_{B_f} \approx \left(\dfrac{2m_A}{m_B}\right) v_{A_i} \approx 0$, since $\left(\dfrac{2m_A}{m_B}\right)$ is very small.

i.e. When a light object A collides with a much more massive object B at rest, the velocity of the light object A is approximately reversed, and the velocity of the massive object B remains approximately at rest. This case will be important when we study the collisions of gas molecules with the walls of a containing vessel in the kinetic theory of gases.

(e) If $m_A \gg m_B$, in addition to the object B is initially at rest (i.e. $v_{B_i} = 0$).

Then, $v_{A_f} \approx v_{A_i}$ and $v_{B_f} \approx 2v_{A_i}$

i.e. The velocity of the massive incident object A is virtually unchanged, but the stationary light object B rebounds with approximately twice the velocity of the incident object A. For example, a cannon ball collides with a stationary golf ball (a small hard ball).

Inelastic collision

Let us consider an isolated system consisting of two objects (non-rotating spheres), A and B, of masses m_A and m_B moving with initial velocities $\vec{v}_{A_i}$ and $\vec{v}_{B_i}$ respectively along the same straight line as shown in figure 12.3 (a). Now, the two objects, A and B, had an inelastic collision and then left (without rotation) with different velocities $\vec{v}_{A_f}$ and $\vec{v}_{B_f}$ respectively, as shown in figure 12.3 (b). Here, the subscripts i and f indicated the velocities referred to before and after the collision. Since the motion is one-dimensional, we can drop vector signs (the overhead arrows) and use only components along the axis, indicating the direction with an algebraic sign.

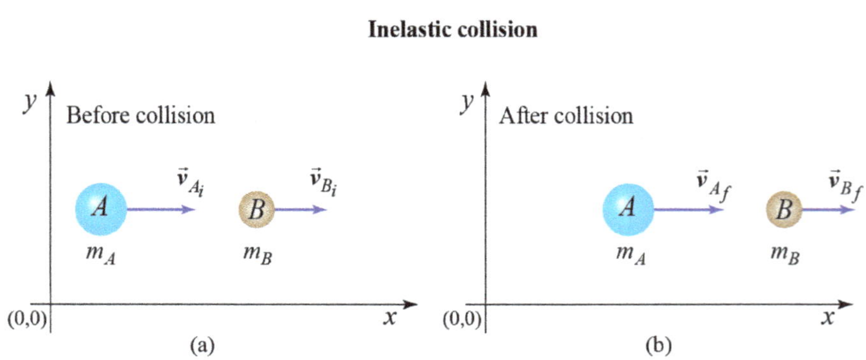

Figure 12.3: Schematic representation of an inelastic collision of two objects of masses m_A and m_B in one dimension (a) before the collision (b) after the collision.

During the collision, momentum is conserved. Therefore, using the law of conservation of linear momentum along the +ve x-axis, we have the following.

$$p_i = p_f$$

i.e. $$m_A v_{A_i} + m_B v_{B_i} = m_A v_{A_f} + m_B v_{B_f} \tag{12.31}$$

Since the collision is inelastic, some of the kinetic energy (KE) of the system is lost during the collision. Therefore, the kinetic energy is not conserved. If KE_i is the initial kinetic energy before the collision and KE_f is the final kinetic energy after the collision, the change in kinetic energy ΔKE can be written as

$$\Delta KE = KE_f - KE_i$$

$$\Delta KE = \left\{ \frac{1}{2} m_A \left(v_{A_f} \right)^2 + \frac{1}{2} m_B \left(v_{B_f} \right)^2 \right\} - \left\{ \frac{1}{2} m_A \left(v_{A_i} \right)^2 + \frac{1}{2} m_B \left(v_{B_i} \right)^2 \right\} \tag{12.32}$$

If we know the values of m_A, , the initial velocities $\vec{v}_{A_i}$, $\vec{v}_{B_i}$, and one of the final velocities, we can find the other final velocity. Hence, we can determine the loss of kinetic energy during the collision.

Perfectly inelastic collision

Let us consider an isolated system consisting of two objects (non-rotating spheres), *A* and *B*, of masses m_A and m_B moving with initial velocities $\vec{v}_{A_i}$ and $\vec{v}_{B_i}$ respectively along the same straight line as shown in figure 12.4 (a). Now, the two objects, *A* and *B*, collide, stick together (perfectly inelastic collision), and then move (without rotation) with some common velocity $\vec{v}_f$ after the collision, as shown in figure 12.4 (b). Here, the subscripts *i* and *f* indicate the velocities referred to before and after the collision. Since the motion is one-dimensional, we can drop vector signs (the overhead arrows) and use only components along the axis, indicating the direction with an algebraic sign.

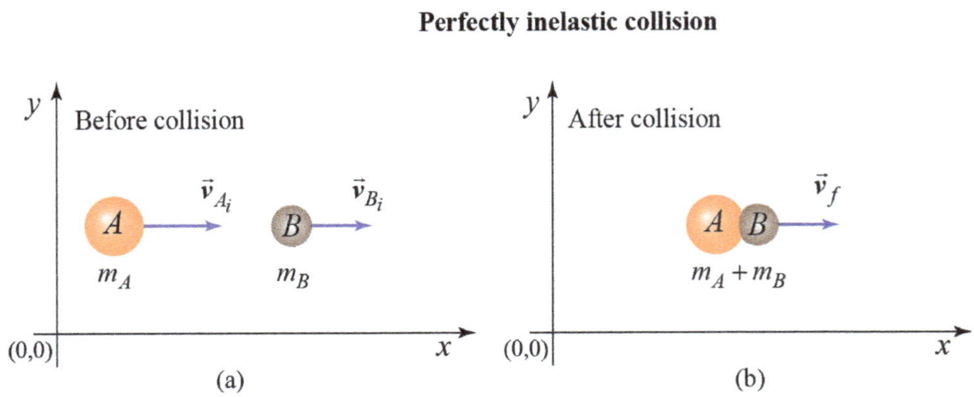

Figure 12.4: Schematic representation of a perfectly inelastic collision of two objects of masses m_A and m_B in one dimension (a) before the collision (b) after the collision, the objects move together

During the collision, momentum is conserved. Therefore, using the law of conservation of linear momentum along the +ve x-axis, we have the following.

$$p_i = p_f$$

i.e. $$m_A v_{A_i} + m_B v_{B_i} = (m_A + m_B) v_f \quad (12.33)$$

Therefore, $$v_f = \frac{m_A v_{A_i} + m_B v_{B_i}}{m_A + m_B} \quad (12.34)$$

If we know the values of m_A, m_B and the initial velocities $\vec{v}_{A_i}, \vec{v}$ we can find the other final velocity of the combined object. Hence, we can determine the loss of kinetic energy during the collision.

Since the collision is inelastic, some of the kinetic energy (KE) of the system is lost during the collision. Therefore, the kinetic energy is not conserved. If KE_i is the total initial kinetic energy before the collision and KE_f is the total final kinetic energy after the collision, the change in kinetic energy ΔKE can be written as

$$= \Delta KE = KE_f - KE_i$$

i.e. $$\Delta KE = \left\{\frac{1}{2}(m_A + m_B)(v_f)^2\right\} - \left\{\frac{1}{2} m_A (v_{A_i})^2 + \frac{1}{2} m_B (v_{B_i})^2\right\} \quad (12.35)$$

Special case

Consider a case where the object with mass m_A and initial velocity $\vec{v}_{A_i}$ along the axis collides inelastically with an object with mass m_B that is initially at rest, i.e. , $\vec{v}_{B_i} = 0$ moving together after collision with the velocity $\vec{v}_f$.

From equation 11.34, the velocity of the combined object $\vec{v}_f$ will be

$$v_f = \left(\frac{m_A}{m_A + m_B} \right) v_{A_i} \tag{12.36}$$

The total initial kinetic energy KE_i before the collision is

$$KE_i = \frac{1}{2} m_A \left(v_{A_i} \right)^2$$

and the total final kinetic energy after the collision is

$$KE_f = \frac{1}{2} (m_A + m_B) \left(v_f \right)^2 = \frac{1}{2} (m_A + m_B) \left(\frac{m_A}{m_A + m_B} \right)^2 \left(v_{A_i} \right)^2$$

i.e. $$KE_f = \frac{1}{2} \left(\frac{m_A^2}{m_A + m_B} \right) \left(v_{A_i} \right)^2$$

If we calculate the ratio of final to initial total kinetic energy, we have

$$\frac{KE_f}{KE_i} = \frac{m_A}{m_A + m_B} \tag{12.37}$$

In the above equation $m_A + m_B > m_A$, hence $KE_f < KE_i$.

If $m_A = m_B$, then $\frac{KE_f}{KE_i} = \frac{1}{2}$. This means that the final total kinetic energy is one-half the initial total kinetic energy.

Activity 12.4

On a frictionless horizontal floor, a ball *P* with a mass of $150\ \text{g}$ is moving at a velocity of $6.00\ \text{m s}^{-1}$ in the positive *x* direction. It elastically collides with another ball *Q* on the same horizontal floor, which has a mass of $300\ \text{g}$. Calculate the velocities of the two balls after the collision.

Activity 12.5

On a frictionless floor, a toy car with a mass of 150 g moves to the right with a velocity of 0.80 m s^{-1}. A small child drops a 35 g piece of clay vertically onto the car. Determine the final velocity of the toy car.

Activity 12.6

Collisions involving atomic and subatomic particles are often regarded as perfectly elastic. In one such collision, a proton travelling to the right at 300 km s^{-1} collides head-on elastically with a stationary alpha particle. The alpha particle is a helium nucleus, having a mass of $6.65 \times 10^{-27} \text{ kg}$.

(a) Calculate the magnitude and direction of the velocity of each particle after the collision.

(b) How much kinetic energy does the proton lose during the collision?

(c) Explain how the collision can be considered elastic even though the proton loses kinetic energy.

12.7 The Ballistic Pendulum

The ballistic pendulum was used to measure the speeds of bullets before electronic timing devices were developed. A large wooden block of mass M is suspended vertically by two cords, as shown in figure 12.5. A bullet of mass m, travelling with a horizontal speed v_i, strikes the pendulum and is embedded in it. If the time of collision is very small, we can assume the cords to remain vertical during the collision. Thus no external horizontal force acts on the system during the collision. Therefore, the system is isolated, and its total linear momentum is conserved.

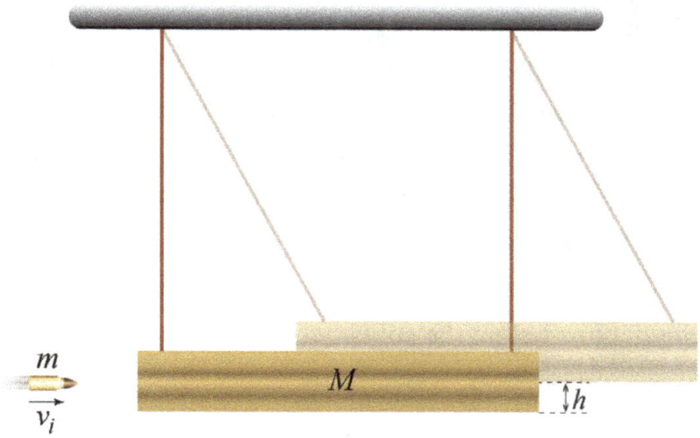

Figure 12.5: Ballistic pendulum: A large wooden block of mass M is suspended vertically by two cords. A bullet of mass m, travelling with a horizontal speed v_i, strikes the pendulum

Let v_f be the speed of the system (wooden block + bullet) after the collision.

The initial momentum of the bullet is mv_i and the momentum of the system just after the collision is $(M+m)v_f$.

Total linear momentum is conserved. Thus,

$$mv_i = (M+m)v_f \tag{12.38}$$

After the collision, the block and the bullet swing up to a height h. The mechanical energy of the system is conserved. Let's take the block's initial level as our reference level of zero gravitational potential energy

Therefore, $$0+\frac{1}{2}(M+m)v_f^2 = 0+(M+m)gh$$

i.e. $$\frac{1}{2}(M+m)v_f^2 = (M+m)gh \tag{12.39}$$

Solving the two equations 12.38 and 12.39 for v_i, we have

$$v_i = \left[\frac{M+m}{m}\right]\sqrt{2gh} \tag{12.40}$$

Thus, if we measure M, m and h, the initial velocity of the bullet can be calculated.

Activity 12.7

The ballistic pendulum, shown in figure 12.5, consists of a large wooden block of mass 5.0 kg hanging from two long cords. A bullet of mass 8.0 g is fired horizontally into the block. The bullet strikes the pendulum and is embedded in it. The time of collision is very small. Subsequently, the block and bullet together swing upward, their centre of mass rising a vertical distance of 6.0 cm before the pendulum comes momentarily to rest at the end of its arc. Determine the speed of the bullet just before the collision.

Questions

NOTE: Although the solutions are given on the last page, it is in your own interest to attempt the questions on your own as an indication of how far you have understood the material of the section.

1. Consider two objects with different masses that have the same kinetic energy. Which one will have the largest magnitude of momentum?

2. In your own words, rephrase each of Newton's three laws of motion as a statement in terms of momentum.

3. As the cricket ball moves with velocity, it has both kinetic energy and momentum.

   (i) Does the cricket ball possess a force that it can apply to any object it comes into contact with?

   (ii) Can the cricket ball transfer more kinetic energy to the bat and the batter than it initially possesses?

   (iii) Can the cricket ball impart more momentum to the bat and batter than it initially carries? Explain each of your answers.

4. State the principle of conservation of momentum and state the condition under which it is valid.

5. State two quantities that are conserved in an elastic collision.

6. A 200 g rubber ball is released from rest at a height of 1.50 m and rebounds from the floor to a height of 0.85 m. Determine the impulse imparted to the ball by the floor.

7. During a cricket match, the wicketkeeper catches a 150 g cricket ball travelling horizontally at a speed of 35.0 m s^{-1}.

   (i) Calculate the impulse exerted on the wicketkeeper by the ball.

   (ii) If the ball's interaction with the wicketkeeper's glove brings it to a complete stop in just 20.0 ms, what is the average force imparted by the ball to the wicketkeeper?

8. A gas molecule undergoes an elastic collision with the walls of a gas cylinder. Before the collision, the molecule moves at a velocity of 350 m s^{-1} perpendicular to the wall.

   (i) Determine the magnitude and direction of the molecule's velocity after the collision.

   (ii) Calculate the change in momentum of the molecule during the collision if it has a mass of $8.0 \times 10^{-26} \text{ kg}$.

   (iii) Explain how the gas molecules exert a force on the container wall using Newton's laws of motion.

9. As shown in the figure, Nathan and Mary were seated on wheeled computer chairs facing each other, initially at rest on a horizontal frictionless floor. Nathan has a larger mass compared to Mary. If they engaged in pushing against each other, answer the following.

   (i) As Nathan and Mary push against each other, will one of them experience a greater magnitude of pushing force, or will they both experience the same magnitude of pushing force? Explain.

(ii) After they finish pushing each other, will one of them experience a greater magnitude of momentum, or will they both experience the same magnitude of momentum? Explain.

(iii) Once they had finished pushing each other, who would have travelled further from their initial position?

10. During a soccer game, a player kicks a $400\ \text{g}$ soccer ball that is initially moving to the left at a speed of $25\ \text{m s}^{-1}$. As shown in the figure, after the kick, the soccer ball moves at 45° upward and to the right with a speed of $35\ \text{m s}^{-1}$. Assume the collision time between the player's foot and the ball is $0.01\ \text{s}$.

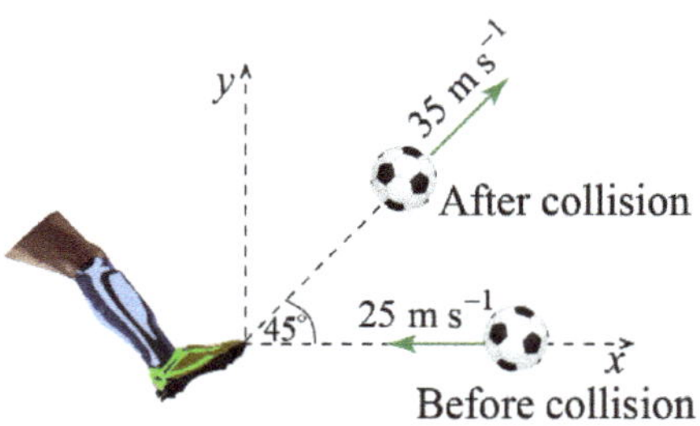

(i) Calculate the impulse of the net force imparted onto the ball.

(ii) Calculate the average net force imparted to the ball and its direction.

(iii) Does the force of gravity of the ball influence the motion of the ball after the kick?

11. In an indoor playground, a child artist with a mass of $40.0\ \text{kg}$ is practising on a trampoline. At a particular moment, her feet reach is a maximum height of $2.5\ \text{m}$ above the trampoline, as shown in the figure. Neglect the air resistance and take $g = 10\ \text{m s}^{-2}$.

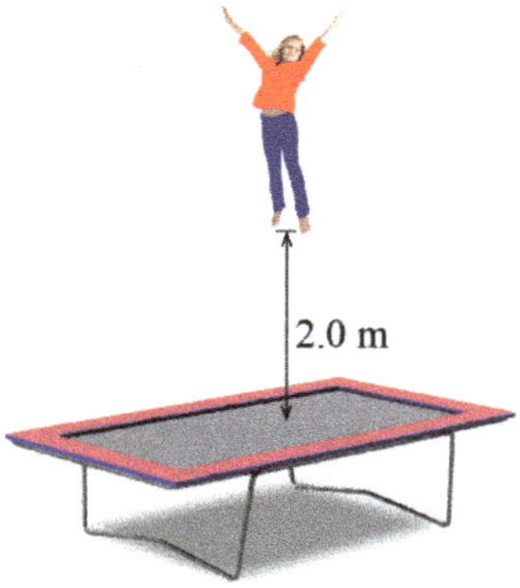

(i) Determine the speed of the child artist as she lands on the trampoline.

(ii) Suppose the child artist rebounds with a speed of $6.0\ \mathrm{m\ s^{-1}}$. Calculate the change in momentum of the child artist.

(iii) Suppose the child artist was in contact with the trampoline for $0.50\ \mathrm{s}$. Calculate the average force exerted by the trampoline on the girl.

(iv) Considering the trampoline behaves as a spring with a spring constant of $6.2 \times 10^4\ \mathrm{N\ m^{-1}}$, determine the amount of compression it will experience.

12. As shown in the figure, a car crash test is conducted to investigate the impact force exerted by a seat belt on a crash test dummy. In this specific crash test, the combined mass of the car and the dummy is $1\ 200\ \mathrm{kg}$, and they collide with a wall.

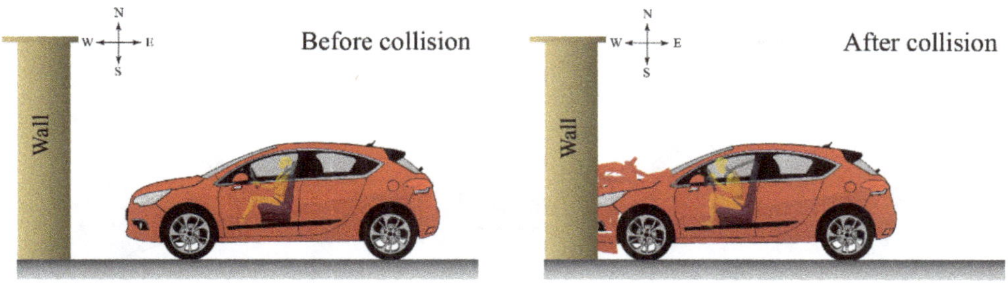

(i) If the car's initial velocity is $17.0\ \mathrm{m\ s^{-1}}$ toward the east, and its final velocity after the collision is $3.00\ \mathrm{m\ s^{-1}}$ toward the west, and the collision lasts for a duration of $0.150\ \mathrm{s}$,

(a) calculate the impulse caused by the collision and the dummy, and

(b) calculate the average net force exerted on the car and the dummy.

(ii) Now, consider a situation in which the car collides with a wall with an initial velocity is $17.0\ \mathrm{m\ s^{-1}}$and comes to a complete stop without rebounding from the wall while maintaining a collision duration of $0.150\ \mathrm{s}$.

(a) Based on the previous results, predict whether the net force on the car in this scenario will be greater or smaller than that of the part (i) (b). Justify your answer.

(b) Calculate the change in momentum of the car and the dummy in this collision.

(c) Calculate the average net force exerted on the car and the dummy in this collision.

(d) The crash test dummy has a mass of $75\ \text{kg}$ and is wearing a seat belt. If the dummy comes to a stop while travelling a distance of $0.5\ \text{m}$ during the collision, determine the average force exerted on the dummy by the seat belt.

13. Falcons often grab their prey from an aerial position, while both the falcon and the prey are in mid-air. In a particular instance, a $1.00\ \text{kg}$ falcon flying at $20.0\ \text{m s}^{-1}$ descends at a 45° angle from behind a $0.350\ \text{kg}$ pigeon flying horizontally at $10.0\ \text{m s}^{-1}$. The falcon successfully captures the pigeon, as shown in the figure.

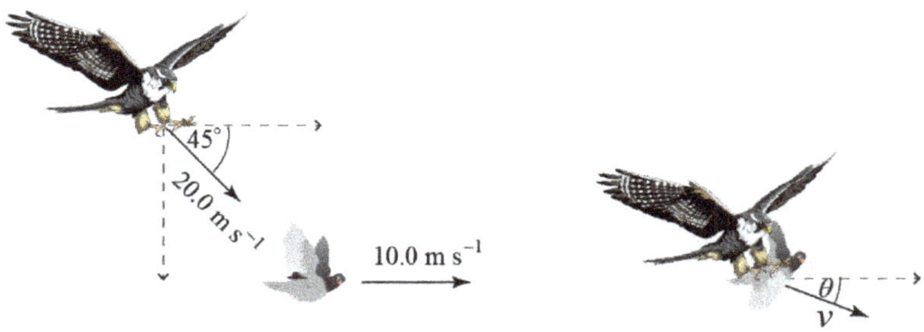

Calculate the magnitude of the velocity v and direction (the angle θ) of the combined system comprising the falcon and pigeon immediately after impact. During the brief collision, neglect the external impulse delivered by the air resistance to the birds.

14. A projectile weighing $10.0\ \text{g}$ is moving towards the east at a velocity of $2.0\ \text{m s}^{-1}$. Suddenly, it explodes into three fragments, as shown in the figure. One of the fragments, weighing $3.0\ \text{g}$, is propelled towards the west with a speed of $10.0\ \text{m s}^{-1}$. Another $3.0\ \text{g}$ fragment moves at an angle of 40° north of east with a velocity of $12.0\ \text{m s}^{-1}$.

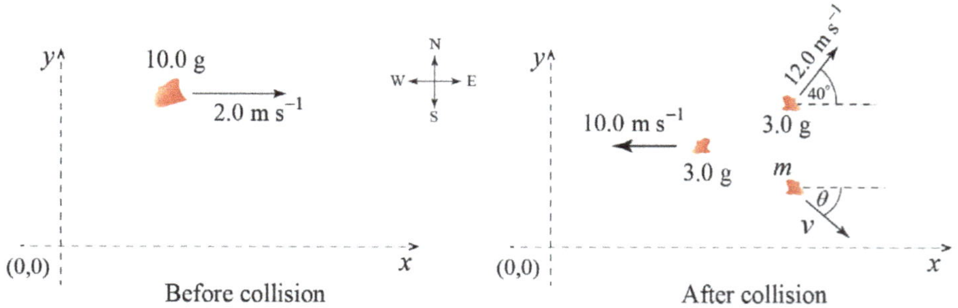

(i) Determine the mass of the third fragment m.

(ii) Determine the magnitude and direction of the velocity v of the third fragment.

15. The figure shows that a 3.00 kg wooden block rests on a table over a large hole. A 55.0 g arrow moving vertically upward at a speed of 40.0 m s^{-1} hits the bottom of the wooden block and remains in the block after the collision. The block and the embedded arrow move vertically. Neglect the air resistance and take $g = 10$ m s^{-2}.

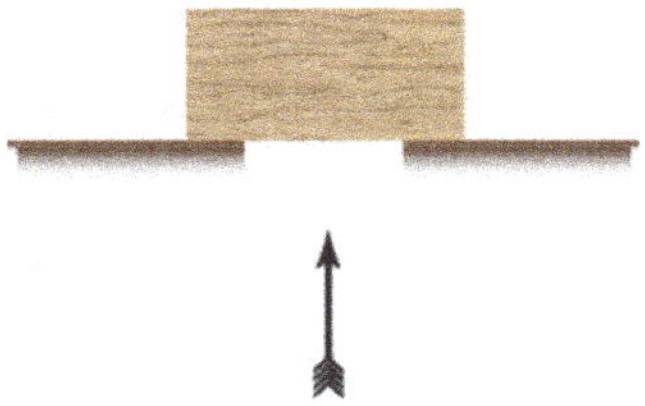

(i) Calculate the kinetic energy possessed by the arrow just before the collision.

(ii) Calculate the momentum of the arrow just before the collision.

(iii) Calculate the momentum of the combined system consisting of the block and the embedded arrow immediately following the collision.

(iv) Calculate the velocity of the combined system consisting of the block and the embedded arrow immediately following the collision.

(v) Calculate the amount of energy lost during the collision.

(vi) How high does the block rise above its initial position?

16. Consider two spherical solid marbles *A* and *B*, with equal masses. These marbles are placed on a horizontal frictionless table, as shown in figure (a).

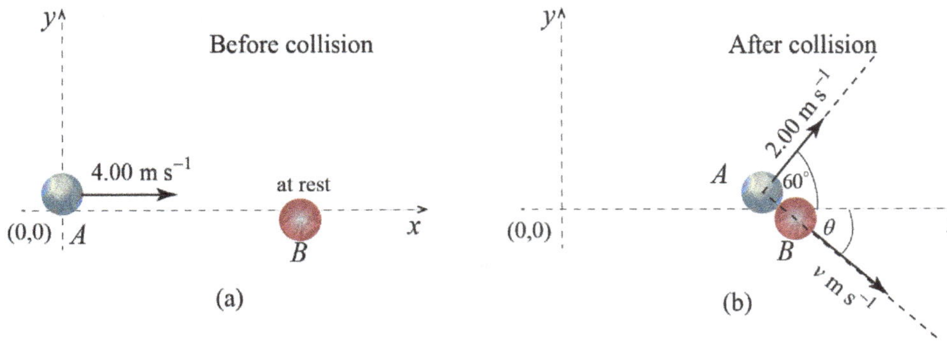

Marble *A* is initially in motion with a velocity of 4.00 m s^{-1} in the positive x direction. It undergoes an elastic collision with marble *B*, which is initially at rest. After the collision, marble *A* moves with a velocity of 2.00 m s^{-1} in a direction 60° to the $+x$ axis and marble *B* moves with a velocity of $v \text{ m s}^{-1}$ in an unknown direction θ, as shown in figure (b). Calculate the magnitude of the final velocity v of marble *B* and the angle θ.

17. A bowling ball is a heavy spherical object used in the sport of bowling. Consider two such bowling balls, *P* and *Q*, with equal masses. These balls are placed on a frictionless horizontal floor. The ball *P* is initially in motion towards the east with a velocity 7.5 m s^{-1}. It collides with ball *Q*, which is initially at rest. After the collision, the ball *P* deflected 30° from its original direction, and the ball *Q* acquired a velocity and deflected through an angle of 45° to the original direction of *P*, as shown in the figure.

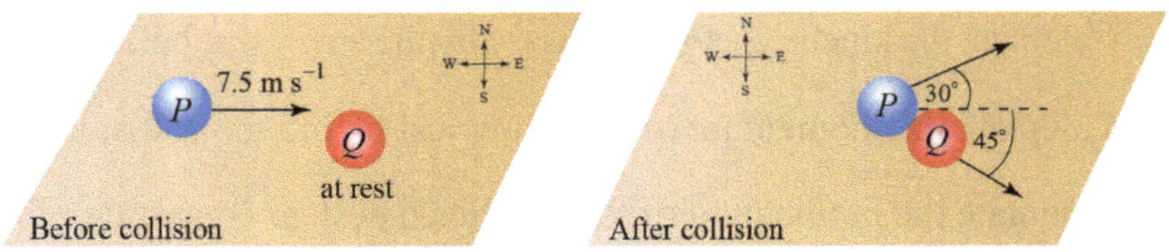

(i) Calculate the speed of each ball after the collision.

(ii) What fraction of the original kinetic energy of ball *P* dissipates during the collision?

18. Consider a situation as shown in the figure, where a car weighing 1200 kg travelling east with a speed of 20 m s^{-1} collides at a T-intersection with a pickup vehicle weighing 2000 kg travelling north at a speed of 18 m s^{-1}. Assume that the vehicles fuse together after the collision and calculate the resultant direction (angle θ) and magnitude of the velocity v of the combined wreckage.

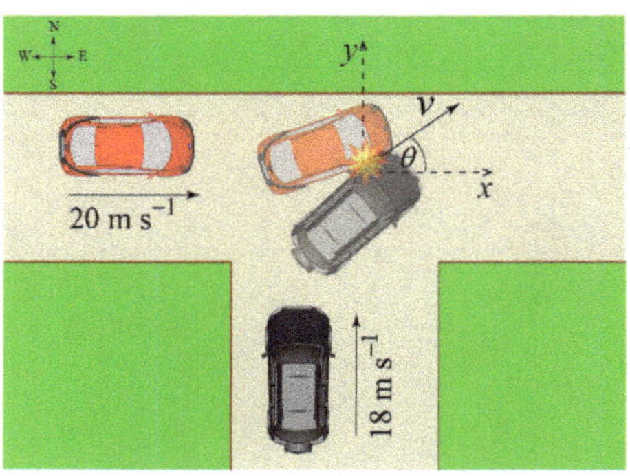

19. A fractured pipe releases a flow of water that lands on a horizontal roof surface. Water moves at 5 m s^{-1} when it hits the roof. The water hits the roof with a mass flow rate of 10 kg s^{-1} without rebounding. Calculate the force of water hitting the roof in newtons. Would your calculated force be greater or smaller if the water rebounded after hitting the roof? Explain.

20. Imagine a situation where a tennis ball of mass m is held just above a basketball of mass m, and its centres are vertically aligned. Both balls are released from rest simultaneously to fall through a distance of h, as shown in the figure. Answer the following questions, neglecting the air resistance and treating the balls as point objects. Take the acceleration due to gravity as g.

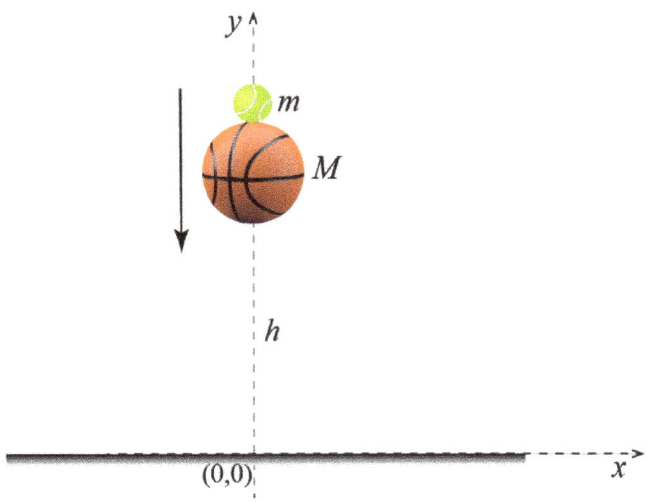

(i) Determine the magnitude of the downward velocity of the basketball upon reaching the ground in terms of g and h.

(ii) What would be the magnitude of the velocity and direction of the tennis ball at the precise moment when the basketball makes contact with the ground? Justify your answer.

(iii) What would be the magnitude of the velocity and direction of the basketball momentarily after it rebounds elastically from the floor? Justify your answer.

(iv) Soon after the basketball rebounds elastically from the floor, the tennis ball collides with the basketball and subsequently rebounds elastically from it. Find an expression for the velocity of the basketball and the tennis ball after the collision in terms of M, m, h and g.

(v) Using the expression you obtained in part (iv), deduce the mass of the tennis ball that can halt the basketball's motion after the collision.

(vi) Using the mass of the tennis ball obtained in part (v), and assuming that the basketball stops its motion shortly after the collision, determine the height that the tennis ball then reaches in terms of h.

Chapter 13

Circular Motion

13.1 Introduction

In Chapter 7, we studied motion along a straight line and introduced concepts such as displacement, velocity, and acceleration. After that, in Chapter 8, we learnt the motion in a plane that involves the movement of objects across a two-dimensional or 2D surface. Projectile motion is a special case of two-dimensional kinematics in which the object is projected into the air while being subject to gravitational force.

In this chapter, we will focus on another important fundamental concept in physics that involves the motion of an object along a circular path with a constant distance from a central point, commonly referred to as circular motion. In a uniform circular motion, the object experiences a continuous change in direction while maintaining a constant speed. This type of motion is governed by principles such as centripetal force, which keeps the object on its circular path, and angular velocity, which measures the rate of rotation around the centre. Circular motion finds diverse applications in fields like mechanics, astronomy, and engineering, including the design of vehicles, machinery, and rotating systems.

13.2 Angular position, displacement, velocity and acceleration

Consider a situation where an object rotates about a point while travelling along a circular path.

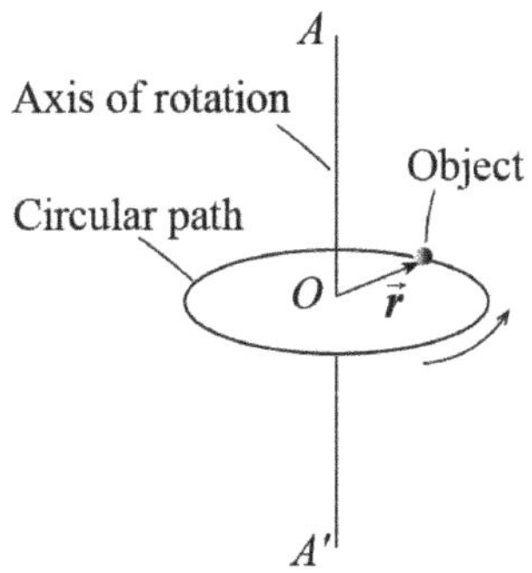

Figure 13.1: An object rotates about the point O while travelling along a circular path. AOA' is called the axis of rotation.

The line that passes through the centre O of the circular path and perpendicular to the plane of rotation AOA' is called the axis of rotation, as shown in figure 13.1. The vector drawn from the centre of the circular path O to the instantaneous location of the rotating object is called the radial vector ($\vec{r}$). Its magnitude is equal to the radius of the circular path, i.e. $|\vec{r}| = r$. The radial vector is constant in magnitude but varies in direction with time.

Angular position

The angular position is the rotational analogue of the linear position. In linear motion, the position is with respect to a fixed point and the displacement is given with respect to that position. In rotational motion, the angular position is the angle between a reference line and the position vector of the rotating object, as shown in figure 13.2. The angular position is denoted by the Greek letter theta (θ).

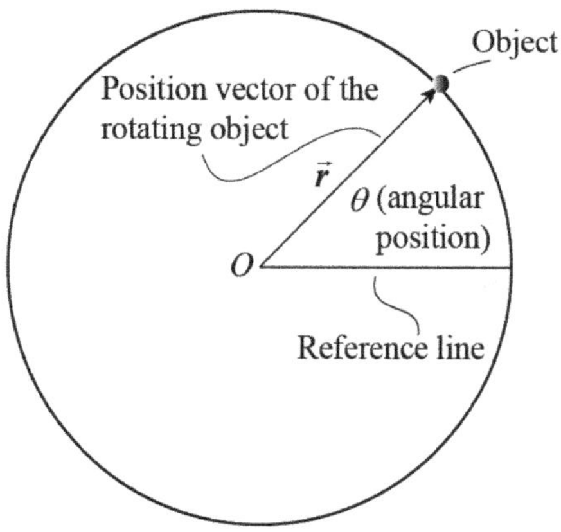

Figure 13.2: The angular position – the angle between a reference line and the position vector of the rotating object

The angular position can be measured in degrees (°), radians (rad), or revolutions. A radian is the angle subtended by an arc length equal to the radius of the circle. A circle has 2π radians, which is equal to 360°. A complete revolution is equivalent to a full rotation of 360° or 2π radians. The angular position can be positive or negative, depending on the direction of rotation. A positive angular position means that the body is rotating in the counterclockwise direction, while a negative angular position means that the body is rotating in the clockwise direction. The angular position is a scalar quantity, meaning that it has magnitude but no direction.

Angular displacement

Angular displacement is the change in the angular position between an object's initial and final positions rotating about a fixed axis. It is measured with respect to a reference line in the plane of motion of the object and the axis of rotation AOA', as shown in figure 13.3.

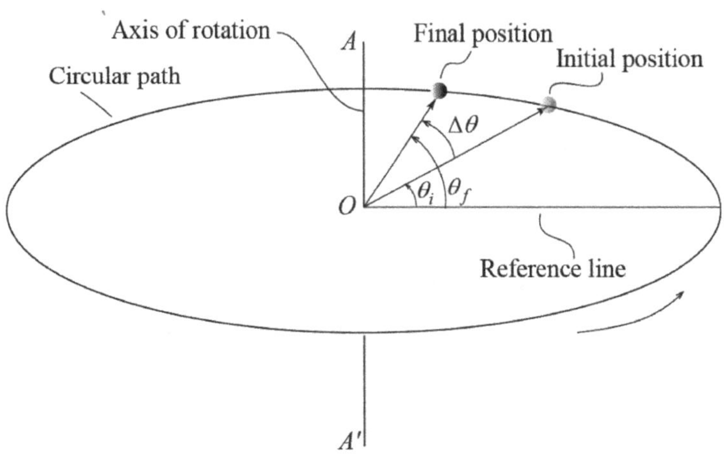

Figure 13.3: Angular displacement - the change in the angular position between the object's initial and final positions

θ_i is the initial angular position,

θ_f is the final angular position, and

$\Delta\theta = \theta_f - \theta_i$ is the magnitude of the change in the angular position between the object's initial and final positions, called the angular displacement. It has both magnitude and direction. The magnitude of the angular displacement is measured in radians, while the direction is indicated by the sense of rotation (clockwise or counterclockwise).

Figure 13.4 shows an angle θ between two radii of a circle, which defines an arc of length s. In other words, θ is the angle subtended by the arc.

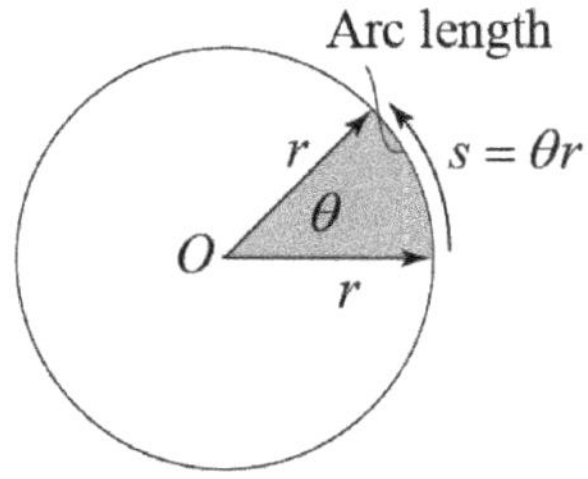

Figure 13.4: Angle θ in radians is the arc length s divided by the radius r.

The length of the arc is proportional to both the radius r of the circle and to the angle subtended (θ). Thus, the angle θ in radians is defined as

$$\theta = \frac{s}{r} \tag{13.1}$$

where θ is the angular displacement in radians,

s is the arc length in metres and

r is the radius of the circular path in metres.

Angular velocity

As shown in figure 13.5, consider an object moving in a circular path. Suppose that its angular position changes from θ_i at the initial time t_i to θ_f at the final time t_f.

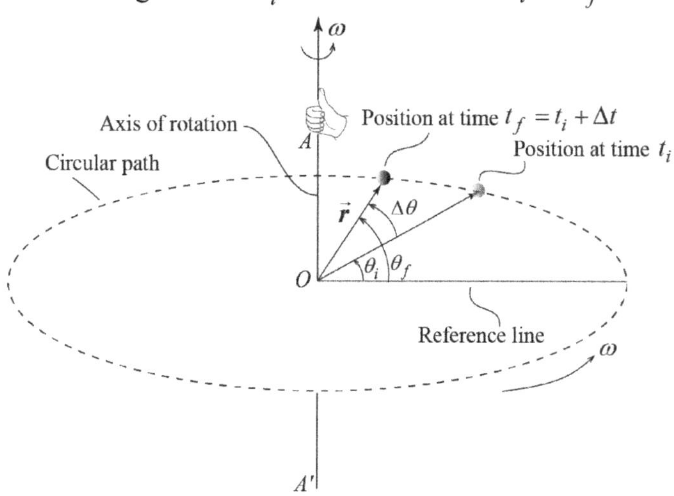

Figure 13.5: An object moving in a circular path - during a very short time Δt, the object's position vector $\vec{r}$ sweeps out the angular displacement $\Delta\theta$

During a very short time Δt, the position vector $\vec{r}$ sweeps out the angular displacement $\Delta\theta$, where $\Delta t = t_f - t_i$ and $\Delta\theta = \theta_f - \theta_i$. Angular velocity is a vector quantity and is defined as the rate of change of the angular displacement with respect to time. It is represented by the Greek letter omega (ω). The direction of angular velocity could be identified by using the right-hand thumb rule, as shown in figure 13.5. Right-hand thumb rule: If the fingers of the right hand are curled in the direction of rotation of the circular path, then the direction of the thumb of the right hand gives the direction of the angular velocity vector.

When an object rotates through an angular displacement $\Delta\theta = \theta_f - \theta_i$ in a time interval $\Delta t = t_f - t_i$, as shown in figure 13.5, the average angular velocity ω_{avg} is defined as the ratio of the angular displacement to the time interval. Thus,

$$\omega_{avg} = \frac{\theta_f - \theta_i}{t_f - t_i} = \frac{\Delta\theta}{\Delta t} \tag{13.2}$$

In the limit $\Delta t \to 0$, ω_{avg} becomes the instantaneous angular velocity ω, and it could be defined as;

$$\omega = \lim_{\Delta t \to 0}\left(\frac{\Delta\theta}{\Delta t}\right) = \frac{d\theta}{dt} \tag{13.3}$$

Since angular displacements can be measured in degrees or radians, angular velocities have units such as degrees/second (deg s^{-1}), radians/second (rad s^{-1}), and degrees/day (deg day^{-1}).

Angular velocity ω is a vector quantity and it can be positive or negative depending on the direction in which the object is rotating.

When an object moves in a circular path with constant speed, we call it a uniform circular motion. Although the speed of the object is constant, the velocity is not since the direction of the velocity is changing. The time for the object to travel completely around the circle is called the period of the motion, denoted by T. Thus, the magnitude of the angular velocity (angular speed) can be written as

$$\omega = \frac{2\pi}{T} \tag{13.4}$$

The frequency f of the motion, which is the number of revolutions per unit time, is defined as

$$f = \frac{1}{T} \tag{13.5}$$

Then, for uniform circular motion

$$\omega = 2\pi f \tag{13.6}$$

Angular acceleration

An object in non-uniform circular motion will have a changing speed and a changing angular velocity. To describe how the angular velocity changes, we define an angular acceleration denoted by α .

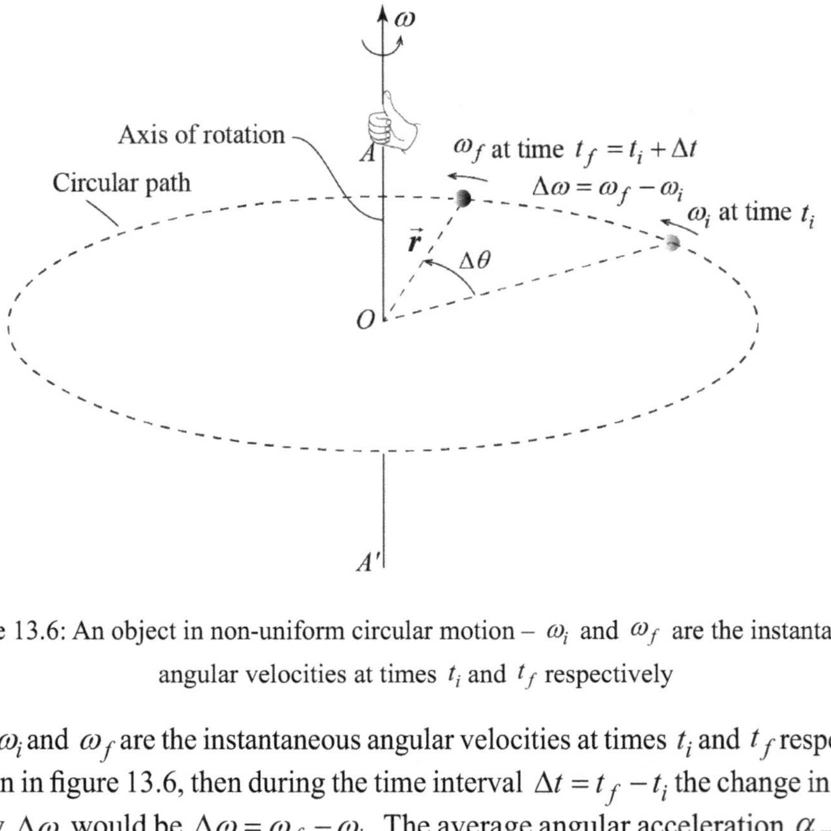

Figure 13.6: An object in non-uniform circular motion – ω_i and ω_f are the instantaneous angular velocities at times t_i and t_f respectively

If ω_i and ω_f are the instantaneous angular velocities at times t_i and t_f respectively, as shown in figure 13.6, then during the time interval $\Delta t = t_f - t_i$ the change in angular velocity $\Delta\omega$ would be $\Delta\omega = \omega_f - \omega_i$. The average angular acceleration α_{avg} of an object moving in a circular path is the change in angular velocity, $\Delta\omega$, divided by the time interval Δt . Thus,

$$\alpha_{avg} = \frac{\omega_f - \omega_i}{t_f - t_i} = \frac{\Delta\omega}{\Delta t} \tag{13.7}$$

In the limit $\Delta t \to 0$, α_{avg} becomes the instantaneous angular acceleration α , and it could be defined as;

$$\alpha = \lim_{\Delta t \to 0} \left(\frac{\Delta \omega}{\Delta t} \right) = \frac{d\omega}{dt} \tag{13.8}$$

If ω is in units of rev s^{-1} or rad s^{-1}, α is in units of rev s^{-2} or rad s^{-2} respectively.

The angular acceleration $\vec{\alpha}$ is also a vector quantity whose direction is taken to be parallel or antiparallel to the direction of the angular velocity $\vec{\omega}$, i.e. it is along the axis of rotation determined by the right-hand thumb rule. If the object is speeding up, then angular acceleration $\vec{\alpha}$ is parallel to $\vec{\omega}$, and it would be taken as positive. If the object is slowing down, then angular acceleration $\vec{\alpha}$ is antiparallel to $\vec{\omega}$, and it would be taken as negative.

Circular motion with constant angular acceleration

Like the three kinematic equations for constant acceleration in linear motion that we studied in Chapter 7, which describes the relationship between displacement, velocity, acceleration, and time, we have three similar equations for constant angular acceleration in circular motion. These three kinematic expressions for the object under constant angular acceleration can be generated from the equations for straight-line by making the following substitutions.

linear displacement s $\rightarrow$ angular displacement $\Delta\theta$

initial velocity u $\rightarrow$ initial angular velocity ω_0

final velocity v $\rightarrow$ final angular velocity ω

constant linear acceleration a $\rightarrow$ constant angular acceleration α

Table 13.1 shows the comparison of linear and circular motions with constant acceleration.

Table 13.1: The comparison of linear and circular motions with constant acceleration.

Straight-line motion with constant linear acceleration	**Circular motion with constant angular acceleration**
$v = u + at$	$\omega = \omega_0 + \alpha t$
$s = ut + \frac{1}{2}at^2$	$\Delta\theta = \omega_0 t + \frac{1}{2}\alpha t^2$
$v^2 = u^2 + 2as$	$\omega^2 = \omega_0^2 + 2\alpha\Delta\theta$

13.3 Relation between Linear and Angular velocity

Consider an object moving in a circular path, as shown in figure 13.7. During a very short time interval Δt, the position vector $\vec{r}$ sweeps out the angular displacement $\Delta\theta$ and it traces an arc length Δs.

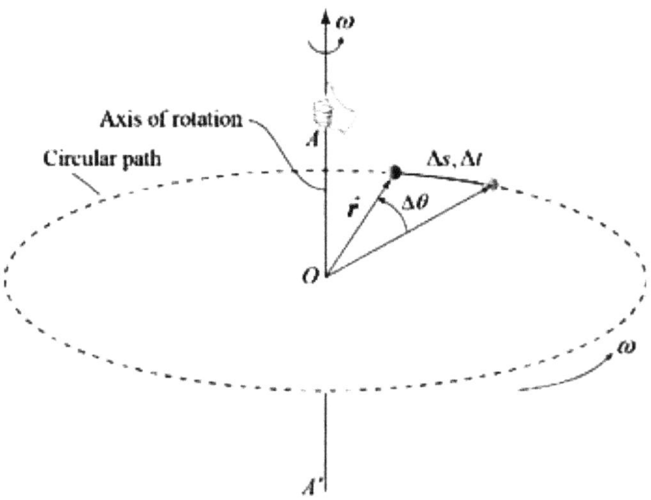

Figure 13.7: An object moving in a circular path - during a very short time Δt, the object's position vector $\vec{r}$ sweeps out the angular displacement $\Delta\theta$ and traces an arc length Δs

From the equation 12.1, we have

$$\Delta\theta = \frac{\Delta s}{r} \tag{13.9}$$

Thus, using the equations 12.2 and 12.7, the average speed of the object is given by

$$v_{avg} = \frac{\Delta s}{\Delta t} = r\frac{\Delta}{\Delta t} = r \quad {}_{avg} \tag{13.10}$$

In the limit $\Delta t \to 0$, the equation 12.8 becomes

$$v = r \lim_{\Delta t \to 0}\left(\frac{\Delta \theta}{\Delta t}\right) = r\omega \tag{13.11}$$

The direction of the object's linear velocity is tangential to its circular path at each point and is denoted as v_t.

13.4 Tangential and Radial acceleration

Consider an object moving in a circular path, and its velocity changes both in direction and in magnitude, as shown in figure 13.8. In this situation, the velocity vector is always tangential to the path, and its acceleration at any instant can be resolved into two components based on an origin O at the centre of the circle as a radial component a_r along the radius of the circle and a tangential component a_t perpendicular to this radius. The tangential acceleration component is parallel to the instantaneous velocity v of the object.

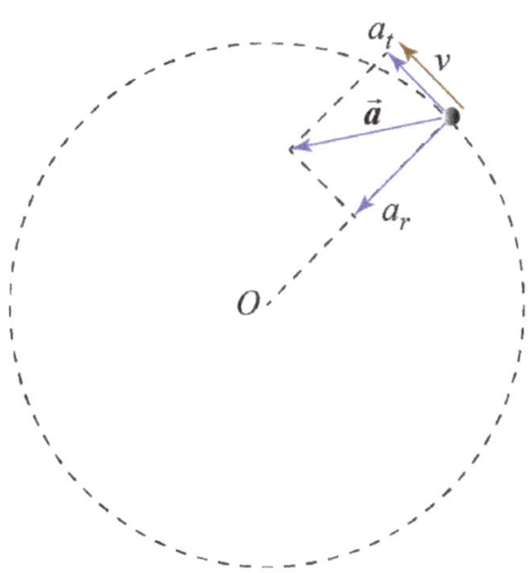

$$\vec{a} = \vec{a}_r + \vec{a}_t$$

Figure 13.8: An object moving in a circular path – the components of the acceleration $\vec{a}$ are a tangential component a_t and a radial component a_r

The total acceleration vector $\vec{a}$ can be written as the vector sum of the component vectors:

$$\vec{a} = \vec{a}_r + \vec{a}_t \tag{13.12}$$

Tangential acceleration

The equation 13.11, i.e. $v = r\omega$, shows that a change in angular velocity ω will produce a change in linear velocity v. These changes in both velocities will result in angular and linear accelerations. Therefore, the relation between two accelerations is derived as follows. From equation 13.11, we have

$$v = r\omega$$

Therefore, we can write the change in velocity Δv as

$$\Delta v = r\Delta\omega, \text{ here } r \text{ is constant}$$

Dividing both sides of the above equation by Δt, and taking the limit both sides as $\Delta t \to 0$, we get

$$\lim_{\Delta t \to 0}\left(\frac{\Delta v}{\Delta t}\right) = r \lim_{\Delta t \to 0}\left(\frac{\Delta\omega}{\Delta t}\right)$$

Using the definitions of instantaneous angular and linear acceleration, we can write the above equation as

$$a = r\alpha$$

This equation gives the relationship between the magnitudes of linear and angular accelerations. The linear acceleration a is also called tangential acceleration a_t. Thus,

$$a_t = r\alpha \tag{13.13}$$

The tangential acceleration component is parallel to the instantaneous velocity v of the object.

Radial acceleration

The radial acceleration component arises from a change in the direction of the velocity vector. As shown in figure 13.9 (a), consider an object moving with constant speed in a circular path with radius r and centre at *O*. Although the magnitude of the velocity remains constant (with constant speed), the direction of the velocity continuously changes as the object moves around the circle. Since acceleration is defined as the rate of change of velocity, a change in the direction of velocity constitutes acceleration, just as a change in the magnitude of the velocity does. Thus, an object revolving in a circle continuously accelerates even when the speed remains constant.

As shown in figure 13.9(a), the object moves from A to B through a distance Δs in a time interval Δt. Let $\vec{v}_A$ and $\vec{v}_B$ be the velocities of the object at points A and B, respectively. The vector change in velocity $\Delta\vec{v} = \vec{v}_B - \vec{v}_A$ during the motion of the object from A to B is shown in figure 13 (b). It can be shown that the vector change in velocity $\Delta\vec{v}$ passes through the centre O. Thus, the acceleration due to this change in velocity is directed towards the centre of the circular path.

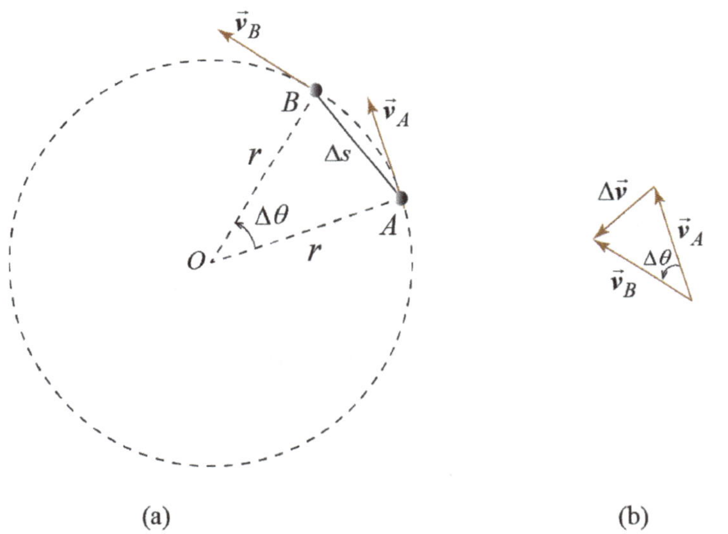

Figure 13.9: (a) An object moves from A to B through a distance Δs in a time interval Δt (b) The vector change in velocity $\Delta\vec{v} = \vec{v}_B - \vec{v}_A$ during the motion of the object from A to B

Since the object is moving with constant speed in a circular path, the magnitudes of the velocities at points A and B are the same. This means that $|\vec{v}_A| = |\vec{v}_B| = v$, where v represents the speed of the object at any point on the circular path during the circular motion.

The two triangles depicted in figure 13.9 (a) and (b) are similar because both are isosceles triangles, and the angles $\Delta\theta$ are the same. Thus, by geometry, the ratios of the corresponding sides of similar triangles are equal. Therefore, we have the following.

$$\frac{|\Delta\vec{v}|}{v_A} = \frac{\Delta s}{r} \quad \Rightarrow \quad \frac{|\Delta\vec{v}|}{v} = \frac{\Delta s}{r} \quad \text{since } |\vec{v}_A| = |\vec{v}_B| = v$$

i.e. $$|\Delta\vec{v}| = \frac{v}{r}\Delta s$$

Therefore, the magnitude of the average acceleration a_{avg} over the time interval t for the object to move from A to B can be written as

$$a_{avg} = \frac{|\Delta\vec{v}|}{\Delta t} = \frac{v}{r}\frac{\Delta s}{\Delta t}$$

The magnitude of the instantaneous acceleration a at point A can be obtained by taking the limit of the above expression as $\Delta t \to 0$. Thus,

$$a = \frac{v}{r}\lim_{\Delta t \to 0}\left(\frac{\Delta s}{\Delta t}\right) = \frac{v^2}{r}$$

Here $\lim_{\Delta t \to 0}\left(\frac{\Delta s}{\Delta t}\right)$ is the speed v at point A. Since this acceleration of the object in uniform circular motion always points toward the centre of the circle, this is referred to as the radial acceleration and is denoted as a_r. Therefore, the magnitude of the radial acceleration a_r of an object in uniform circular motion having a radius r and moving with constant speed v is given by

$$a_r = \frac{v^2}{r} \tag{13.14}$$

Equation 12.14 tells us that the magnitude of the radial acceleration a_r is proportional to the square of the speed, i.e. $a_r \propto v^2$, and inversely proportional to the radius, i.e. $a_r \propto \frac{1}{r}$.

Since this radial acceleration of an object in uniform circular motion is always directed toward the centre of the circle, it is sometimes called centripetal acceleration. The word centripetal means "seeking the centre", which is derived from Latin words.

In a uniform circular motion, the angular speed ω can be written as $\omega = \frac{2\pi}{T}$ (equation 12.4), where T is the period of the motion. Since the speed v of the object in uniform circular motion is equal to the circumference of the circular path $(2\pi r)$ divided by the period T, we have

$$\omega = \frac{2\pi}{T} = 2\pi\left(\frac{v}{2\pi r}\right) = \frac{v}{r} \quad \Rightarrow \quad v = r\omega \tag{13.15}$$

Thus, the centripetal acceleration of an object in uniform circular motion in terms of angular speed ω, and radius r can be written as

$$a_r = \frac{v^2}{r} = \frac{r^2\omega^2}{r} = r\omega^2 \quad \Rightarrow \quad a_r = r\omega^2 \tag{13.16}$$

Non-uniform Circular Motion

In a non-uniform circular motion, we still get the radial component of acceleration, i.e. $a_{rad} = \frac{v^2}{r}$, which is always perpendicular to the instantaneous velocity and directed toward the centre of the circle. Since in non-uniform circular motion, the speed v of the object has different values at different points in the motion, the radial component of acceleration a_{rad} is not constant.

Furthermore, since in non-uniform circular motion, the speed v of the object has different values at different points in the motion, we get another component of acceleration that is parallel to the instantaneous velocity, which is the tangential acceleration component a_t discussed earlier. Therefore, as seen in equation 12.12, the resultant acceleration vector $\vec{a}$ can be written as the vector sum of the component vectors, i.e. $\vec{a} = \vec{a}_r + \vec{a}_t$. Thus, the magnitude of $\boldsymbol{a}$ is

$$a = \sqrt{a_r^2 + a_t^2} \tag{13.17}$$

It should be noted that in a uniform circular motion, where v is constant, i.e. $a_t = 0$ and the acceleration is always completely radial, as discussed earlier.

Activity 13.1

If an object moves in a circular path of radius 150 mm with 300 rev min^{-1}, calculate

(i) its angular velocity in rad s^{-1}.

(ii) the linear speed of a point on the circular path.

(iii) the time for one revolution.

Activity 13.2

Imagine a CD rotating at a rate of 400 revolutions per minute (rpm). What are the angular and linear speeds at a location 4.0 cm away from the CD's centre?

13.5 Angular Momentum of a Particle

As shown in figure 13.10, let us consider a single particle at a position $\vec{r}$ that rotates about an axis passing through the origin O in a $x-y$ plane. If m is the mass of the particle and $\vec{v}$ is the instantaneous velocity of the particle, then the linear momentum of the particle will be $\vec{p} = m\vec{v}$. The magnitude of the angular momentum L of the particle is defined as the product of its linear momentum p and the perpendicular distance d from its linear momentum vector $\vec{p}$ to point O.

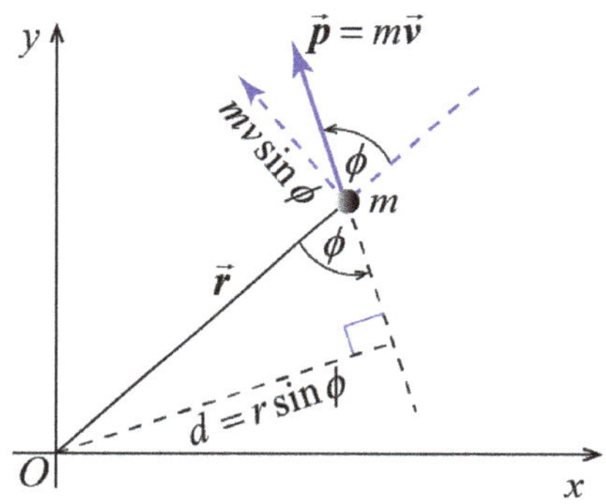

Figure 13.10: A particle with mass m rotating in the $x-y$ plane. The angular momentum of the particle has a magnitude $L = mvd$

Thus, the angular momentum of the particle can be expressed mathematically as

$$L = pr\sin\phi = mv\left(r\sin\phi\right) = mvd \tag{13.18}$$

The angular momentum L is a vector quantity, and the right-hand thumb rule determines its direction. If the vector $\vec{r}$ rotates onto the vector $\vec{p}$, and the fingers of the right hand are aligned with this rotation, then the thumb of the right hand indicates the direction of angular momentum L. According to figure 13.10, the direction of the angular momentum L will be perpendicular to the x-y plane pointing outward.

For the specific case of a particle of mass m executing a circular orbit of radius r, with instantaneous velocity v and instantaneous angular velocity ω, the magnitude of the particle's angular momentum is $L = mvr$.

But we know that $v = r\omega$. Thus,

$$L = mvr = mr^2\omega \quad (13.19)$$

13.6 Centripetal Force

In previous sections, we learned that when an object travels in a circle, even at constant speed, it undergoes acceleration. In this case, the acceleration does not act to increase or decrease the magnitude of the velocity vector but rather to change its direction. Newton's first law of motion says that an object's velocity will stay the same unless there is a force acting on it. Since an object travelling in a circle, even at constant speed, has centripetal acceleration, there must be a force causing this acceleration. This force is called the centripetal force and acts towards the centre of the circle.

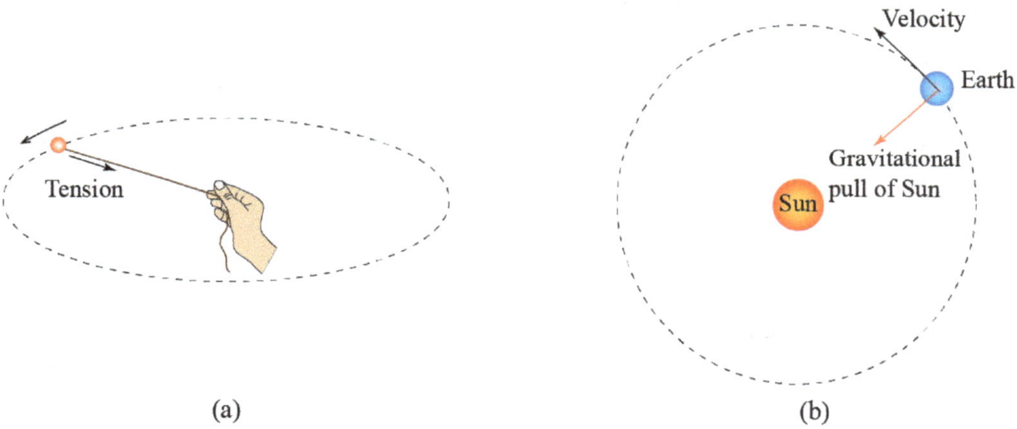

Figure 13.11: (a) Spinning a ball attached to a string in a horizontal circular motion (b) Earth orbiting the Sun

To demonstrate centripetal force, think of the following two different situations.

(i) Imagine a situation where a ball attached to a string is being *spun* in a horizontal circular motion, as shown in figure 13.11 (a). To maintain its circular path, you need to exert a force by pulling the string. This pulling force applied by the string on the ball is an external force that continuously alters the ball's velocity as it traces a horizontal circle. The force on the ball is directed towards the centre of the circle, and we describe the net force acting towards the centre of the circle as a centripetal force. If the string was released suddenly, the tension in the string would disappear, causing the ball to shoot off tangentially from the circle.

(ii) Similarly, as Earth orbits the Sun (figure 13.11 (b)), it constantly changes velocity. According to Newton's first law, there must be an external force acting on it. That force is the Sun's gravitational pull, and it is directed towards the centre of the circular orbit. We refer to this force, responsible for keeping Earth on its circular path, as the centripetal force. If the force resulting from the Sun's gravitational pull disappeared, the Earth would travel off in a straight line.

The word centripetal in this text is used to describe a force that is making something travel along a circular path. It does not tell us what causes this force, which might be gravitational, electrostatic, magnetic, frictional, etc.

When travelling in a car or bus, or riding on a merry-go-round, you often feel pushed toward the outside that goes around a circular path. This is due to the force called centrifugal force. Centrifugal means "fleeing from the centre".

The centrifugal force is an apparent force that acts on objects moving in a circular path. It is directed away from the centre of rotation. The centrifugal force is not a real force, but rather a fictitious force that arises when describing motion in a rotating reference frame. Since the centrifugal force is not a real force, such an outward force cannot be included in the free-body diagram.

Fictitious forces may not be real, but they can have real effects. The concept of centrifugal force can be applied in rotating devices, such as centrifuges (used to separate different components of a mixture based on their density. For example, they are used to separate blood cells from plasma), washing machines (when the drum of the washing machine spins fast, it causes the clothes to rotate in a circular path. This circular motion pushes the clothes outwards towards the walls of the drum, forces the water droplets trapped in clothes out from it through the small holes in the walls of the washing machine), etc.

Activity 13.3

In each of the scenarios below, state what provides the resultant force causing centripetal acceleration.

(i) A car is going round a bend on a flat, rough road.

(ii) The weight on the end of a swinging pendulum.

(iii) The Moon orbiting the Earth.

Activity 13.4

A ball of mass 200 g is attached to the end of a string 1.00 m long. The ball moves in a horizontal circle, as shown in figure 13.10 (a). Assume the string remains horizontal during the motion.

(i) If the string can withstand a maximum tension of 20.0 N, what is the maximum speed at which the ball can move before the string breaks?

(ii) Suppose the ball travels in a circle of a larger radius at the same speed obtained in part (i). Is the string more likely or less likely to break?

Activity 13.5

Determine the period of revolution of a satellite with mass m that orbits the Earth in a circular path of radius $8000\ \text{km}$. Take the gravitational force F_g between the Earth and the satellite as $F_g = \frac{Gmm_E}{r^2}$, where G is the gravitational constant, m_E is the mass of the Earth and r is the distance of the satellite from the centre of the Earth. Assume $m_E = 5.98 \times 10^{24}$ kg and $G = 6.674 \times 10^{-11}$ N m^2 kg^{-2}.

13.7 Motion on an unbanked and banked roadway

An unbanked roadway refers to a flat or level road surface without inclination. When a vehicle travels on a curved unbanked roadway, as shown in figure 13.12(a), it encounters the centripetal force required to keep the vehicle moving in a circle, and it is provided by the friction force acting between the tyres and the road surface. The vehicle's radial acceleration $\vec{a}_r$ is toward the centre of the circular path. This frictional force acts sideways toward the centre of the vehicle's circular path, as shown in figure 13.12(b). Further, the normal $\vec{N}$ and the static friction $\vec{f}_s$ forces represent the total normal and frictional forces acting on all four tyres. The frictional force might also have a tangential component. If the vehicle's speed v is constant, then the component of the frictional force will be negligibly small.

Figure 13.12: (a) A vehicle travelling on a curve at constant speed on an unbanked roadway. The vehicle's radial acceleration is toward the centre of the circular path (b) A head-on view of the same vehicle. The centre of the circular path is to the left, as shown here

As long as the vehicle tyres roll without slipping, there is no relative motion between the bottom of the tyres and the road, so the force of static friction $\vec{f}_s$ acts. If the car starts to slip, then the smaller force of kinetic friction will begin to act, since the lower part of the tyre slides along the pathway.

If the vehicle moves too fast, the frictional force acting sideways toward the centre of its circular path may not be enough to keep it on the road, and the vehicle may slip. This is more likely to happen on wet or icy roads, where the coefficient of friction is lower.

Suppose v_{max} is the maximum that the vehicle can have around the curve at which it is on the verge of skidding outward. Then, at this point, the friction force f_s has its maximum value $f_{s,max} = \mu_s N$, where μ_s is the coefficient of static friction between the rubber of the tyres and the road, N is the magnitude of the normal force and its value $N = mg$, and m is the mass of the vehicle.

For uniform circular motion, we have $f_{s,max} = \mu_s N = m\frac{v_{max}^2}{r}$, where r is the radius of the curve. Thus, the maximum speed that the vehicle can reach and still make the turn successfully would be

$$v_{max} = \sqrt{\frac{\mu_s Nr}{m}} = \sqrt{\frac{\mu_s mgr}{m}} = \sqrt{\mu_s gr}$$

From this we see that the maximum speed does not depend on the mass of the vehicle, which is why curved highways do not need multiple speed limits to cover the various masses of vehicles.

However, banked circular path roads are designed to help vehicles navigate curves safely without slipping. The outer edge of a banked roadway is raised higher than the inner edge, which creates horizontal components N_x and N_y of the normal force $\vec{N}$ acting on the vehicle, as shown in figure 13.13 (b). The normal force $\vec{N}$ shown in figure 13.13 (a) represents the total normal force acting on all four tyres.

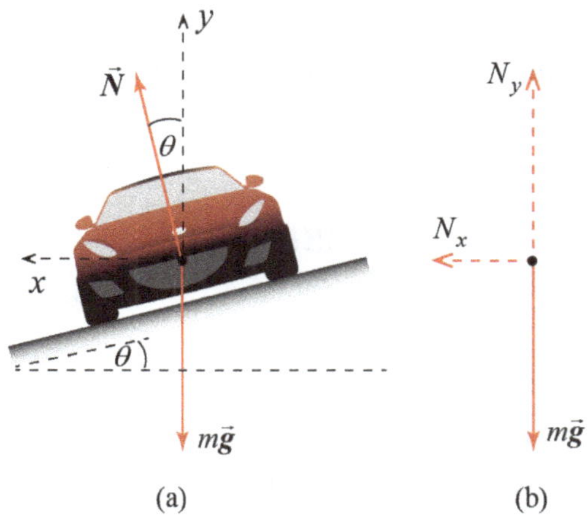

Figure 13.13: (a) Head-on view of a vehicle travelling on a curve at constant speed on a banked roadway. The vehicle's acceleration is toward the centre of the circular path. The vehicle moves at just the right speed so that the frictional force is zero (b) FBD for the vehicle with the normal force represented by its components N_x and N_y

The horizontal component of the normal force N_x, as shown in figure 13.13 (b), helps to provide the centripetal force required to keep the vehicle moving in a circle. The angle θ at which a roadway is banked is designed to be equal to the angle of the centripetal force required for a vehicle to travel safely around the curve at a given speed, as shown in figure 13.13 (a). This means that a vehicle travelling at the design speed on a banked roadway will not need to rely on friction to keep the vehicle moving in a circular path on the road.

Activity 13.6

Consider a car travelling on an unbanked curve at the recommended speed 10 m s^{-1} indicated at the road sign, as shown in figure 13.13.

(i) If the radius of curvature of the path is 25 m, determine the radial acceleration required for the recommended speed of 10 m s^{-1}.

(ii) If the coefficient of static friction between the rubber of the tyres and the road is 0.70, does the car skid as it goes around the curve?

(iii) What happens if the driver ignores the highway speed limit sign and travels at $17\ \text{m s}^{-1}$?

(iv) Determine the safe speed that the car needs to maintain to travel around the curve if the road surface is wet from a recent rainstorm and the coefficient of static friction between the wet road and the rubber of the tyres is $\mu_s = 0.40$.

(v) What banking angle would be required for a car to safely travel along the curved roadway in wet conditions at a speed of $13\ \text{m s}^{-1}$?

Take the acceleration due to gravity g as $10\ \text{m s}^{-2}$.

Questions

NOTE: Although the solutions are given on the last page, it is in your own interest to attempt the questions on your own as an indication of how far you have understood the material of the section.

1. How would you convert an angle from degrees to radians?
2. What is the definition of angular speed?
3. What is meant by the period and frequency of a rotation?
4. In a clock, determine the angular speed of the second hand.
5. What is meant by centripetal acceleration? Write down the equation for centripetal acceleration in terms of angular speed. Define all the symbols you use.
6. An object moves around a circular track of radius 0.3 m at 2 revolutions per second. Calculate the

   (i) angular speed ω.

   (ii) period T.

   (iii) linear speed v of the object.
7. If the Moon completes a full orbit around the Earth in approximately 28 days at an average distance of about 3.84×10^{8} m, determine its respective angular and linear velocities.

8. A fan rotating with $\omega = 100 \text{ rad s}^{-1}$, is switched off. After $2n$ rotation, its angular velocity becomes 50 rad s^{-1}. Calculate the angular velocity of the fan after n rotations.

9. As shown in the figure, a pot maker rotates the potter's wheel from rest to 240 rpm in a time of 0.90 s.

   (i) Determine the angular acceleration of the wheel during this time, assuming constant angular acceleration.

   (ii) During this time interval, how many revolutions does the wheel make?

   (iii) Determine the tangential and radial components of the acceleration of a point 10 cm from the rotation axis when the wheel is spinning at 200 rpm.

10. Consider a car travelling on a flat road and approaching a patch of oil on a bend. Can the car safely and smoothly go around the bend when it hits the oil patch? Explain. What might happen if the driver tries to turn the steering wheel when the car is in the oil patch?

11. Part of the track of a roller coaster ride in which a truck loops a loop is shown in the figure below. When the truck is at the top of the loop, as shown, there is no reaction force between the truck wheels and the track. The diameter of the loop on the track is 10.0 m.

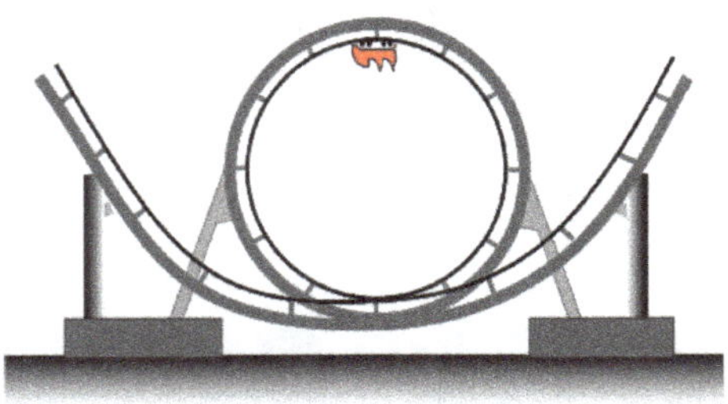

(i) Explain what provides the centripetal force to keep the truck moving in a circle.

(ii) Calculate the speed of the truck. Take the acceleration due to gravity g as $9.8\ \text{m s}^{-2}$.

12. A small ball of mass $m = 50\ \text{g}$ is suspended from a string of length $l = 1.00\ \text{m}$ with negligible mass. The ball revolves with constant speed v in a horizontal circle of radius $r = 0.15\ \text{m}$ as shown in the figure. Since the string sweeps out the surface of a cone, the system is referred to as a conical pendulum. Take the acceleration due to gravity g as $10\ \text{m s}^{-2}$.

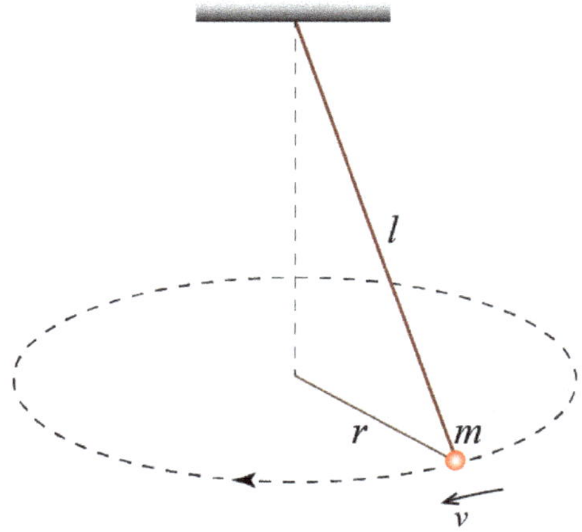

(i) Draw a free-body diagram for the ball.

(ii) Calculate the tension in the string.

(iii) Calculate the speed v of the ball.

(iv) Calculate the period of the motion.

13. Communications satellites are launched in such a way that they remain above the same point on the Earth's surface and move in a circular orbit above the equator.

(i) Determine the time it takes for the communication satellite to complete one orbit around Earth.

(ii) Draw a free-body diagram of the satellite in orbit.

(iii) Apply Newton's second law to the satellite and find its altitude above the Earth's surface.

Take the gravitational force F_g between the Earth and the satellite as $F_g = \frac{Gmm_E}{r^2}$ where G is the gravitational constant, m is the mass of the satellite, m_E is the mass of the Earth and r is the distance of the satellite from the centre of the Earth. Assume $m_E = 5.98\times10^{24}$ kg, $R_E = 6.38\times10^{6}$ m and $G = 6.674\times10^{-11}$ N m^2 kg^{-2}.

14. An aeroplane is flying in a horizontal circle at a speed of 120 m s^{-1}. As shown in the figure below, if its wings are tilted at an angle $\theta = 40°$ to the horizontal, determine the radius of the circle in which the plane is flying. Assume that the required force is provided entirely by an "aerodynamic lift" that is perpendicular to the wing surface. Take the acceleration due to gravity g as 10 m s^{-2}.

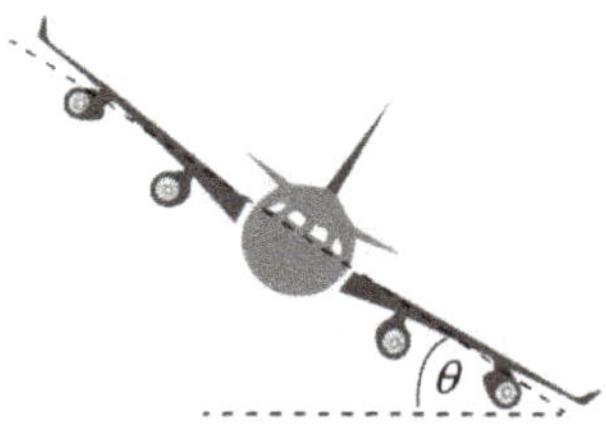

15. In the process of planning a curved roadway, a roadworks engineer has decided to design a banked roadway, as shown in the figure, to enhance safety and prevent skidding when a vehicle drives along the curve. The recommended speed for the road is to be 35 km h^{-1} and the radius of the curve is 40.0 m.

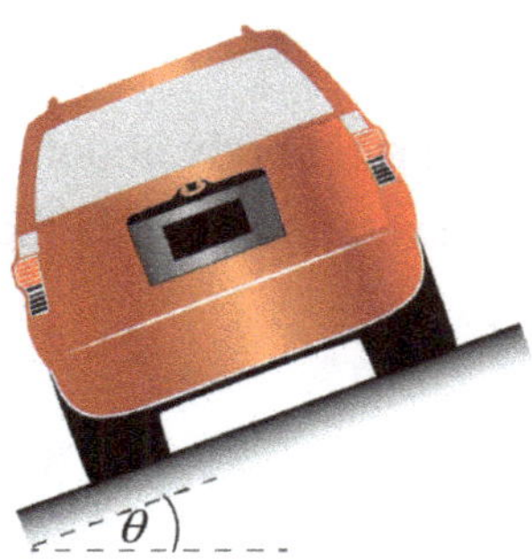

(i) Draw a free-body diagram of the vehicle.

(ii) Determine the angle θ to the horizontal at which the curved road should be banked.

Chapter 14

Rotational Motion

14.1 Introduction

So far, our study of motion has focused almost exclusively on the particle model, in which an object is represented as a mass at a single point in space, and has examined only the motion of translation, in which an object moves along a straight or curved line. In this chapter, we will expand our study of motion to include rotational motion, which is when an object turns about an axis. To do this, we need to consider extended objects whose size and shape matter. Therefore, this chapter will be based on the rigid-body model. A rigid body is an extended object whose size and shape do not change as it moves. A rigid body cannot be stretched, compressed, or deformed. All points on the body have the same angular velocity and angular acceleration. Rotational motion is an extension of circular motion that we have studied in the previous chapter. Therefore, we will begin this chapter with a brief summary of Chapter 13.

14.2 Brief Review of Rotational Kinematics

We now examine the rotation of a rigid body about a fixed axis. Figure 14.1 shows a rigid body of arbitrary shape in rotation about a fixed axis (the z axis), called the axis of rotation or the rotation axis.

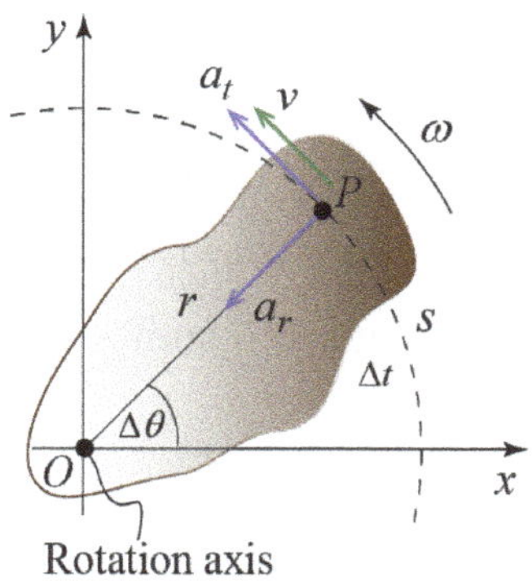

Figure 14.1: A rigid body rotates about a fixed axis (the *z* axis) through *O*, the point *P* on the rigid body experiences a tangential component of translational acceleration a_t and a radial component of translational acceleration a_r.

In pure rotational motion, every point of the rigid body moves in a circle whose centre lies on the axis of rotation, and every point on the rigid body moves through the same angle during a particular time interval. As shown in figure 14.1, if $\Delta\theta$ is the angular displacement during the time interval Δt, from the knowledge that we have learned from the circular motion in Chapter 13, we can write,

the instantaneous angular velocity	$\omega = \dfrac{d\theta}{dt}$
the instantaneous angular acceleration	$\alpha = \dfrac{d\omega}{dt}$
the linear velocity (in the tangential direction)	$v = r\omega$
the tangential acceleration of the point P	$a_t = r\alpha$
the radial or centripetal acceleration at point P	$a_r = \dfrac{v^2}{r} = r\omega^2$

If the rigid body rotates about a fixed axis with constant acceleration, then we have the rotational kinematics as given in Table 14.1.

Table 14.1: Rotational kinematics for constant angular acceleration

$\omega = \omega_0 + \alpha t$
$\Delta\theta = \omega_0 t + \dfrac{1}{2}\alpha t^2$
$\omega^2 = \omega_0^2 + 2\alpha\Delta\theta$

The period of revolution T for the motion of each point and for the rigid body itself is given by

$$T = \frac{2\pi}{\omega}$$

The frequency f of the motion, which is the number of revolutions per unit time, would be

$$f = \frac{1}{T}$$

Then, for uniform rotational motion

$$\omega = 2\pi f$$

Activity 14.1

A flywheel on a motor increases its rate of motion uniformly from 120 rev min^{-1} to 300 rev min^{-1} in 10 s. Calculate (i) its angular accleration (ii) its angular displacement in this time.

14.3 Kinetic Energy of Rotation and Moment of Inertia

Consider the rotation of a rigid body about a fixed axis, as shown in figure 14.2. The rigid body contains many particles with their associated masses, and each of these particles of the rotating rigid body has a certain kinetic energy. Thus, the rigid body that consists of mass in motion has kinetic energy. The kinetic energy due to rotation is called rotational kinetic energy and will be denoted by K_{rot}.

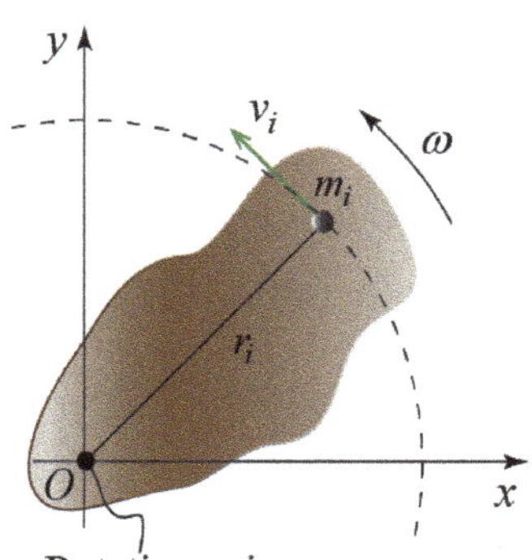

Figure 14.2: A rigid body rotating about an axis through O. i^{th} particle of mass m_i rotates about the axis with the radius r_i and speed v_i

Figure 14.2 shows a rigid body that rotates with angular velocity ω. The th particle of the rigid body of mass m_i rotates in a circle of radius r_i and moves with speed $v_i = r_i\omega$. The kinetic energy of the ith particle is $\frac{1}{2}m_i v_i^2 = \frac{1}{2}m_i r_i^2 \omega^2$. We can then add up the kinetic energies of all the particles to find the rotational kinetic energy K_{rot} of the rigid body as a whole. Thus,

$$K_{rot} = \frac{1}{2}m_1 v_1^2 + \frac{1}{2}m_2 v_2^2 + \frac{1}{2}m_3 v_3^2 + \ldots \qquad (14.1)$$

where particles of mass $m_1, m_2, m_3, \ldots$ at distances $r_1, r_2, r_3, \ldots$ from the axis of rotation O, moving with speed of $v_1, v_2, v_3, \ldots$ respectively. Since $v_i = r_i\omega$, the equation 12.1 can be rewritten as

$$K_{rot} = \frac{1}{2}m_1 r_1^2 \omega^2 + \frac{1}{2}m_2 r_2^2 \omega^2 + \frac{1}{2}m_3 r_3^2 \omega^2 + \ldots = \frac{1}{2}\left(\sum m_i r_i^2\right)\omega^2 \qquad (14.2)$$

The quantity $\sum m_i r_i^2$ is called the rigid body's moment of inertia, and it is denoted by I.

i.e. $$I = m_1 r_1^2 + m_2 r_2^2 + m_3 r_3^2 + \ldots = \sum m_i r_i^2 \qquad (14.3)$$

Thus, the rotational kinetic energy of the rigid body is

$$K_{rot} = \frac{1}{2}I\omega^2 \qquad (14.4)$$

The units of the moment of inertia are kg m^2. The moment of inertia of a rigid body depends on (i) the location of the axis of rotation, (ii) the shape of the body and (iii) the distribution of its mass.

Table 14.2 gives the moments of inertia for a number of uniform rigid bodies with various geometrical shapes about specific axes.

Table 14.2: Moment of Inertia of uniform rigid bodies with various geometrical shapes

Rigid Body	Axis of Rotation	Moment of Inertia	Rigid Body	Axis of Rotation	Moment of Inertia
Thin hollow cylindrical shell	About the central axis 	MR^2	Long, thin rod	About the axis through center 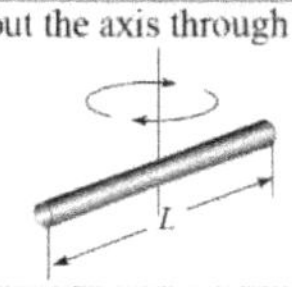	$\frac{1}{12}ML^2$
Thin hollow hoop	About the central axis 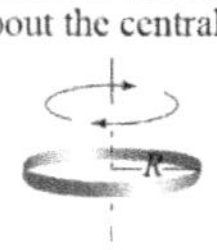	MR^2	Long, thin rod	About the axis through end 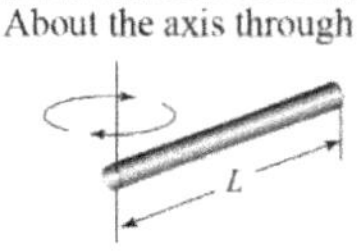	$\frac{1}{3}ML^2$
Solid cylinder	About the central axis 	$\frac{1}{2}MR^2$	Solid cylinder	About the central diameter 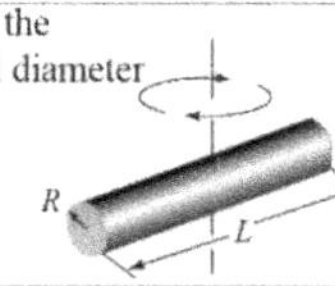	$\frac{1}{4}MR^2+\frac{1}{12}ML^2$
Solid disk	About the central axis 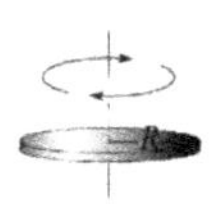	$\frac{1}{2}MR^2$	Solid sphere	About any diameter 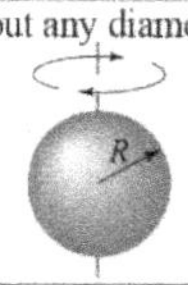	$\frac{2}{5}MR^2$
Hollow cylindrical shell	About the central axis 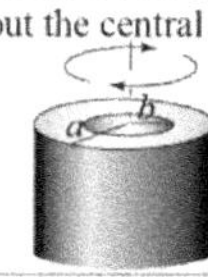	$\frac{1}{2}M\left(a^2+b^2\right)$	Thin spherical shell	About any diameter 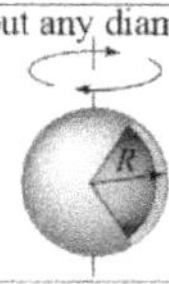	$\frac{2}{3}MR^2$
Hollow disk	About the central axis 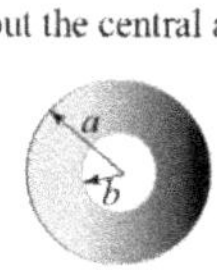	$\frac{1}{2}M\left(a^2+b^2\right)$	Rectangular slab	Along edge of slab 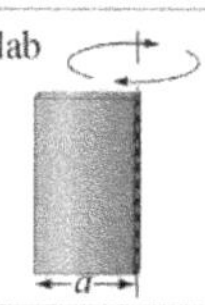	$\frac{1}{3}Ma^2$
Rectangular slab	Perpendicular to slab through **centre** 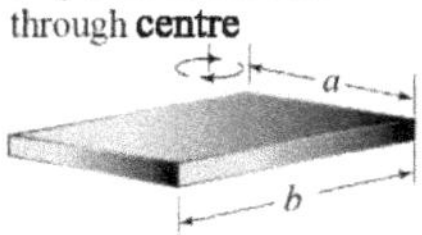	$\frac{1}{12}M\left(a^2+b^2\right)$	Rectangular slab	Parallel to edge of slab through **centre** 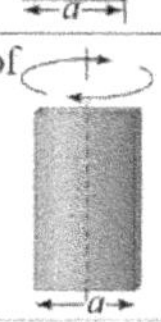	$\frac{1}{12}Ma^2$

So far, we have discussed the moment of inertia of any object about an axis through its centre of mass. If you know the moment of inertia of an object about its centre of mass, then we can find the moment of inertia of the same object rotating about an arbitrary axis through O, which is parallel to the axis through its centre of mass and displaced from it by a distance d, as shown in figure 14.3.

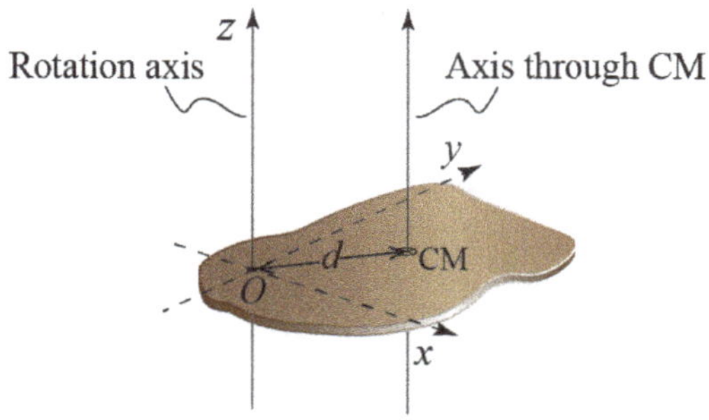

Figure 14.3: The moment of inertia of an object rotating about an arbitrary axis, which is parallel to the axis through its centre of mass (CM) and displaced from it by a distance d

The moment of inertia of the object rotating about the arbitrary axis through O, which is parallel to the axis through its centre of mass and displaced from it by a distance d, can be shown as

$$I = I_{cm} + Md^2 \tag{14.5}$$

where M is the mass of the object and I_{cm} is the moment of inertia of the object about its centre of mass. Equation 14.5 is called the parallel-axis theorem (the proof of this theorem is beyond the scope of this chapter).

14.4 Rigid body under a net Torque

In Chapter 10, we learned about the moment of a force, known as a moment and also known as a torque with the symbol τ. It describes the ability of a force to rotate a rigid body around a specific fixed rotation axis or pivot point. This means that torque can cause the rotation of a rigid body, as we generally use torque to rotate a door.

Now consider a rigid body undergoing pure rotational motion about a fixed rotation axis, as shown in figure 14.4. The forces $\vec{F}_1, \vec{F}_2, \vec{F}_3, ...$ are the external forces acting on particles of masses $m_1, m_2, m_3, ...$ that are part of the rigid body and acting at a distance $r_1, r_2, r_3, ...$ respectively from the fixed rotation axis as shown. These forces exert torques $\tau_1, \tau_2, \tau_3, ...$ about the axis of rotation.

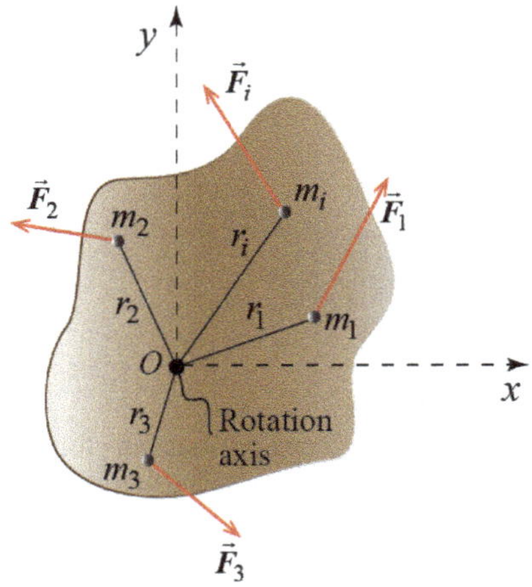

Figure 14.4: A rigid body rotating about a fixed rotation axis through O. Each particle of mass m_i rotates about the axis with the same angular acceleration α

Thus, the net torque on the rigid body is the sum of the torques, which is given by:

$$\tau_{net} = \sum_i \tau_i \tag{14.6}$$

Let us concentrate on the ith particle that is acted on by force $\vec{F}_i$ and undergoes circular motion with radius r_i. In Chapter 13, we learned that the radial component of force $\vec{F}_i$ is responsible for the centripetal acceleration of the circular motion of the ith particle, while the tangential component of force $\vec{F}_i$, i.e. $(F_i)_t$, causes the particle to accelerate or slow down with a tangential acceleration $(a_i)_t$. We also know that $(a_i)_t = r_i c$, where α is the angular acceleration. Further, all particles in the rigid body have the same angular acceleration. i.e. α is the angular acceleration of the entire rigid body.

Newton's second law for the ith particle will give

$$(F_i)_t = m_i (a_i)_t = m_i r_i \alpha \tag{14.7}$$

If we multiply both sides of the equation 14.6 by r_i we get

$$(F_i)_t r_i = m_i r_i^2 \alpha \tag{14.8}$$

Here, the radius r_i and tangential component of force $(F_i)_t$ are perpendicular to each other. Also, $(F_i)_t r_i$ is the torque τ_i about the axis of the ith particle (refer to section 10.4 of Chapter 10). Thus,

$$\tau_i = m_i r_i^2 \alpha \tag{14.9}$$

Now, if we insert the above result into equation 4.5, we get

$$\tau_{net} = \sum_i \tau_i = \sum_i m_i r_i^2 \alpha = \left(\sum_i m_i r_i^2 \right) \alpha = I\alpha$$

$$\tau_{net} = I\alpha \tag{14.10}$$

where the quantity $\sum_i m_i r_i^2$ is the moment of inertia Δ of the rigid body about the fixed rotation axis.

Equation 14.9, i.e. $\tau_{net} = I\alpha$, sometimes referred to as Newton's second law for rotational motion, is the fundamental equation of rigid-body dynamics.

In the absence of a net torque, i.e. $\tau_{net} = 0$, the rigid body either does not rotate, i.e. $\omega = 0$, or rotates with constant angular velocity, i.e. ω = constant.

Activity 14.2

A wheel is rotating initially at 90 rev min^{-1}. What torque is required to bring it to rest in 5.0 rev if the wheel has moment of inertia 0.80 kg m^2?

14.5 Work and Power in Rotational Motion

We have already learned in Chapter 11 that the work is the transfer of energy from one object to another by the application of force. In rotational motion, work is done when a torque is applied to an object, causing it to rotate about a fixed axis. For example, when you pedal a bicycle, you apply forces to a rotating body and do work on it. We can express the work in rotational motion in terms of torque and angular displacement.

Consider a situation where a tangential force having a constant magnitude F_t acts at the rim of a circular rotating platform with radius r, as shown in figure 14.5 (for example, imagine that you are applying a constant tangential force at the rim of a merry-go-round platform). When the platform rotates through an angular displacement $\Delta\theta$ about a fixed axis through a centre O during a small time interval Δt and displace the point at which the force acts through a distance Δs .

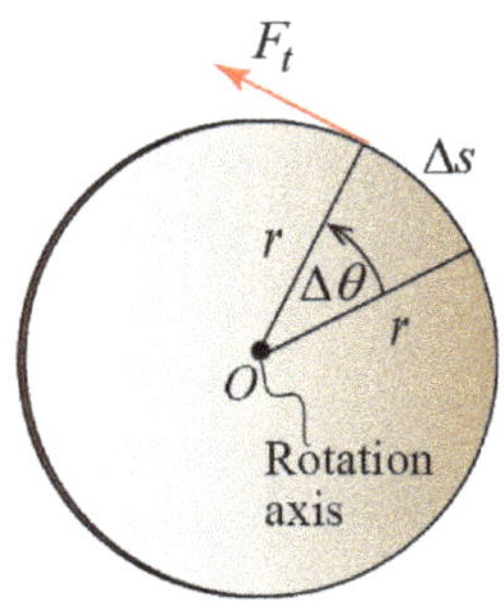

Figure 14.5: The work done by a tangential force acting on a circular rotating platform with radius r

By definition, the work ΔW done by the force F_t is $\Delta W = F_t \Delta s$. If θ is measured in radians, $\Delta s = r\Delta\theta$, thus we have

$$\Delta W = F_t r \Delta\theta \tag{14.11}$$

Now, $F_t r$ is the torque τ due to the force Δt acting on the platform as shown in figure 14.5. Thus, $\vec{r}$.

Therefore we can define the work done on a body by a constant torque τ equals the product of the torque τ and the angular displacement $\Delta\theta$ of the body, i.e.

$$\Delta W = \tau \; \Delta\theta \tag{14.12}$$

This equation is the rotational analogue of the work done by a constant force with the component in the direction of displacement (see Chapter 11). If is expressed in newton-metres (N m) and θ in radians, then the work is expressed in joules.

In the above scenario, we have not considered the radial component of the force. If we assume that such a radial component is acting on the body at the point of application of force, it would do no work because the displacement of the point of application has no radial component. Furthermore, the radial component of the force would also not contribute to the torque about the axis of rotation. Therefore, equation 14.12 is correct for any force, even if it does have a radial component.

Power in rotational motion measures the rate at which work is done, or energy is transferred within a rotational system. It is a representation of how quickly the rotational work is being performed. Power can be calculated by dividing the work done by the time taken to perform the work. Accordingly, i.e. we divide both sides of equation 14.12 by the time interval Δt during which the angular displacement occurs, we get the following result.

$$\frac{\Delta W}{\Delta t} = \tau \frac{\Delta \theta}{\Delta t}$$

Here, $\frac{\Delta W}{\Delta t}$ is the rate of doing work, i.e. the power P, and $\frac{\Delta \theta}{\Delta t}$ is the body's angular velocity ω. Therefore, we can define the power P as a product of torque τ and angular velocity ω.

i.e. $$P = \tau\omega \tag{14.13}$$

Equation 14.13 is the analogue of the relation that we developed in Chapter 11.

Table 14.3 summarises the various equations we have discussed pertaining to rotational motion together with the analogous expressions for linear motion.

Table 14.3: Useful Equations in Rotational and Linear Motion

Rotational Motion About a Fixed Axis		**Linear Motion**	
Angular speed	$\omega = \frac{d\theta}{dt}$	Linear speed	$v = \frac{dx}{dt}$
Angular acceleration	$\alpha = \frac{d\omega}{dt}$	Linear acceleration	$a \quad \frac{dv}{dt}$
Net torque	$\tau_{net} = \sum \tau_{ext} = I\alpha$	Net force	$F_{net} = \sum F_{ext} = ma$
If α is constant	$\omega = \omega_0 + \alpha t$ $\Delta\theta = \omega_0 t + \frac{1}{2}\alpha t^2$ $\omega^2 = \omega_0^2 + 2\alpha\Delta\theta$	If a is constant	$v = u + at$ $s = ut + \frac{1}{2}at^2$ $v^2 = u^2 + 2as$
Rotational kinetic energy	$K_{rot} = \frac{1}{2}I\omega^2$	Kinetic energy	$K = \frac{1}{2}mv^2$
Work	$\Delta W = \tau\ \Delta\theta$	Work	$\Delta W = F_x dx$
Power	$P = \tau\omega$	Power	$P = Fv$

Activity 14.3

An electric motor applies a constant torque of 15.0 N m to a grindstone attached to its shaft. Take the moment of inertia of the grindstone as $I = 2.5 \text{ kg m}^2$. If the system starts from rest, calculate

(i) the work done by the motor in 10.0 s.

(ii) the kinetic energy at the end of 10.0 s.

(iii) the average power delivered by the motor during 10.0 s .

14.6 Rolling Motion of a Rigid Object

The rolling motion is a combination of translational and rotational motions. If a rigid object moves while rotating about a point of contact with a surface without slipping, we call the motion a pure rolling motion. In this section, we will consider only objects that roll without slipping.

Consider a circular object such as a wheel or a sphere of radius R rolling without slipping on a horizontal surface, as shown in figure 14.6. As shown, concentrate exactly on one revolution of its motion. The point at the bottom follows the curve to the top and back to the bottom.

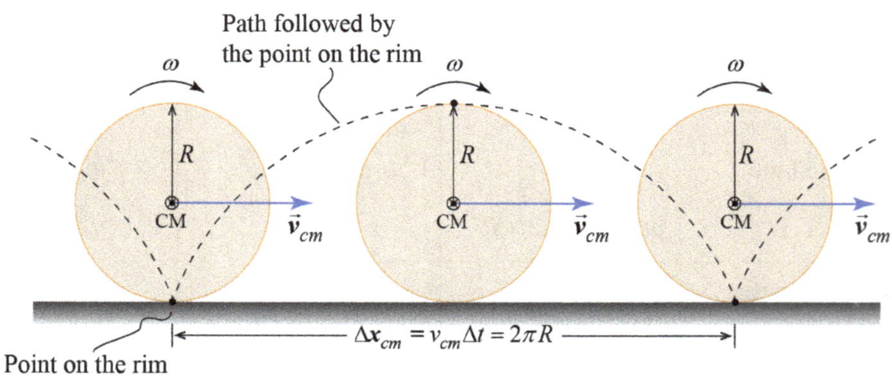

Figure 14.6: A circular object rolling without slipping through one revolution

Since the object does not slip, the centre of mass (CM) moves forward exactly one circumference, i.e. $\Delta x_{cm} = 2\pi R$. This distance travelled could be written in terms of the velocity of the centre of mass v_{cm} . If T is the time taken by the object to make one complete revolution (i.e. the rotation period), then we can write $\Delta x_{cm} = v_{cm}T$. These two expressions for Δx_{cm} we obtained come from two perspectives of the motion: one looking at the rotation and the other looking at the translation of the centre of mass. However, both distances are the same. Thus, these two expressions must be equal.

i.e. $$\Delta x_{cm} = 2\pi R = v_{cm}T \tag{14.14}$$

If we divide the equation 14.14 by T, we can write the centre of mass velocity as

$$v_{cm} = \left(\frac{2\pi}{T}\right)R = R\omega \tag{14.15}$$

Here $\frac{2\pi}{T}$ is the angular velocity ω. Equation 14.15 is the basic link between the translation and rotation of an object that rolls without slipping.

Now, we look at the rolling object very closely, as shown in figure 14.7.

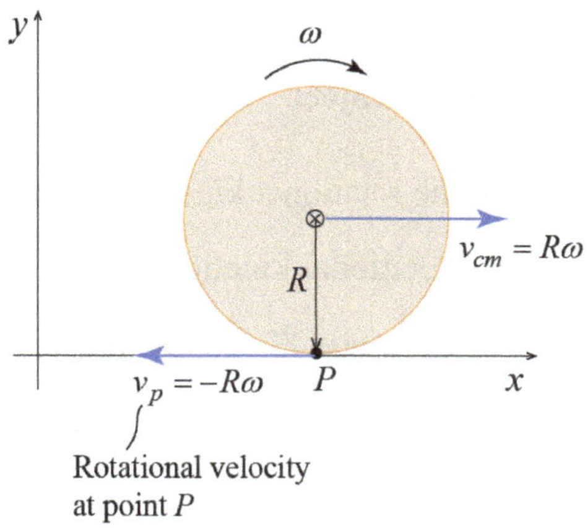

Figure 14.7: The motion of the point *P* on the bottom of a rolling object

The point *P* at the very bottom of the rolling object is the point of contact between the object and the surface. This point *P* is moving around the centre of the object at angular velocity ω. This means that the rotational velocity at *P* is $v_P = -R\omega$. This velocity is relative to the rolling object, i.e. relative to the centre of mass. The negative sign indicates that at point *P*, the motion is opposite to the $+ve$ x direction. At the same time, the velocity of the centre of mass is $v_{cm} = R\omega$, and is in the direction of $+ve\ x$. Thus, by adding these two velocities (i.e. v_p and $v =_{\iota}$), we could find the velocity at point *P*, which gives the value zero. This means that a rolling object's lowest point (i.e., the point on the bottom) is instantaneously at rest. In other words, to roll without slipping, the bottom point, i.e. the point touching the surface, must be at rest.

Figure 14.8 shows the velocity vectors at the top, centre, and bottom of a rolling object. These are found by adding the rotational velocity vectors to the centre of mass velocity. You can see the velocities at the bottom $v_{bottom} = 0$, the top $s = ut + \frac{1}{2}at^2$ and the centre of mass $v_{cm} = R\omega$.

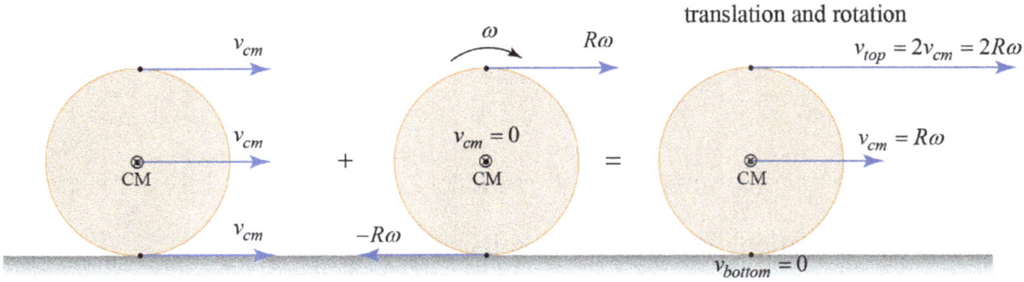

Figure 14.8: Rolling without slipping is a combination of translation and rotation

14.7 Kinetic Energy of a Rolling Object

In Section 14.3, we discussed the rotational kinetic energy of a rigid body in pure rotational motion and find the rotational kinetic energy as $K_{rot} = \frac{1}{2}I\omega^2$ (refer to equation 14.4). In this section, we would like to find the kinetic energy of an object that rolls without slipping, a combination of rotational and translational motion, as shown in figure 14.9.

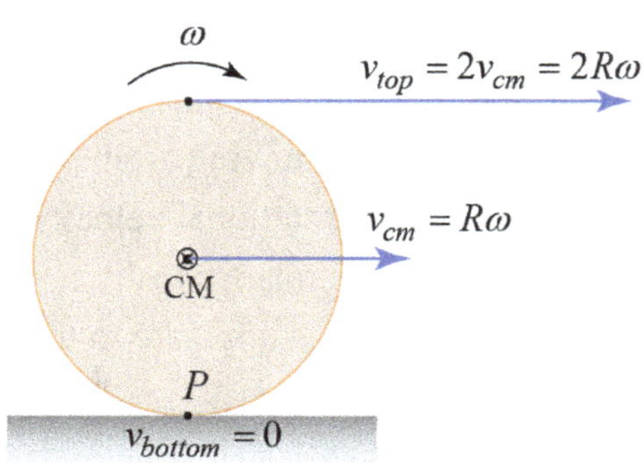

Figure 14.9: Rolling motion without slipping can be considered as an instantaneous rotation about point P

As we have seen in the previous section, when an object rolls without slipping, the bottom point P (figure 14.9) is instantaneously at rest. This type of rolling motion without slipping can also be visualised as an instantaneous rotation about the bottom point P. Suppose that we think of an axis through P as an axis of rotation. Then, the motion of the object is a pure rotation about point P. Thus, the kinetic energy of pure rotation can be written as

$$K = K_{rot-about\ P} = \frac{1}{2} I_P \omega^2 \tag{14.16}$$

where I_P is the moment of inertia for rotation about point P and ω is the angular velocity of the rotating object. We can use the parallel-axis theorem (refer to section 14.3) to write Δs in terms of the moment of inertia I_{cm} about the centre of mass, the mass of the object Δt and the distance $\vec{r}$ between the parallel axis through the centre of mass and the axis through P. Thus,

$$\Delta\theta \tag{14.17}$$

From equations 14.6 and 14.17, we have

$$K = \frac{1}{2} I_{cm} \omega^2 + \frac{1}{2} M (R\omega)^2 \tag{14.18}$$

In Section 14.6, we have observed that the basic link between the translation and rotation of an object that rolls without slipping is $v_{cm} = R\omega$. Thus, the kinetic energy of a rolling object is

$$K_{roll} = \frac{1}{2} I_{cm} \omega^2 + \frac{1}{2} M v_{cm}^2 = K_{rot} + K_{cm} \tag{14.19}$$

The term $\frac{1}{2} I_{cm} \omega^2$ represents the rotational kinetic energy of the rigid object about its centre of mass, denoted by K_{rot} , and the term $\frac{1}{2} M v_{cm}^2$ denoted by K_{cm} represents the kinetic energy the rigid object would have if it were just translating through space without rotating. This K_{cm} is sometimes called the translational kinetic energy of its centre of mass and is denoted by K_{trans}.

Therefore, equation 14.19 states that the total kinetic energy of a rolling object K_{roll} is the sum of the rotational kinetic energy K_{rot} about the centre of mass and the translational kinetic energy K_{cm} of the centre of mass.

Activity 14.4

A 15.0 kg mass cylinder rolls about its horizontal axis through its centre of mass smoothly without slipping on a horizontal surface. At a given time, its centre of mass is travelling at a velocity of 12.0 m s^{-1}. Calculate the following:

(i) The translational kinetic energy of its centre of mass,

(ii) The rotational kinetic energy about the horizontal axis through its centre of mass and

(iii) Its total energy.

14.8 Angular Momentum

Every rotational quantity that we have seen in the previous sections is the analogue of some quantity related to the motion of a particle. Likewise, in rotational motion, the analogue of the linear momentum of a particle is called the angular momentum. It is a vector quantity and is denoted by $\vec{L}$.

When a rigid body with the moment of inertia v with respect to a specified axis rotates with angular velocity ω about that axis, the angular momentum L of the body with respect to the axis is defined as the product of the moment of inertia v about the axis and the angular velocity ω.

i.e. $$L = I\omega \tag{14.20}$$

The direction of angular momentum depends on the direction of ω and is identified using the right-hand thumb rule (refer to Chapter 13 Section 13.2). Thus, in vector form, equation 14.20 can be written as

$$\vec{L} = I\vec{\omega} \tag{14.21}$$

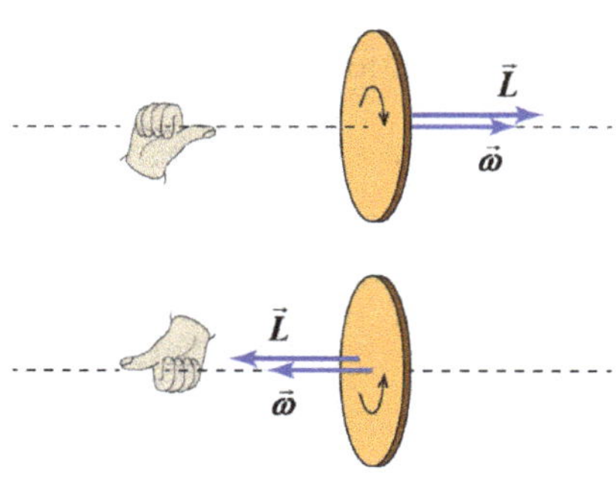

Figure 14.10: The right-hand rule for angular quantities

Right-hand thumb rule: If the fingers of the right hand are curled in the direction of rotation of the circular path, then the direction of the thumb of the right hand gives the direction of the angular velocity $\vec{\omega}$ and the angular momentum $\vec{L}$, as shown in figure 14.10. Usually, the sign of the angular momentum is positive for counterclockwise rotation and negative for clockwise rotation. The unit of angular momentum is $\text{kg m}^2\text{s}^{-1}$.

During the rotational motion of a rigid body if the rotational axis does not change, then the moment of inertial I about that axis will be constant. Suppose α_{avg} is the average angular acceleration of the rotational motion of a rigid body, then

$$I\alpha_{avg} = I\frac{\Delta\omega}{\Delta t} = \frac{\Delta I\omega}{\Delta t} = \frac{\Delta L}{\Delta t}$$

In the limit $\Delta t \to 0$, α_{avg} becomes the instantaneous angular acceleration α , and it could be defined as;

$$I\alpha = I\lim_{\Delta t\to 0}\left(\frac{\Delta\omega}{\Delta t}\right) = \lim_{\Delta t\to 0}\left(\frac{\Delta I\omega}{\Delta t}\right) = \frac{dL}{dt}$$

Refer to Chapter 13 Section 13.2 and equation 13.8.

From equation 14.10 we have $\tau_{net} = I\alpha$. Thus, we have

$$\tau_{net} = \sum\tau = I\alpha = \frac{dL}{dt} \tag{14.22}$$

This means that the rate of change of angular momentum of a rigid body with respect to any axis equals the sum of the torques of the forces acting on it with respect to that axis. In vector form the equation 14.22 can be written as

$$\vec{\tau}_{net} = \sum\vec{\tau} = \frac{d\vec{L}}{dt} \tag{14.23}$$

It should be noted that $\vec{\tau}$ and $\vec{L}$ are measured about the same axis.

We know that a rigid body consists of a collection of particles. As shown in Figure 14. 11, let us consider $m_1, m_2, m_3, \ldots, m_i \ldots$ are the masses of the collection of particles at a distance $r_1, r_2, r_3, \ldots, r_i \ldots$ from the axis of rotation through point O. When the object rotates with angular velocity ω , each particle that comprises the rigid body will also be rotating with the same angular velocity ω . Therefore, the magnitude of the velocity v_1 of the particle at a distance r_1 can be written as $v_1 = r_1\omega$, and the magnitude of the angular momentum L_1 of this particle is $L_1 = m_1 v_1 r_1$ (Refer to Chapter 13 Section 13.5 and equation 13.19). This can be rewritten as

$$L_1 = m_1 v_1 r_1 = m_1 \left(r_1\omega\right) r_1 = m_1 r_1^2 \omega$$

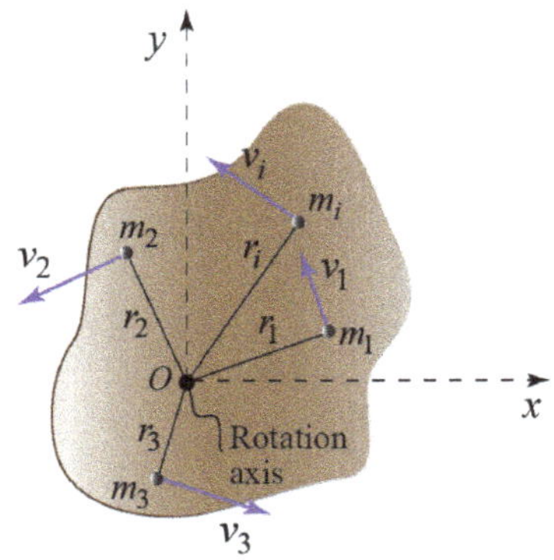

Figure 14.11: A rigid body consists of a collection of particles having masses $m_1, m_2, m_3, ...m_i ...$ at a distance $r_1, r_2, r_3, \cdots, r_i ...$ from the axis of rotation through point O and have corresponding velocities, $v_1, v_2, v_3, ..., v_i ..$ respectively

Since all the particles that comprise the rigid body rotate with the same angular velocity Δt, the total angular momentum of all the particles can be written as

$$a = \frac{v}{r} \lim_{\Delta t \to 0} \left(\frac{\Delta s}{\Delta t} \right) = \frac{v^2}{r}$$

This agrees with our previous definition (equation 14.20) of the angular momentum of a rigid body.

Activity 14.5

A uniform solid disk of mass 3.00 kg has a radius 10.0 cm and spins horizontally with a frequency of 1.00 rev s^{-1} about its vertical axis, passing through its centre. What is its angular momentum?

Activity 14.6

An object is rotating with angular frequency ω about a fixed axis, having a moment of inertia I. Suppose the angular momentum of the rotating object about the fixed axis is $L = I\omega$, show that the rotational kinetic energy of the object as $K_{rot} = \dfrac{L^2}{2I}$.

14.9 Conservation of Angular Momentum

In Chapter 12, Section 12.4, we learnt that the total linear momentum of a system of particles remains constant if the system is isolated, that is, if the net external force acting on the system is zero. We have an analogous conservation law in rotational motion, called the principle of conservation of angular momentum. This law of conservation of angular momentum is also based on the concept of an isolated system and is stated as follows.

The total angular momentum of a system is constant in both magnitude and direction if the net external torque acting on the system is zero, that is, if the system is isolated.

This principle follows directly from equation 14.23 and can be related to if

$$\vec{\tau}_{net} = \sum \vec{\tau} = \frac{d\vec{L}}{dt} = 0 \tag{14.24}$$

then $\Delta\vec{L} = 0$ (14.25)

That is, the total angular momentum $\vec{L}$ of the system is constant. Therefore, from equation 14.25 for an isolated system, this can be stated as

$$\vec{L} = \text{Constant}$$

In the case of a rigid body rotating about an axis, we have from equation 14.21

$$\vec{L} = I\vec{\omega}$$

where $\vec{L}$ is the angular momentum about the rotation axis and I is the moment of inertia about the same axis. The angular velocity of the body is given by $\vec{\omega}$. If no external torques act on the body, any change in either I or $\vec{\omega}$ must result in an equal but opposite change in the other so that the product $I\vec{\omega}$ remains constant. This means that when I decreases, angular velocity $\vec{\omega}$ increases and when v increases, angular velocity $\vec{\omega}$ decreases so that the product $I\vec{\omega}$ remains constant.

The conservation of angular momentum has many applications in the real world. For example, it is used to explain the motion of planets in orbit, the spin of gyroscopes, and the way ice skaters spin.

Application of angular momentum

(i) Suppose a person sits on a stool that is free to rotate with as little friction as possible. He is made to rotate about the vertical axis with a modest initial angular speed ω while holding two dumb-bells/weights in his outstretched hands, as shown in figure 14.12 (a). His angular momentum vector $\vec{L}$ lies along the vertical rotation axis, pointing upward. No net external torque acts on the system consisting of the person, stool, and dumb-bells. Thus, the angular momentum of that system about the axis of rotation must remain constant, no matter how the person manoeuvres the dumb-bells.

Suppose that he suddenly pulls his arms towards his chest, as shown in figure 14.12 (b). This action reduces his moment of inertia from its initial value I to a smaller value I' because he moves mass closer to the axis of rotation. Since the angular momentum remains constant, his angular speed of rotation will increase to ω'. The person can then slow down by extending his arms once more, moving the dumb-bells outward.

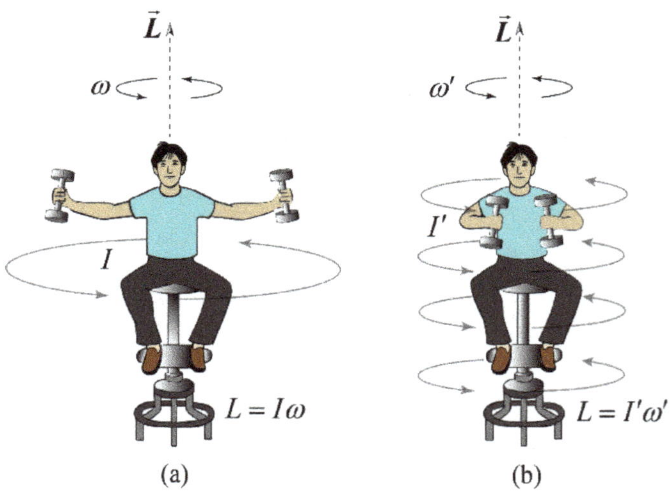

Figure 14.12: A man sits on a stool that is free to rotate with as little friction as possible. While holding two weights in his hands, (a) he is made to rotate, with his arms outstretched, (b) he suddenly brings his arms towards his chest

(ii) As shown in figure 14.13, a diver leaps from the diving springboard doing a forward somersault dive. When leaving the springboard, the diver has a certain angular momentum $\vec{L}$ about a horizontal axis through her centre of mass. The direction of the angular momentum is represented by the tail $\otimes$ of an arrow that is perpendicular to the plane of figure 14.13. There are no external forces acting on her except gravity, and gravity does not exert a torque about her centre of mass. Therefore, her angular momentum remains constant in both magnitude and direction throughout the dive. Since she wishes to do a somersault before reaching the water level below, she pulls her arms and legs toward the centre of her body, as shown in figure 14.13. By doing this, the diver considerably decreases her moment of inertia about the same axis through the centre of mass. As her total angular momentum remains constant, her angular speed increases, and she can perform a rotation or somersault.

At the end of the dive, she extends her limbs again, increasing her moment of inertia. Therefore, the angular velocity drops back to the initial value so that she can enter the water with little splash. Note that her centre of mass continues to move along a parabolic path under the influence of gravity.

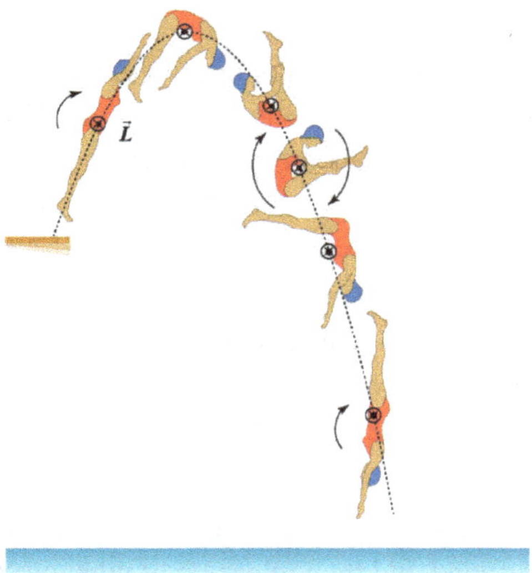

Figure 14.13

(iii) Similarly, ice skaters and ballet dancers can increase or decrease the angular speed of a spin about a vertical axis by pulling in or extending their limbs, respectively. Figure 14.14 shows the rotational motion of an ice skater. To initiate the spin, the skater begins by gliding with her arms extended and subsequently initiates the rotation of her body around a vertical axis by exerting force against the ice with one of her skates. The push of the ice against the skate provides the external torque that gives the skater her initial angular momentum.

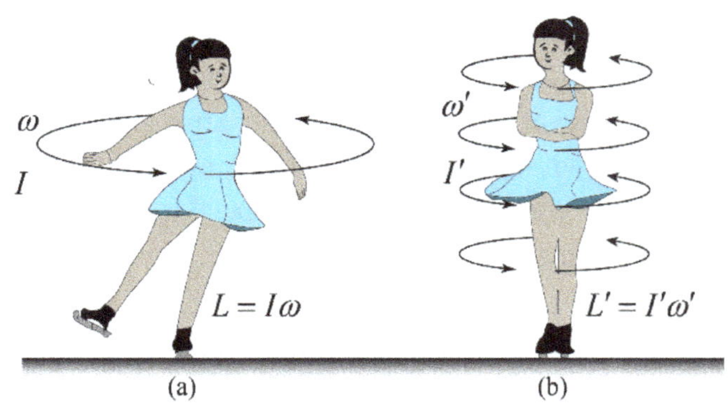

Figure 14.14: An ice skater uses her moment of inertia to control her spin. Angular momentum $\vec{r}$ is conserved. (a) I is large and ω is small (b) I' is smaller than I and ω' is larger than ω . i.e. Her angular velocity is higher in (b) than in (a)

Initially, her arms and the leg that is not in contact with the ice are extended away from her body, as shown in figure 14.14 (a). The mass of the arms and legs, when extended, contribute more to her moment of inertia I than they do when held close to the body. Therefore, the skater pulls her arms and legs close and straightens her body, as shown in figure 14.14 (b), to decrease her moment of inertia to a value I'. Since no external torque acts on the skater, the angular momentum cannot change. Thus, with a decrease in moment of inertia, the angular speed must increase, and she spins faster. Again, extending her arms increases her moment of inertia, and her angular velocity drops until she can gracefully exit the spin.

(iv) You would have noticed that if you hold a cat upside down in your hands at a height above the ground and then release it, it manages to land on its feet upon the ground. Why and how is this possible? If you film the cat during its fall and project the film in slow motion, you will see what is happening. As it falls, the cat throws its tail and its limbs in a given direction. Angular momentum is generated about a horizontal axis. Since the total angular momentum of the cat must remain constant, its body rotates in an opposite sense to compensate for this angular momentum. You will realise that the cat has no knowledge of the principle of conservation of angular momentum! But nature works according to such principles.

Activity 14.7

Consider an ice skater who is spinning at 0.70 rev s^{-1} with her arms extended, as shown in figure 14.14 (a). Her moment of inertia of her arms when extended is 2.35 kg m , and when her arms are close to her body is 0.35 kg m^2.

(i) Calculate her angular velocity in revolutions per second after she pulls her arms close to her body.

(ii) What is her rotational kinetic energy before and after she does this?

14.10 Precession of Gyroscopes and Tops

In Section 14.8, we learnt that the rate of change of angular momentum of a rigid body with respect to any axis equals the sum of the torques of the forces acting on it with respect to that axis. This could be written as

$$\vec{\tau}_{net} = \sum \vec{\tau} = \frac{\Delta \vec{L}}{\Delta t} \qquad (14.26)$$

Equation 14.26 shows that in the time interval Δt, the non-zero torque produces a change in angular momentum $\Delta \vec{L}$, which is in the same direction $\vec{\tau}_{net}$. It should be noted that the torque vector is always perpendicular to the axis of rotation and thus $\Delta \vec{L}$ is always perpendicular to $\vec{L}$.

Precession describes a change in the direction of the axis of a rotating object. Precession is observed in rotating machines such as gyroscopes. Figure 14.15 shows a typical horizontal gyroscope with the flywheel spinning with an angular velocity ω in a vertical plane that is supported at only one end of its axle.

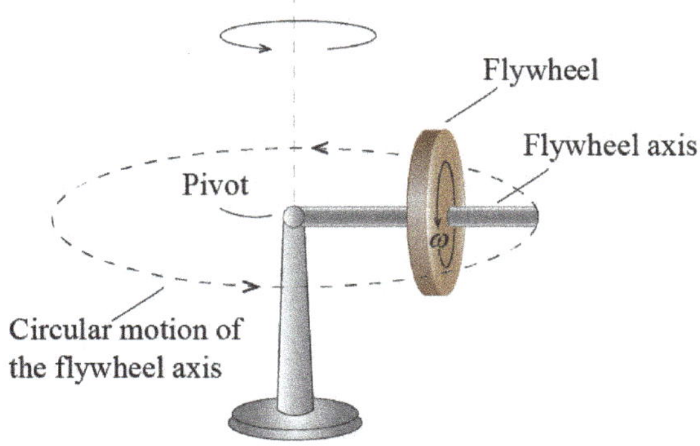

Figure 14.15: A typical horizontal gyroscope

Let us lift the flywheel of the stationary gyroscope of mass m until its axis is horizontal, as shown in figure 14.16. Two forces will act on the gyroscope: one is the weight of the flywheel $m\vec{g}$ acting downward at the centre of mass (assume that the axle is massless), and the other will be the normal force $\vec{n}$ of the support pushing upward at the pivot point O. The normal force does not exert torque at the pivot point because it acts at the pivot point. Therefore, the net torque ($\vec{\tau}$) on the gyroscope is entirely due to the weight of the flywheel. At this stage, if we leave the stationary gyroscope, then the free end of the axis simply drops because of gravity. This means that the weight of the flywheel mg acting downward at the centre of mass and the distance d from the pivot to the centre of the disk will create a resulting torque, which causes the free end of the flywheel axis to fall along a path determined by the right-hand rule.

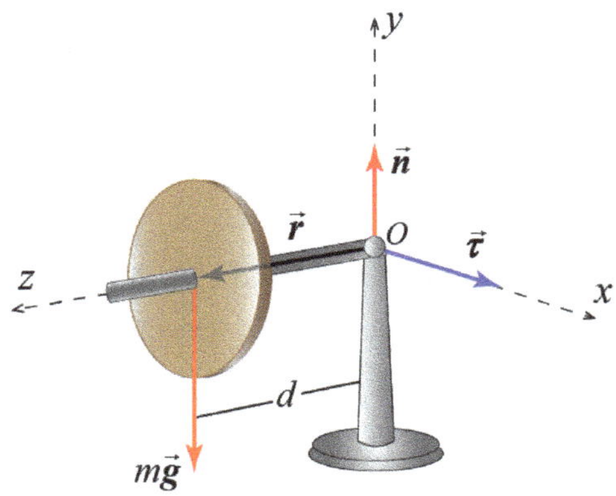

Figure 14.16: A nonspinning gyroscope causes it to fall over

The vector $\vec{r}$ is the position vector of the centre of the disk from the pivot point O and $|\vec{r}| = d$ is the distance from the pivot to the centre of the disk. Vectors $\vec{r}$ and the weight F_g are perpendicular, and by using the right-hand rule, we see that the torque (τ) points along the x-axis. The magnitude of the τ is given by

$$\tau = mgd \tag{14.27}$$

Suppose that after lifting the stationary gyroscope's flywheel until its axis is horizontal and if we start rotating the flywheel before we let it go, as shown in figure 14.17 (a), then the situation will be different.

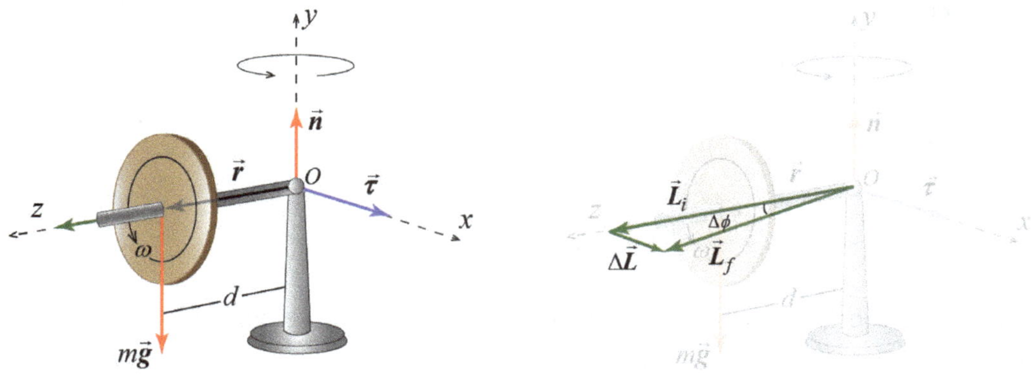

(a) (b)

Figure 14.17: A spinning gyroscope, (a) the torque τ changes the direction but not the magnitude of the angular momentum L (b) the torque results in a change in angular momentum $\Delta\vec{L}$ in a direction parallel to the torque vector. The gyroscope axle sweeps out an angle $d\phi$ in a time interval dt

In this case, before we let it go, the flywheel's initial angular momentum $\vec{L}_i$ is along its axis of rotation. Once we let the flywheel go, then it will be acted upon by a torque due to the weight and normal forces. Since the normal force exerts no torque on the pivot point because it acts at the pivot point, the net torque on the gyroscope is entirely due to the weight of the flywheel. Thus, as mentioned earlier in equation 14.27, the magnitude of the τ can be written as $\tau = mgd$. This torque results in a change in angular momentum $\Delta\vec{L}$ in a direction parallel to the torque vector. The gyroscope axle sweeps out an angle $\Delta\phi$ in a time interval Δt, as shown in figure 14.7 (b). From the vector diagram shown in figure 14.7 (b), after a time Δt, the angular momentum $\vec{L}_f$ will be $\vec{L}_f = \vec{L}_i + \Delta\vec{L}$. This means that the gyroscope axle has turned through a small angle $\Delta\phi$ during the time interval Δt. Therefore, for a small value of $\Delta\phi$, we can write

$$\Delta\phi = \frac{|\Delta\vec{L}|}{|\vec{L}|} \tag{14.28}$$

Furthermore, using equation 14.26, the change in angular momentum ΔL in a short time interval Δt can be written in terms of τ as

$$\Delta\vec{L} = \vec{\tau}\ \Delta t \tag{14.29}$$

Using equations 14.21, 14.27, 14.28 and 14.29, the rate at which the axis moves, i.e. $\frac{\Delta\phi}{\Delta t}$ which is defined as the precession angular velocity, denoted by Ω can be written as

$$\Omega = \frac{\Delta\phi}{\Delta t} = \frac{|\Delta\vec{L}|}{|\vec{L}|\Delta t} = \frac{\tau}{L} = \frac{mgd}{I\omega} \tag{14.30}$$

From equation 14.30, we see that the precession angular velocity Ω is inversely proportional to the angular velocity ω of spin about the axis. Therefore, as the flywheel slows down, the precession angular velocity increases.

We shift our attention to the rotational movement of the spinning top. Most likely, during your childhood, you would have had the opportunity to interact with a spinning top, which must have brought you a great deal of enjoyment. Let us now try to understand the actual behaviour of such a top in terms of the physical principles you have learnt.

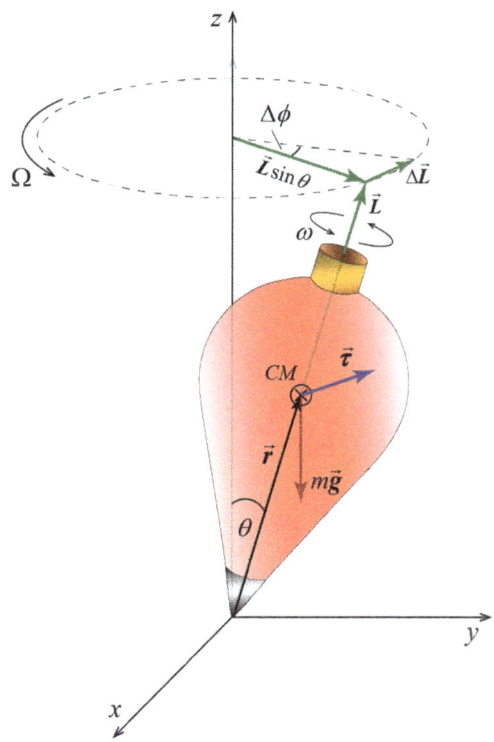

Figure 14.18: Precession of spinning top

When you spin a top on a flat surface, you will see that its upper end slowly revolves about the vertical direction, a process called precession. As the spin of the top slows, you will observe that this precession is getting faster and faster.

Figure 14.18 shows a spinning top, with its basal pivot at the origin of the rectangular coordinates *xyz*. While spinning about its own axis, it is also precessing about the *z*-axis. If m is the mass of the top, the weight $m\vec{g}$ of the top acts through its centre of mass, which has the position vector $\vec{r}$ from the pivot point *O*. The circle around the z-axis shows the path around which the axis of the top precesses. Suppose that the initial angular momentum of spin is $\vec{L}$ whose direction is determined from the sense of rotation of the top, using the right-hand rule which you have already studied. The force of gravity acting on the center of mass produces a torque $\vec{\tau}$ in the direction perpendicular to $\vec{L}$. The magnitude of the torque acting on the top due to gravity is $mgr\sin\theta$, where r is the magnitude of the position vector $\vec{r}$ and θ is the inclination of the spin axis of the top to the vertical. By our usual vectorial convention, this torque acts in a direction perpendicular to both $\vec{\omega}$ and $\vec{r}$, which is out of the paper. This changes the direction of the angular momentum vector $\vec{L}$ according to $\Delta\vec{L} = \vec{\tau}\ \Delta t$. Magnitude of the angular momentum vector $\vec{L}$ does not change, but changes its direction. Furthermore, $\Delta\vec{L}$ is in the direction of the torque vector $\vec{\tau}$. The top precesses around a vertical axis, since the torque is always horizontal and perpendicular to $\vec{L}$.

When the top precesses around a vertical axis, the spin axis of the top will turn through a small angle $\Delta\phi$ during the time interval Δt, as shown in figure 14.18. Therefore, for a small value of $\Delta\phi$, we can write

$$\Delta L = \left(mgr\sin\theta\right)\Delta t \tag{14.31}$$

also, we have

$$\Delta\phi = \frac{\Delta L}{L\sin\theta} \tag{14.32}$$

Therefore, $\frac{\Delta\phi}{\Delta t}$ is the precession angular velocity, denoted by 10 can be written as

$$\Omega = \frac{\Delta\phi}{\Delta t} = \frac{mgr}{L} = \frac{mgr}{I\omega} \tag{14.33}$$ \

From equation 14.33, we see that the precession angular velocity Ω is independent of θ the inclination of the spin axis of the top to the vertical and inversely proportional to L the spin angular momentum of the top. Thus, as G decreases, i.e. as the spinning rate m of the top slows down, its rate of precession increases. This result (for both gyroscope and top) is valid only when $\Omega \ll \omega$. Otherwise, a much more complicated motion is involved.

Activity 14.8

The disk of the gyroscope, having a mass of $400\ \text{g}$ and a radius of $4.00\ \text{cm}$, spins at a rate of $25.0\ \text{rev s}^{-1}$. The gyroscope is supported on a single pivot with its centre of mass at a horizontal distance of $5.00\ \text{cm}$ from the pivot. If the gyroscope precesses in a horizontal plane with negligible friction, determine the precession angular velocity.

Questions

NOTE: Although the solutions are given on the last page, it is in your own interest to attempt the questions on your own as an indication of how far you have understood the material of the section.

1. In what direction is the Earth's angular velocity vector as it rotates daily about its axis?

2. Suppose that you are standing on the edge of a large freely rotating turntable. What happens if you walk towards the centre?

3. If a stationary, rigid object experiences the application of only two external forces that are equal in magnitude but opposite in direction, under what condition does the object start to rotate?

4. We learned that momentum and angular momentum are conserved. However, in reality, most moving or rotating objects eventually slow down and stop. Explain.

5. Many helicopters have a small propeller attached to the tail that rotates in a vertical plane. What is the role of this propeller, and why is it attached to the tail rather than another part of the helicopter?

6. Animals that can run fast always have thin legs. Their leg muscles are concentrated close to the hip joint, while the lower leg consists mostly of tendons. Using the concept of rotational inertia, explain how this helps them run fast.

7. A door closer is used in a door to close automatically after being opened. If you wish to place a wedge-shaped door stopper on a slippery floor to hold the door open, what is the best place to put it? Should it be placed close to the hinge or far from it? Explain.

8. In a children's park, a playground merry-go-round has a radius of 2.00 m and a moment of inertia of 250 kg m^2 and turns with negligible friction about a vertical axle through its centre. The merry-go-round is initially at rest, and a child tangentially applies a force 25.0 N to the edge of the merry-go-round, as shown in the figure.

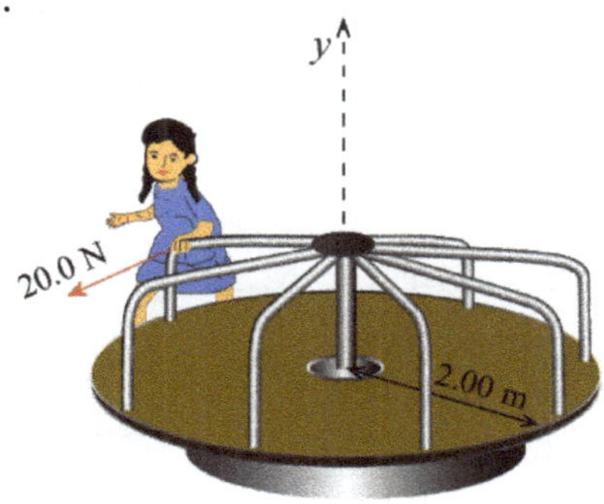

   (i) Calculate the merry-ground's angular velocity after this 20.0 s interval.

   (ii) Determine the work done by the child on the merry-go-round.

   (iii) Calculate the average power supplied by the child during this time.

9. A cyclist maintains a constant pedalling speed of 75.0 revolutions per minute on a sports bicycle that has wheels with a diameter of 62.0 cm, as shown in the figure.

The chain interacts with a front sprocket having a diameter of 15.0 cm and a rear sprocket with a diameter of 7.50 cm .

(i) Calculate

(a) the chain's speed relative to the bicycle frame.

(b) the angular speed of the bicycle wheels.

(c) the speed of the bicycle relative to the road.

(ii) If the length of the pedal cranks (not shown in the figure) is 17.5 cm , determine the linear speed of the pedals.

10. As shown in the figure, a wheel with radius R , mass M , and moment of inertia I is mounted on a frictionless horizontal axle. Attached to the wheel is a light weight cord that is tied to a bucket of mass m .

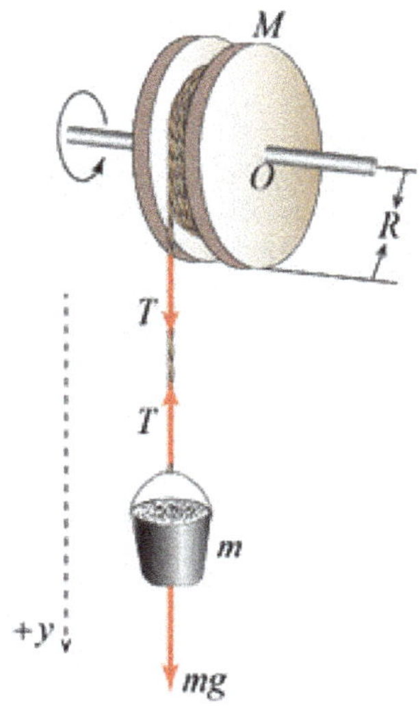

When the wheel is released from the resting state, the object experiences a downward acceleration of magnitude a , the cord unwinds from the wheel without slipping, and the wheel undergoes angular acceleration α .

(i) Does the Earth exert the gravitational force on the wheel, and the normal force exerted by the axle on the wheel will contribute to the torque taken about its axis of rotation? Explain.

(ii) Show that the angular acceleration α of the wheel as $\alpha = \dfrac{mgR}{mR^2 + I}$.

(iii) Show that the magnitude of the downward acceleration a of the bucket as $a = \dfrac{mgR^2}{mR^2 + I}$.

(iv) Show that the tension T in the cord would be $T = \dfrac{mgI}{mR^2 + I}$.

11. In a mechanical workshop, there is a set-up where an electric motor turns a flywheel through a drive belt that joins a pulley to the motor and a pulley that is rigidly attached to the flywheel, as illustrated in the figure.

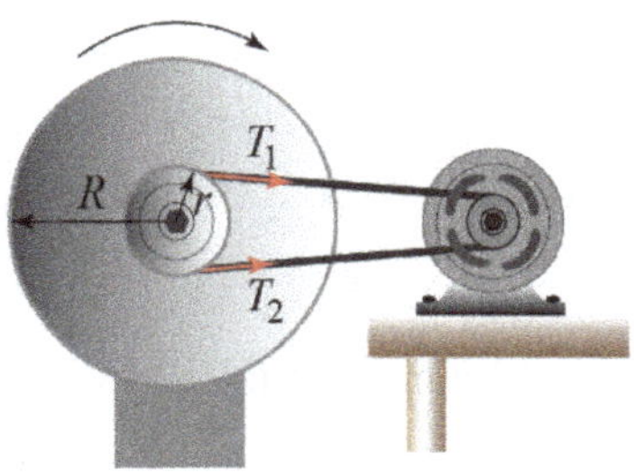

The flywheel is a solid disk with a mass of $75.0\ \text{kg}$ and a radius R of $60.0\ \text{cm}$. It turns on a frictionless axle. Its pulley has a negligible mass and a radius of r of $20.0\ \text{cm}$. It starts from rest and experiences a uniform acceleration due to the torque from the belt, resulting in a clockwise angular acceleration of $1.50\ \text{rad s}^{-2}$.

(i) Calculate the moment of inertia of the flywheel.

(ii) Calculate the net torque on the flywheel.

(iii) If the tension T_1 of the upper segment of the belt is $125\ \text{N}$, calculate the tension T_2 in the lower segment of the belt using the result obtained in part (ii).

(iv) How long does the flywheel take to reach its final operating speed of $100\ \text{rev min}^{-1}$?

(v) Through how many revolutions does it turn while accelerating?

12. The figure shows that various rigid bodies (solid sphere, solid cylinder, thin-walled hollow sphere, thin-walled hollow cylinder) roll without slipping from rest down an inclined plane of height h. Which rigid body rolls down the incline fastest? Explain.

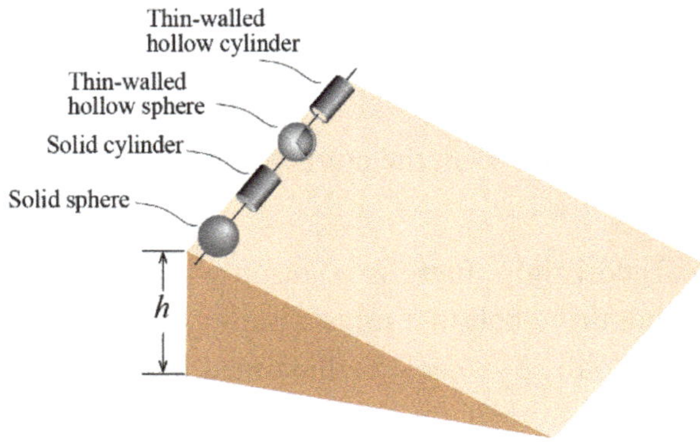

(i) Show that after descending a distance h, the speed of the centre of mass of the said rigid bodies is independent of their respective mass M and radius R.

(ii) Which rigid body rolls down the incline fastest? Explain.

The moment of inertia of uniform rigid bodies is given in table 14.2.

13. A yo-yo is made using two uniform disks, each with mass m and radius R, connected by a light axle of radius r. A light, thin string is wound several times around the axle and then held stationary while the yo-yo is released from rest, dropping as the string unwinds, as shown in the figure. Take g as the gravitational acceleration.

(i) Show that the linear acceleration a of the centre of mass and the angular acceleration α of the yo-yo is given by

$$a_{cm} = \frac{2gr^2}{2r^2 + R^2} \qquad \text{and} \qquad \alpha = \frac{2gr}{2r^2 + R^2}$$

(ii) Determine the tension in the string in terms of m, R, r and g.

(iii) Now consider the similar yo-yo having a rotational inertia of 1000 g cm^2 and a mass of 125 g. Its light axle radius is 3.5 mm, and its string is 120 cm long. The yo-yo rolls from rest down to the end of the string. Take $g = 10 \text{ m s}^{-2}$.

(a) Determine the radius R of the yo-yo disk.

(b) Calculate the magnitude of its linear acceleration using the results obtained in part (i).

(c) How long does it take to reach the end of the string?

(d) As it reaches the end of the string, calculate its (α) linear speed, (β) translational kinetic energy and (γ) rotational kinetic energy.

14. A spinning wheel, while rotating on its axis of rotation, has an angular momentum of $225 \text{ kg m}^2 \text{ s}^{-1}$. After applying a constant braking torque for 3.0 s, it slows and receives a new angular momentum of $100 \text{ kg m}^2 \text{ s}^{-1}$. Calculate the torque applied.

15. By considering the Earth as a uniform solid sphere with radius 6.38×10^{6} m and mass of 5.97×10^{24} kg calculate the magnitude of the

(i) angular momentum of the Earth in a circular orbit around the sun.

(ii) angular momentum of the Earth due to its rotation around an axis through the north and south poles.

Assume that the Earth is in a circular orbit around the sun with radius 1.50×10^{11} m. The Earth completes one rotation on its axis in 24 hours $= 86\ 400$ s and one orbit in 1 year $= 3.156\times10^{7}$ s. Since the radius of the Earth is much less than its orbit radius, it is very reasonable to model it as a particle for this calculation.

16. A flywheel with a moment of inertia $_0$ rotates without friction at an angular velocity ω_0 on a frictionless, vertical shaft of negligible rotational inertia. A second flywheel, which is at rest, having a moment of inertia $3I_0$ and initially not rotating, drops onto the first disk, as shown in the figure below. Due to friction between the flywheel surfaces, they very quickly reach the same angular velocity ω, and after which they spin together.

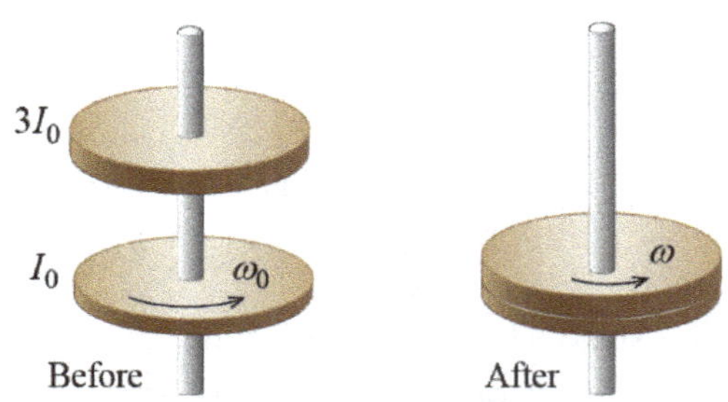

(i) Show that $\omega = \frac{1}{4}\omega_0$.

(ii) Determine the ratio of the final to the initial rotational kinetic energy. Hence, what fraction of the initial rotational energy is lost in the coupling of the flywheels?

17. A large wooden turntable in the shape of a flat disk with a radius of 2.00 m and a total mass of 86.0 kg initially rotates at 5.00 rad s^{-1} about a vertical axis through its centre. Subsequently, a flying bird with a mass of 7.0 kg gently lands on the outermost edge of the turntable (treat the bird as a particle).

    (i) Calculate the total moment of inertia of the turntable and the bird after the bird lands on the outermost edge of the turntable.

    (ii) Determine the final angular velocity of the turntable after the landing of the bird.

18. A uniform 12.0 kg solid wooden door of 0.90 m wide is mounted on a vertical hinge so that it swings without friction about a vertical axis. A 18.0 g bullet travelling horizontally with a speed of 500 m s^{-1} strikes the exact centre of the door perpendicularly opposite the hinges and embedding itself. The door swings open when the bullet strikes. The figure below shows the overhead view of a bullet striking a door. Take the moment of inertia of a door of mass M and width d rotating about its vertical hinged edge as $I = \frac{1}{3}Md^2$.

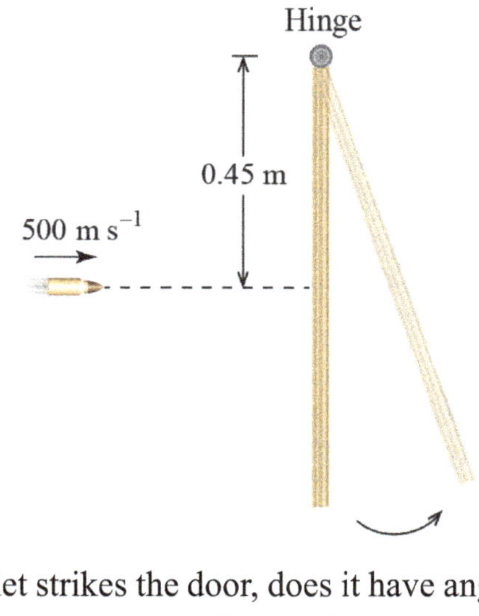

    (i) Before the bullet strikes the door, does it have angular momentum relative to the door's axis of rotation? If yes, calculate the angular momentum of the bullet about the central hinge just before it strikes the door. If not, explain why there is no angular momentum.

(ii) Does the torque created by the weight of the door act as an external torque, causing the door to rotate? If so, can we apply the conservation of angular momentum? Explain.

(iii) Calculate the angular speed of the door just after the bullet strikes it and embeds itself.

(iv) Determine the total kinetic energy of the door and bullet just after it strikes and embeds itself. Is the kinetic energy conserved in the impact? Justify your answer.

19. A knife-sharpening grindstone in the form of a solid disk has a mass of 15.0 kg and a diameter of 30.0 cm. When it starts rotating counterclockwise at 85 rad s^{-1}, a knife is pressed against the rim with a normal force of 50.0 N, and the grindstone comes to rest in 10.0 s. Ignore friction in the bearings on the rotating disk axle.

(i) Draw the free-body diagram for the grindstone when the knife is pressed against its rim.

(ii) Calculate the angular acceleration of the grindstone.

(iii) Determine the coefficient of friction between the knife and the grindstone.

20. A gyroscope is precessing about a vertical axis as shown in he figure.

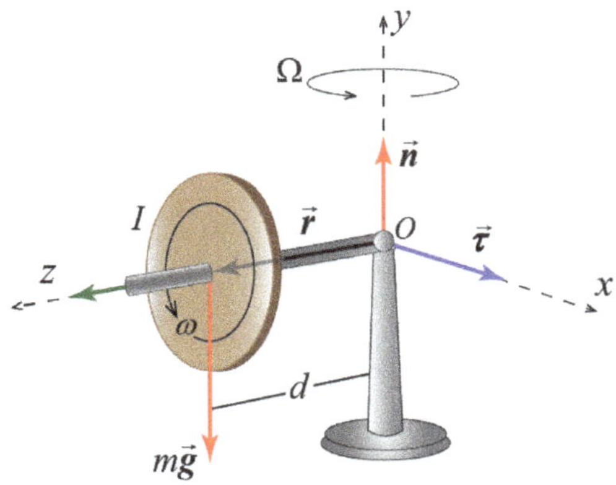

(i) Write down the expression for the precession angular speed Ω of the gyroscope in terms of the angular speed ω of the spinning flywheel, the mass m of the flywheel, the moment of inertia I about the axis of the spinning flywheel, the distance d from the pivot to the centre of gravity and the Earth's gravitational acceleration g.

(ii) Explain what happens to the precession angular speed if the following changes in the variables are made, with all other variables remaining the same.

(a) When the angular speed ω of the spinning flywheel is doubled.

(b) When the weight mg is doubled.

(c) When the moment of inertia I about the axis of the spinning flywheel is doubled.

(d) When the distance d from the pivot to the centre of gravity is doubled.

(iii) What happens if all four variables mentioned in (a) through (d) in part (ii) are doubled?

(iv) A certain gyroscope precesses at a rate of 0.6 rad s^{-1} when used on Earth. What would be its precession rate if it were taken to the Moon, where the acceleration due to the Moon's gravity is $\frac{g}{6}$? Assume the gyroscope m, d, I and ω are same on Earth and Moon.

Solution to Activities

Chapter 2

Activity 2.1

$$\frac{x^3x}{x^{-4}\left(x^2\right)^3} = \frac{x^3x^1}{x^{-4}x^6} = x^3x^1x^4x^{-6} = x^{(3+1+4-6)} = x^2$$

when $x = 2$, $\dfrac{x^3x}{x^{-4}\left(x^2\right)^3} = x^2 = 2^2 = 4$

Activity 2.2

(i) $3x^2 + 4 = 31$

$$x^2 = \frac{31-4}{3} = \frac{27}{3} = 9 = 3^2$$

$$x = 3$$

(ii) $x = u_0t + \frac{1}{2}at^2$

$$\frac{1}{2}at^2 = x - u_0t$$

$$a = \frac{2(x - u_0t)}{t^2}$$

Activity 2.3

The quadratic formula: $x = \dfrac{-b \pm \sqrt{b^2 - 4ac}}{2a}$

For $x^2 + x = 1 \Rightarrow x^2 + x - 1 = 0$, hence, $a = 1$, $b = 1, c = -1$

$$x = \frac{-1 \pm \sqrt{1^2 - 4 \times 1 \times (-1)}}{2 \times 1} = \frac{-1 \pm \sqrt{5}}{2} = \frac{-1 \pm 2.24}{2}$$

Since x is non-negative value,

$$x = \frac{-1 + 2.24}{2} = \frac{1.24}{2} = 0.62$$

Activity 2.4

(i) $5x - 4y = 1$ and $5x - 3y = 2$

When subtracting the 2nd equation from the 1st equation, we get

$$(5x - 4y) - (5x - 3y) = 1 - 2$$
$$-y = -1$$
$$y = 1$$

Substituting this in the 1st equation, we get

$$5x - 4 \times 1 = 1$$
$$5x = 5$$
$$x = 1$$

(ii) $3x + 2y = 12$ and $2x - y = 1$

Multiply the 2nd equation by 2 and then add it to the 1st equation, we get

$$(3x + 2y) + 2(2x - y) = 12 + 2$$
$$3x + 2y + 4x - 2y = 14$$
$$7x = 14$$
$$x = 2$$

Substituting this to the 1st equation, we get

$$3 \times 2 + 2y = 12$$
$$2y = 12 - 6$$
$$y = 3$$

Activity 2.5

Since x and y are directly proportional to each other, we have $\frac{x_1}{y_1} = \frac{x_2}{y_2} = k$.

When substituting the values, we get, $\frac{2}{8} = \frac{x}{12}, \quad \Rightarrow \quad x = 3$

Activity 2.6

According to Boyle's Law, if the temperature of an ideal gas is kept constant, its pressure, P, is inversely proportional to its volume, V. Thus, $P_1V_1 = P_2V_2$.

Here, $P_1 = 12\times10^4$ Pa, $V_1 = 8\times10^{-3}$ m^3 and $P_2 = 24\times10^4$ Pa. Hence,

$$V_2 = \frac{P_1V_1}{P_2} = \frac{12\times10^4\times8\times10^{-3}}{24\times10^4} \text{ m}^3 = 4\times10^{-3} \text{ m}^3$$

Activity 2.7

Let F be the force of gravity on an object that is at a distance d from the Earth's centre. Hence, $F \propto \frac{1}{d^2} \quad \Rightarrow \quad F = \frac{k}{d^2} \quad \Rightarrow \quad Fd^2 = k$.

When $d = 6.38\times10^6$ m, $F = 700$ N, thus, $k = 700\times\left(6.38\times10^6\right)^2$ N m^2.

When the astronaut is 6000 km $\left(= 6\times10^6 \text{ m}\right)$ from the Earth's surface, then

$$d = 6.38\times10^6 + 6\times10^6 = 12.38\times10^6 \text{ m}$$

$$F = \frac{k}{d^2} = \frac{700\times\left(6.38\times10^6\right)^2}{\left(12.38\times10^6\right)^2} = 185.91 \text{ N} \approx 186 \text{ N}$$

Hence,

Activity 2.8

$$N = N_0e^{-\alpha t} \quad \Rightarrow \quad \frac{N}{N_0} = e^{-\alpha t}$$

Take the natural logarithm of both sides of the equation and solve for t

$$\ln\left(\frac{N}{N_0}\right) = -\alpha t\text{, thus, } t = -\frac{1}{\alpha}\ln\left(\frac{N}{N_0}\right).$$

Activity 2.9

$$\sqrt{1.02} = (1.02)^{\frac{1}{2}} = (1+0.2)^{\frac{1}{2}} \approx 1+\frac{1}{2}\times 0.02 = 1.01$$

Activity 2.10

When $r = 0.20$ m and $h = 0.40$ m,

The total surface area $S = 2\pi r^2 + 2\pi rh = 2\pi\left(0.2^2 + 0.2\times 0.4\right) = 2\pi\times 0.12 = 0.75\ \text{m}^2$

The volume $V = \pi r^2 h = \pi\times 0.2^2\times 0.4 = 0.05\ \text{m}^3$.

Chapter 3

Activity 3.1

(i) All three digits are significant in the measurement 5.64 g, so it has 3 significant figures.

(ii) The first zero in the measurement 0.120 g is not significant since it is used only to place the decimal point. The digits 1, 2 and the final zero are significant, so it has 3 significant figures.

(iii) The first two zeros in the measurement 0.013 m are not significant since it is used only to place the decimal point. Digits 1 and 3 are significant, so there are 2 significant figures.

(iv) Both 8 and 3 are significant in the measurement 8.300×10^3 s. Further, the two zeros are also significant since they come after the decimal point. Therefore, it has 4 significant figures.

Activity 3.2

$$358.32\ \text{m} - 5.824\ \text{m} = 352.492 = 352.49$$

The subtraction obtained in the calculation cannot be more precise than the least precise of the two quantities. Thus, all these digits obtained in the calculation cannot be written in the answer. The result of the subtraction would be reported as

$$358.32\ \text{m} - 5.824\ \text{m} = 352.49$$

there are 5 significant figures in the result.

The density, $d = \dfrac{\text{mass}}{\text{volume}} = \dfrac{25.318}{18.2} = 1.3910989011\ \text{g cm}^{-3} = 1.39\ \text{g cm}^{-3}$.

The smallest number of significant figures in the calculation is the dimension $18.2\ \text{cm}^3$, having only three significant figures. Thus, the result should have only three significant figures and the density would be recorded as $1.39\ \text{g cm}^{-3}$.

Chapter 4

Activity 4.1

Convert metres to miles and seconds to hours

$$65.0\ \text{mi h}^{-1} = \left(65.0\ \frac{\cancel{\text{mi}}}{\text{h}}\right)\left(\frac{1.609\ \text{km}}{1\ \cancel{\text{mi}}}\right) = 104.6\ \text{km h}^{-1}$$

$$100.0\ \text{km h}^{-1} = \left(100.0\ \frac{\cancel{\text{km}}}{\text{h}}\right)\left(\frac{1\ \text{mi}}{1.609\ \cancel{\text{km}}}\right) = 62.2\ \text{mi h}^{-1}$$

Yes, the driver exceeding the speed limit ($100.0\ \text{km h}^{-1}$) and should slow down.

Activity 4.2

Volume of 2 cubic inches

$$= 2\text{ in}^3 = \left(2\text{ in}^3\right)\left(2.54\ \frac{\text{cm}}{1\text{ in}}\right)^3 = \left(2\ \cancel{\text{in}^3}\right)(2.54)^3\left(\frac{\text{cm}^3}{\cancel{\text{in}^3}}\right)^3 = 32.77\text{ cm}^3$$

Activity 4.3

Consider the equation $s = ut + \frac{1}{2}at^2$

Dimension of s, $[s] = \text{L}$

Dimension of ut, $[ut] = [u]\times[t] = \text{LT}^{-1}\text{T} = \text{L}$

Dimension of at^2, $\left[at^2\right] = [a]\times\left[t^2\right] = \text{LT}^{-2}\text{T}^2 = \text{L}$

Therefore, dimensions $[s] = [ut] = \left[at^2\right]$. Hence, we have the same dimensions on both sides of the equation $s = ut + \frac{1}{2}at^2$ and it is dimensionally correct.

Activity 4.4

Expression for the accelertion a with a dimensionless constant of proportionality k is,

$$a = kr^{\alpha}v^{\beta}$$

Dimension of r, $[r] = \text{L}$

Dimension of v, $[v] = \text{LT}^{-1}$

Dimension of a, $[a] = \text{LT}^{-2}$

When substituting the dimension to the equation $a = kr^{\alpha}v^{\beta}$, we get

$$\text{LT}^{-2} = \text{L}^{\alpha}\left(\text{LT}^{-1}\right)^{\beta} = \text{L}^{\alpha+\beta}\text{T}^{-\beta}$$

Equating the exponents of L and T, we get,

$1 = \alpha + \beta$ and $-2 = -\beta$

Hence, $\alpha = -1$ and $\beta = 2$

Hence, expression for acceleration a is $a = kr^{-1}v^2 = k\dfrac{v^2}{r}$

Chapter 5

Activity 5.1

The first result is more precise than the second because the spread in the first set of measurements is between 2.4 cm and 2.6 cm, whereas the spread in the second set of measurements is between 2.3 cm and 2.7 cm. That is, the measurements of the first experiment are less uncertain than those of the second.

Activity 5.2

(iii) only the accuracy of the readings.

Activity 5.3

(ii) precise but not accurate.

Activity 5.4

$$49 \text{ MSD} = 50 \text{ VSD}\,.$$

This means,

$$1 \text{ VSD} = \frac{49}{50} \text{ MSD}$$

Since one MSD is equal to 0.5 mm, we have

$$1 \text{ VSD} = \frac{49}{50} \times 0.5 \text{ mm}$$

The least count (LC) is defined as the difference between the length of one MSD and the length of one VSD. i.e.,

$$1 \text{ LC} = 1 \text{ MSD} - 1 \text{ VSD}$$

Therefore, the least count of the measuring instrument is

$$\text{LC} = \left(0.5 - \frac{49}{50} \times 0.5\right) \text{ mm} = 0.5\left(1 - \frac{49}{50}\right) \text{ mm} = 0.01 \text{ mm}$$

Activity 5.5

Measured value = MSR + CSR × LC − zero error

The final reading $= 12.5\text{ mm} + 16 \times 0.01\text{ mm} - 0.02\text{ mm} = 12.64\text{ mm}$

Chapter 6

Activity 6.1

(i)

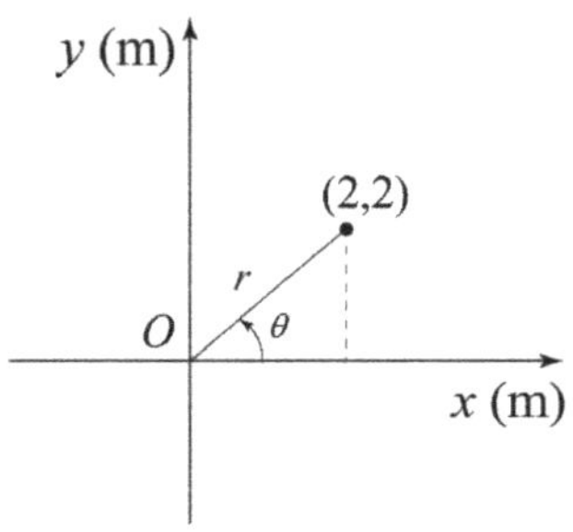

$$r = \sqrt{2^2 + 2^2} = 2\sqrt{2}\text{ m}, \qquad \tan\theta = \frac{2}{2} = 1, \qquad \Rightarrow \qquad \theta = \frac{\pi}{4} = 45^\circ$$

Thus the polar coordinates are $(2\sqrt{2}\text{ m}, 45^\circ)$

(ii)

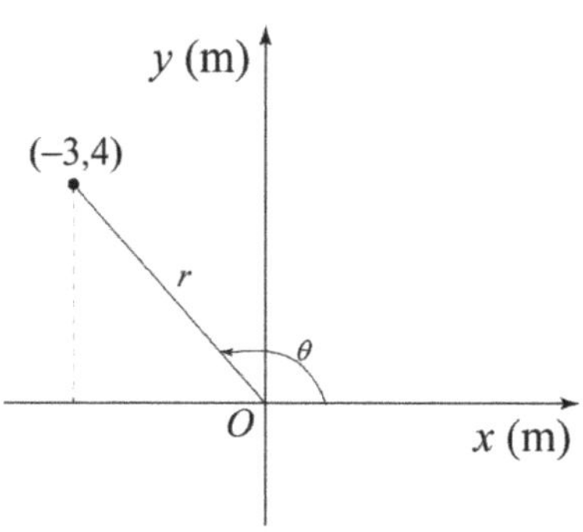

$r = \sqrt{(-3)^2 + 4^2} = 5$ m, $\tan(\pi - \theta) = \frac{4}{3}$, $\pi - \theta = 0.93$ radians,

$$\theta = \pi - 0.93 = 2.21 \text{ radians} = 126.62^\circ$$

Thus the polar coordinates are $(5 \text{ m}, 126.62^\circ)$.

Activity 6.2

(i) The vector diagram for the entire trip is a right angled triangle, and we can find the length of the hypotenuse by using the Pythagorean theorem.

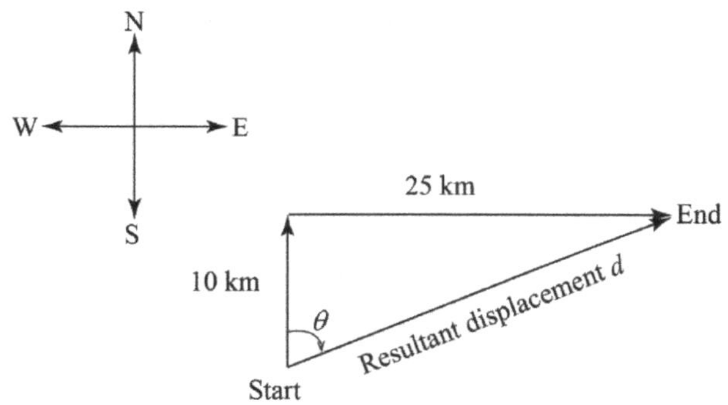

The length d of the resultant displacement vector (the distance from the starting point):

$d = \sqrt{(10.0)^2 + (25.0)^2}$ km $= 26.9$ km (Magnitude of displacement vector)

$\tan\theta = \frac{25}{10} = 2.5$, thus $\theta = 68.2^\circ$ (direction of displacement vector)

(ii) The magnitude of the resultant displacement is the distance d 26.9 km, and the direction is $\theta = 68.2^\circ$ east of north.

Activity 6.3

We could start drawing the vector diagram with either the 3 N force or the 4 N force (it does not matter which). Let start with the 3 N force.

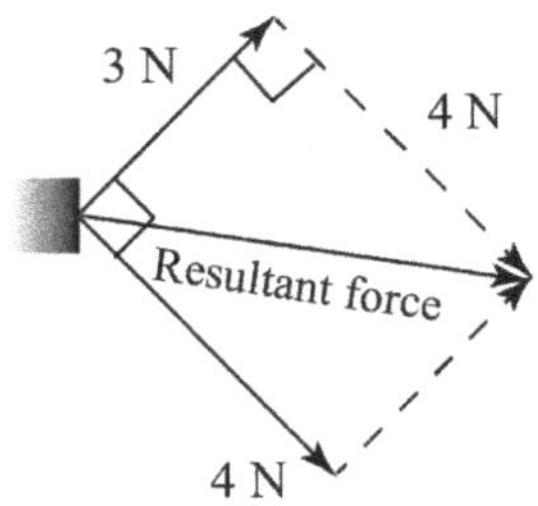

The resultant force R is $R=\sqrt{3^2+4^2}\text{ N}=5\text{ N}$

Activity 6.4

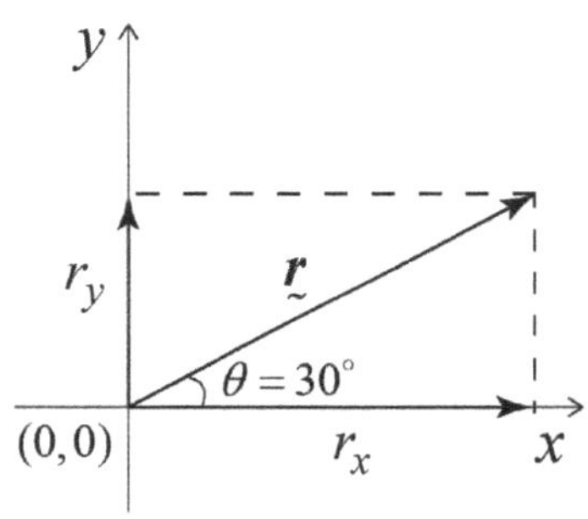

The x and the y components of the displacement vector $\underset{\sim}{r}$ lying in the xy plane are given by $r_x=r\cos\theta$ and $r_y=r\sin\theta$. In this case, $r=20.0\text{ m}$ and $\theta=30$. Thus,

$$r_x=20.0\cos 30^\circ\text{ m}=17.3\text{ m} \quad\text{and}\quad r_y=20.0\sin 30^\circ\text{ m}=10.0\text{ m}$$

Activity 6.5

The x and y component of the vector $\vec{P}$ is $P_x=30.0\text{ m}$ and $P_y=50.0\text{ m}$ respectively.

(i) The magnitude of the vector $\vec{P}$ is

$$P=\sqrt{P_x^2+P_y^2}=\sqrt{(30.0)^2+(50.0)^2}\text{ m}=58.3\text{ m}$$

(ii) The angle θ between the direction of the vector $\vec{P}$ and the positive direction of the axis.

$$\tan\theta = \frac{P_y}{P_x} = \frac{50.0}{30.0} \quad \Rightarrow \quad \theta = 59^\circ$$

The vector is in the first quadrant.

Activity 6.6

The x and the y components of the position vector having the polar coordinates (r, θ) is given by $r_x = r\cos\theta$ and $r_y = r\sin\theta$. Thus

$$r_x = 15.0\cos 150^\circ \text{ m} = -13.0 \text{ m} \quad \text{and} \quad r_y = 15.0\sin 150^\circ \text{ m} = 7.5 \text{ m}$$

Activity 6.7

Use coordinates for which the $+x$ axis is horizontal parallel to the floor and the + direction is vertically upward.

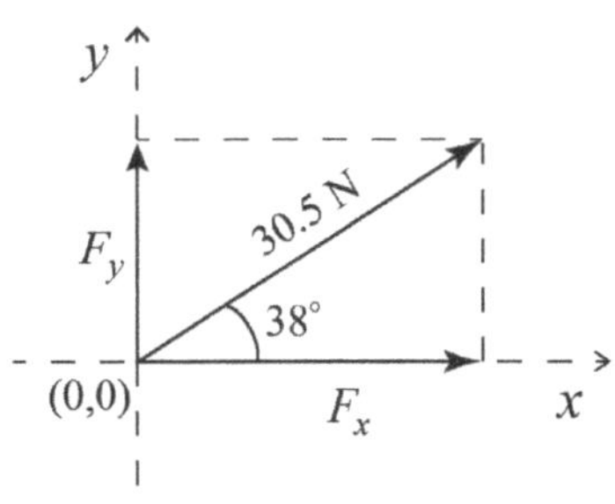

(i) The force given horizontally to pull the box forward

$$F_x = 30.5\cos 38^\circ \text{ N} = 24.0 \text{ N}$$

(ii) The force given vertically upward

$$F_y = 30.5\sin 38^\circ \text{ N} = 18.8 \text{ N}$$

Activity 6.8

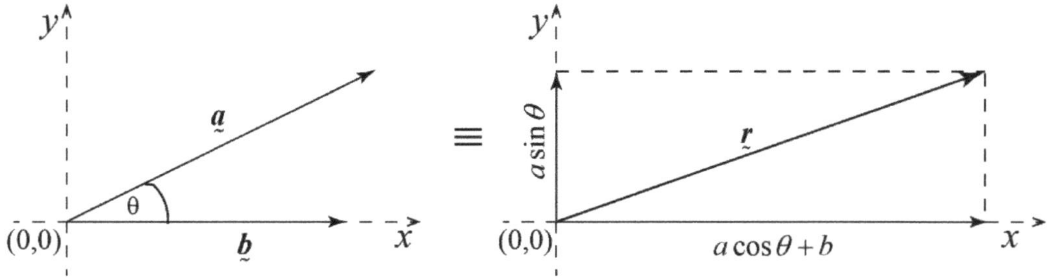

The sum of the components of vectors $\underset{\sim}{a}$ and $\underset{\sim}{b}$ along the x axis is: $a\cos\theta + b$.

Similarly the sum of the components along the y axis is: $a\sin\theta + 0 = a\sin\theta\, a\sin\theta$

Thus, if r is the length of the resultant vector $\underset{\sim}{r}$, then

$$\begin{aligned} r &= \sqrt{(a\cos\theta + b)^2 + (a\sin\theta)^2} \\ &= \sqrt{a^2\cos^2\theta + 2ab\cos\theta + b^2 + a^2\sin^2\theta} \\ &= \sqrt{a^2\left(\cos^2\theta + \sin^2\theta\right) + 2ab\cos\theta + b^2} \\ &= \sqrt{a^2 + b^2 + 2ab\cos\theta} \end{aligned}$$

Activity 6.9

The distance that the student would have to walk due north is $4.0\sin 30^\circ = 2.0$ km

The distance that the student would have to walk due east is $4.0\cos 30^\circ = 3.5$ km

Activity 6.10

Let $\underset{\sim}{r}$ be the resultant vector and its x, y components be r_x and r_y respectively.

(i) The resultant vector along the x axis:

$$\begin{aligned} r_x &= -|\underset{\sim}{a}|\sin 45^\circ + |\underset{\sim}{b}|\sin 45^\circ + 0 + |\underset{\sim}{d}|\cos 20^\circ \\ &= -20.0\sin 45^\circ + 35.0\sin 45^\circ + 20.0\cos 20^\circ = 29.4 \text{ unit} \end{aligned}$$

The resultant vector along the y axis:

$$r_y = |\underset{\sim}{\boldsymbol{a}}|\cos 45^\circ + |\underset{\sim}{\boldsymbol{b}}|\cos 45^\circ - |\underset{\sim}{\boldsymbol{c}}| - |\underset{\sim}{\boldsymbol{d}}|\sin 20^\circ$$

$$= 20.0\cos 45^\circ + 35.0\cos 45^\circ - 10.0 - 20.0\sin 20^\circ = 22.1 \text{ unit}$$

(ii) The magnitude of the resultant vector $\underset{\sim}{\boldsymbol{r}}$:

$$r = \sqrt{r_x^2 + r_y^2} = \sqrt{(29.4)^2 + (22.1)^2} = 36.8 \text{ unit}$$

Direction of the resultant vector $\underset{\sim}{\boldsymbol{r}}$:

$$\tan\theta = \frac{22.1}{29.4} \quad \Rightarrow \quad \theta = 36.9^\circ$$

Chapter 7

Activity 7.1

When $t = 2.0$ s, $x = 10t^2 = 10\times(2.0)^2 = 40.0$ m.

When $t = 5.0$ s, $x = 10t^2 = 10\times(5.0)^2 = 250.0$ m.

Thus the average velocity $= \dfrac{x_f - x_i}{t_f - t_i} = \dfrac{250.0 - 40.0}{5.0 - 2.0} = \dfrac{210.0}{3.0} = 70.0 \text{ m s}^{-} =$

Activity 7.2

(i) The total distance (say d) covered by the runner is $d = 300 + 100 = 400$ m

(ii) The total displacement (say Δx) of the runner is $\Delta x = x_f - x_i = 300 - 100 = 200$ m east or $+x$ direction.

(iii) Average velocity (say $v_{avg,x}$) is $v_{avg,x} = \dfrac{\Delta x}{\Delta t} = \dfrac{200}{320} = 0.625 \text{ m s}^{-1}$

(iv) Average speed (say v_{avg}) is $v_{avg} = \dfrac{\Delta x}{\Delta t} = \dfrac{800}{320} = 2.50 \text{ m s}^{-1}$

Activity 7.3

$v_{x,f} = 40 \text{ m s}^{-1}$, $v_{x,i} = 20 \text{ m s}^{-1}$.

Thus change in velocity $\Delta v_x = v_{x,f} - v_{x,i} = (40 - 20) \text{ m s}^{-1}$, the change in time $\Delta t = 5 \text{ s}$

Therefore, the average acceleration $a_{avg,x} = \dfrac{\Delta v_x}{\Delta t} = \dfrac{20}{5} \text{ m s}^{-2} = 4 \text{ m s}^{-2}$.

This means that the car's velocity increased by 4 metres per second every second, on average, over the 5-second period.

Activity 7.4

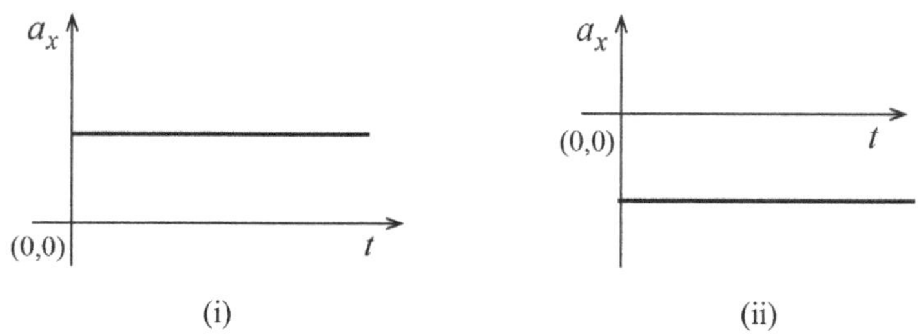

(i) Acceleration is positive and constant.

(ii) Acceleration is negative and constant.

Activity 7.5

Take the origin of coordinates to be at the initial position of the car. i.e. at $t = 0$, $u = 52.2 \times \dfrac{1000}{3600} \text{ m s}^{-1} = 14.5 \text{ m s}^{-1}$, $a = 3.0 \text{ m s}^{-2}$.

(i) The acceleration is constant during the 6.0 s time interval

The velocity after 6.0 s is,

$$v = u + at = 14.5 + 2.0 \times 6.0 = 26.5 \text{ m s}^{-1} = 26.5 \times \frac{3600}{1000} \text{ km h}^{-1} = 95 \text{ km h}^{-1}$$

(ii) The distance the car has travelled during the time interval 6.0 s is,

$$s = ut + \frac{1}{2}at^2 = 14.5 \times 6.0 + \frac{1}{2} \times 2 \times (6.0)^2 \text{ m} = 123 \text{ m}$$

Activity 7.6

The fishing boat starts from rest. Therefore, at $t = 0$, $u = 0$. The acceleration of the boat $a = 2.50 \text{ m s}^{-2}$, the final velocity 15.0 m s^{-1}.

(i) Use the equation $v = u + at$ to determine the time t, thus

$$15.0 = 0 + 2.50 \times t \qquad \Rightarrow \quad t = \frac{15.0}{2.50} = 6.00 \text{ s}.$$

(ii) Use the equation $v^2 = u^2 + 2as$ to determine the distance travelled. Thus,

$$(15.0)^2 = 0^2 + 2 \times 2.50 \times s \qquad \Rightarrow \quad s = \frac{(15.0)^2}{2 \times 2.50} = 45 \text{ m}.$$

Activity 7.7

Let the touching point on the runway be the origin, where $x = 0$. The initial velocity $u = 70.0 \text{ m s}^{-1}$. After it has travelled 1000 m its velocity (final velocity) $v = 7.0 \text{ m s}^{-1}$. Travel time t is not given. Thus we use the equation $v^2 = u^2 + 2as$. Therefore,

$$(7.0)^2 = (70.0)^2 + 2a \times 1000$$

$$\Rightarrow \quad a = \frac{(7.0)^2 - (70.0)^2}{2 \times 1000} = -2.4 \text{ m s}^{-2}$$

The negative sign of the answer indicates that the plane's velocity which is in the direction of motion, North is decreasing. The acceleration has a magnitude of 2.4 m s^{-2} and its direction is opposite the direction of motion, i.e. South.

Activity 7.8

(i) Accelerates uniformly for first 10 s from rest, then travels at constant velocity for the next 15 s, and then decelerates for the next 25 s and comes to rest.

Acceleration for the first 10 s is $= \dfrac{20 - 0}{10 - 0} = 2 \text{ m s}^{-2}$

Next 15 s travels at a constant velocity, thus zero acceleration

Next, 25 s the acceleration is $= \dfrac{0-20}{50-25} = -0.8 \text{ m s}^{-2}$, i.e. negative acceleration, i.e. means the car decelerates.

Next 4 s, the velocity is zero, and the car does not move, acceleration is zero.

(ii) The total distance (d) travelled = the area under the curve

Thus, $d = \dfrac{1}{2} \times (20+50) \times 20 = 700 \text{ m}$

(iii)

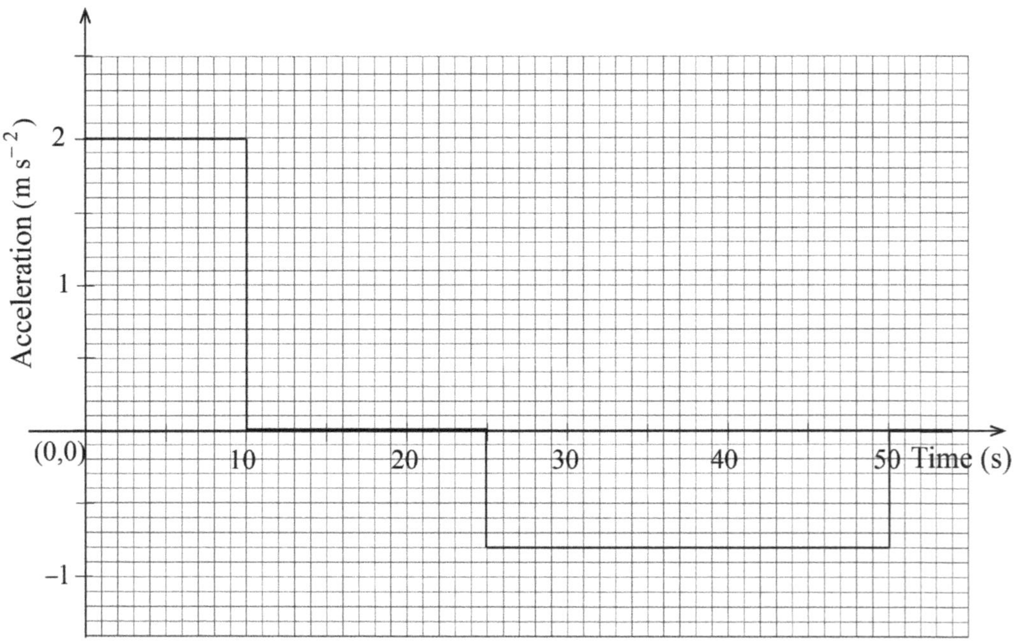

Activity 7.9

Since the motion is downward, we choose a vertical y axis and the upward direction to be positive. Let the edge of the tall building be the origin. Further the ball is dropped, thus the magnitude of the initial velocity $u = 0$. Motion is under the acceleration of gravity and is downward direction, i.e. in the negative y direction.

Hence, $a_y = -g = -9.8 \text{ m s}^{-2}$, use the equation $s = ut + \dfrac{1}{2}at^2$ in the downward direction.

$$y = 0 \times t + \frac{1}{2} \times (-9.8) \times t^2 \quad \Rightarrow \quad y = -4.9t^2$$

Use the equation, $v = u + at$ in the downward direction.

$$v_y = 0 + (-9.8) \times t \quad \Rightarrow \quad v_y = -9.8t$$

Therefore, the position and velocity

at $t = 1$ s

$$y_1 = -4.9 \text{ m}, \; v_{1y} = -9.8 \text{ m s}^{-1}$$

at $t = 2$ s

$$y_2 = -19.6 \text{ m}, \; v_{2y} = -19.6 \text{ m s}^{-1}$$

at $t = 3$ s

$$y_3 = -44.1 \text{ m}, \; v_{3y} = -29.4 \text{ m s}^{-1}$$

at $t = 4$ s

$$y_4 = -78.4 \text{ m}, \; v_{4y} = -39.2 \text{ m s}^{-1}$$

The values of position y and the velocity v_y are negative because we have chosen the positive y axis to be upward direction.

Activity 7.10

(i) The motion is upward, and we choose a vertical y axis and the upward direction to be positive. Let the hand of the person be the origin. Further, the ball is thrown upward, the magnitude of the initial velocity $u = 14.0 \text{ m s}^{-1}$. Motion is under the acceleration of gravity and is downward direction, i.e. in the negative y direction.

Initial velocity $u = 14.0 \text{ m s}^{-1}$, the maximum height $s = h_{max}$, velocity at the maximum height $v = 0$, Apply the equation $v^2 = u^2 + 2as$, in the upward direction, from the initial point to the point of maximum height. Thus,

$$0 = (14.0)^2 + 2 \times (-9.8) \times h_{max} \quad \Rightarrow \quad h_{max} = 10.0 \text{ m} \;\; h_{max} = 10.0 \text{ m}$$

(ii) Let t_{max} be the time to reach the maximum height. Apply the equation $v = u + at$ in the upward direction, from the initial point to the point of maximum height.

$$0 = 14.0 + (-9.8) \times t_{max} \qquad \Rightarrow \quad t_{max} = 1.43 \text{ s}$$

(iii) Let t be the time to return to the person's hand. Apply the equation $s = ut + \frac{1}{2}at^2$ from the initial point to the point of reaching the person's hand. Thus the displacement is zero.

$$0 = 14.0 \times t + \frac{1}{2}(-9.8)t^2 \qquad \Rightarrow \quad t = 2.86 \text{ s} \text{ or } \quad t = 0$$

$t = 0$ corresponds to the initial position. Thus, the time taken to return to the person's hand is 2.86 s.

(iv) Let v be the velocity when it returns to the person's hand. Apply the equation $v^2 = u^2 + 2as$ from the initial point to the point of reaching the person's hand. Thus the displacement is zero.

$$v^2 = (14.0)^2 + 2 \times (-9.8) \times 0 \qquad \Rightarrow \quad v = \pm 14.0 \text{ m s}^{-1},$$

The velocity at the initial position is $v = 14 \text{ m s}^{-1}$. Thus, when it returns the velocity of the ball at the person's hand is $v = -15 \text{ m s}^{-1}$, the negative sign indicates the direction of the velocity is downward.

(v) Let t' be the time to pass the position 7.0 m above the person's hand. Apply the equation $s = ut + \frac{1}{2}at^2$ from the initial point to the point of the position 7.0 m above the person's hand.

$$7.0 = 14.0 \times t + \frac{1}{2} \times (-9.8) \times t^2 \qquad \Rightarrow \quad 4.9t^2 - 14.0t + 7.0 = 0$$

This is a quadratic equation in t. Thus

$$t = \frac{14.0 \pm \sqrt{(14.0)^2 - 4 \times 4.9 \times 7.0}}{2 \times 4.9} = 2.21 \text{ s} \quad \text{ or } 0.65 \text{ s}$$

Here both solutions are valid. At the position 7.0 m above the person's hand, when it goes up, at $t = 0.65$ s it will pass, and at $t = 2.21$ s when it returns, it will pass.

(f)

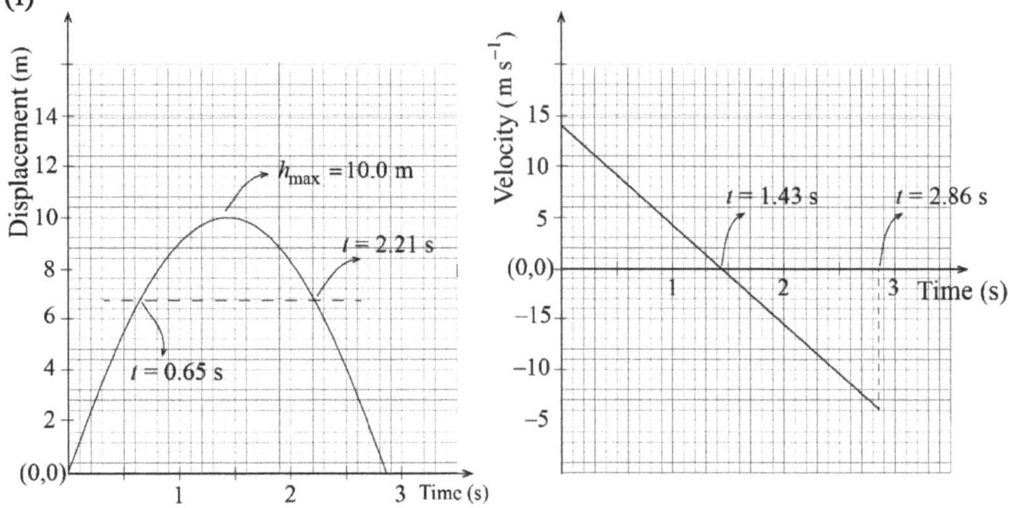

Activity 7.11

Let A be the coconut that falls from the tall tree and let B be the coconut from the shorter tree. Let the origin be at the ground. Thus $y_A = 2y_B$ and intial velocities of both coconunts will be zero. Use the equation $v^2 = u^2 + 2as$.

(i) Use the equation $v^2 = u^2 + 2as$. Let v_B the velocity of the coconut from the shorter when it reaches the ground. Thus,

$$v_A^2 = v^2 = 0 + 2\times(-g)\times y_A \text{ and } v_B^2 = 0 + 2\times(-g)\times y_B.$$

Therefore, $\left(\frac{v_A}{v_B}\right)^2 = \frac{y_A}{y_B} = 2 \Rightarrow |v_B| = \frac{v}{\sqrt{2}}$. Hence, the speed v_B is $\frac{v}{\sqrt{2}}$.

(ii) Use the equation $s = ut + \frac{1}{2}at^2$. Let t_A be the time taken for the coconut from the taller tree to reach the ground. Thus,

$$y_A = 0 + \frac{1}{2}(-g)\times t_A^2 \text{ and } y_B = 0 + \frac{1}{2}\times(-g)\times t^2.$$

Therefore, $\left(\frac{t}{t_A}\right)^2 = \frac{y_B}{y_A} = \frac{1}{2} \Rightarrow t_A = \sqrt{2}t$. Hence, the time t_A is $t_A = \sqrt{2}t$.

Activity 7.12

Let the velocity of the motorcycle rider with respect to the Earth be $\underset{\sim}{v}_{M,E} = 65 \text{ km h}^{-1} \rightarrow$ along the eastward direction. i.e. $\underset{\sim}{v}_{E,M} = 65 \text{ km h}^{-1} \leftarrow$ along the westward direction. Let the velocity of the wind with respect to the Earth be $\underset{\sim}{v}_{W,E} = 20 \text{ km h}^{-1} \leftarrow$, along the westward direction. Let the velocity of the wind felt by the motorcycle rider be $\underset{\sim}{v}_{W,M}$. Therefore,

$$\underset{\sim}{v}_{W,M} = \underset{\sim}{v}_{W,E} + \underset{\sim}{v}_{E,M}$$

$$\Rightarrow \quad \underset{\sim}{v}_{W,M} = (20 + 65) \text{ km h}^{-1} \leftarrow \quad = 85 \text{ km h}^{-1} \leftarrow$$

Therefore, the velocity of the wind felt by the motorcycle rider is

$$\Rightarrow \quad \underset{\sim}{v}_{W,M} = 85 \text{ km h}^{-1} \leftarrow \text{ along the westward direction.}$$

Chapter 8

Activity 8.1

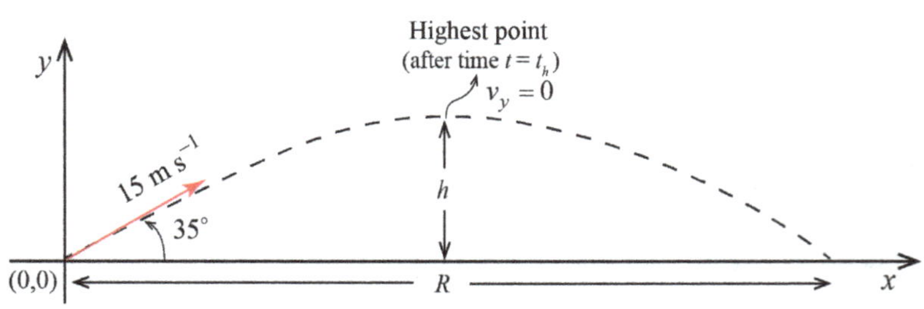

(i) Use the equation $v^2 = u^2 + 2as$ for the projectile motion in the vertical direction from the initial point to the highest point.

$$0 = (15.0 \sin 35.0°)^2 + 2(-10)h$$

$$\Rightarrow \quad h = \frac{\left(15.0 \sin 35.0^\circ\right)^2}{2 \times 10} = 3.70 \text{ m}$$

Therefore, the maximum height that the ball reaches above the ground is 3.7 m.

(ii) Use the equation $s = ut + \frac{1}{2}at^2$ for the projectile motion in the vertical direction from the initial point to the point where the ball hit the ground.

$$0 = \left(15.0 \sin 35.0^\circ\right)t_f + \frac{1}{2}(-10)t_f^2$$

$$\Rightarrow \quad 5t_f^2 - 8.6t_f = 0 \quad \Rightarrow \quad t_f = 0 \quad \text{or} \quad t_f = \frac{8.6}{5} = 1.72 \text{ s}$$

Therefore, the time it takes for the football to hit the ground is 1.72 s (Note that $t_f = 0$ is for the solution for the initial point).

(iii) Use the equation $s = ut + \frac{1}{2}at^2$ for the projectile motion in the horizontal direction from the initial point to the point where the ball hit the ground.

$$R = \left(15.0 \cos 35.0^\circ\right) \times 1.72 + 0 = 21.1 \text{ m}$$

Therefore, the horizontal distance the ball travels before hitting the ground is 21.1 m

(iv) At the highest point, there is no vertical component to the velocity. There is only a horizontal component to the velocity along the direction of +ve x axis and its magnitude is $15.0 \cos 35^\circ = 12.3 \text{ m s}^{-1}$.

(v) The acceleration is constant throughout the motion including at the highest point with the magnitude of 9.8 m s^{-2} (in this question it has been taken as 10 m s^{-2}) and acts downwards direction.

Activity 8.2

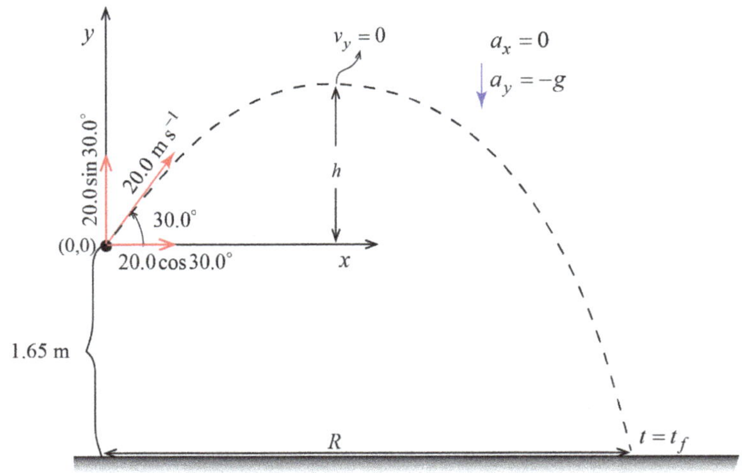

(vi) Use the equation $v^2 = u^2 + 2as$ for the projectile motion in the vertical direction from the initial point to the highest point.

$$0 = (20.0 \sin 30.0^\circ)^2 + 2(-10)h \qquad \Rightarrow \quad h = \frac{(20.0 \sin 30.0^\circ)^2}{2 \times 10} = 5.0 \text{ m}$$

Therefore, the maximum height that the stone reaches above the ground is $(1.65 + 5.0)$ m $= 6.65$ m

(vii) Use the equation $s = ut + \frac{1}{2}at^2$ for the projectile motion in the vertical direction from the initial point to the point where the stone hit the ground.

$$-1.65 = (20.0 \sin 30.0^\circ)t_f + \frac{1}{2}(-10)t_f^2$$

$$\Rightarrow \quad 5t_f^2 - 10.0t_f - 1.65 = 0 \quad \Rightarrow \quad t_f = \frac{10.0 \pm \sqrt{(-10.0)^2 - 4 \times 5 \times (-1.65)}}{2 \times 5}$$

$$\Rightarrow \quad t_f = 2.15 \text{ s}$$

(viii) Use the equation $s = ut + \frac{1}{2}at^2$ for the projectile motion in the horizontal direction from the initial point to the point where the stone hit the ground.

$$R = \left(20.0\cos 30.0^\circ\right) \times 2.15 + 0 = 37.2 \text{ m}$$

Activity 8.3

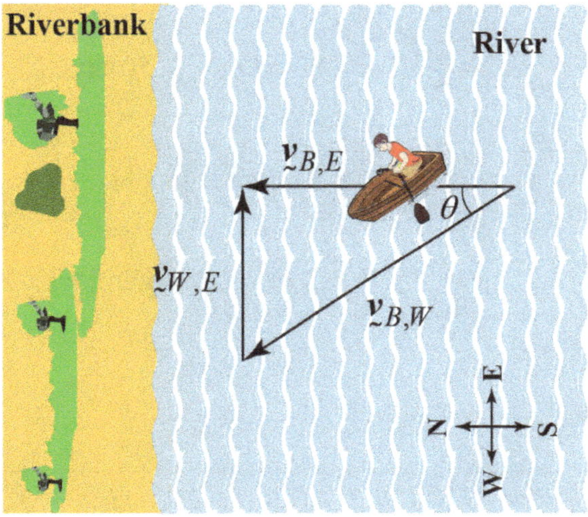

Let, $\underset{\sim}{v}_{B,W}$ be the velocity of the boat relative to water, $\underset{\sim}{v}_{W,E}$ be the velocity of the water relative to Earth, $\underset{\sim}{v}_{B,E}$ be the velocity of the boat relative to Earth.

We have $\left|\underset{\sim}{v}_{B,W}\right| = v_{B,W} = 2.5 \text{ m s}^{-1}$, $\left|\underset{\sim}{v}_{W,E}\right| = v_{W,E} = 1.2 \text{ m s}^{-1}$.

According to the vector diagram, $\underset{\sim}{v}_{B,E} = \underset{\sim}{v}_{B,W} + \underset{\sim}{v}_{W,E}$

Therefore, using Pythagorean theorem, the magnitide of the velocity of boat relative to the Earth will be $\left|\underset{\sim}{v}_{B,E}\right| = v_{B,E} = \sqrt{v_{B,W}^2 - v_{W,E}^2}$, thus,

$$v_{B,E} = \sqrt{2.5^2 - 1.2^2} = 2.8 \text{ m s}^{-1} = 2.2 \text{ m s}^{-1}, \theta = \tan^{-1}\left(\frac{v_{W,E}}{v_{B,E}}\right) = \tan^{-1}\left(\frac{1.2}{2.2}\right) = 28.6^\circ$$

The boat is moving with the magnitide of the velocity 2.2 m s^{-1} in the direction 28.6° west of north relative to the Earth. The boat must be aimed towards upstream.

Chapter 9

Activity 9.1

(i) $\sum F_x$ is the sum of the x components and $\sum F_y$ is the sum of the y components.

Thus,
$$R_x = \sum F_x = 200\cos 60^\circ + 150\cos 30^\circ = 229.9 \text{ N}$$
$$R_y = \sum F_y = 200\sin 60^\circ + 150\sin 30^\circ = 273.2 \text{ N}$$

(ii) $\left|\vec{R}\right| = R = \sqrt{R_x^2 + R_y^2} = \sqrt{229.9^2 + 273.2^2} = 357 \text{ N}$

$\theta = \tan^{-1}\left(\frac{R_y}{R_x}\right) = \tan^{-1}\left(\frac{273.2}{229.9}\right) = 49.9^{\circ\,o}$ counterclockwise from the x axis.

Activity 9.2

$R = \sqrt{F_1^2 + F_2^2 + 2F_1F_2\cos\theta} = \sqrt{20.0^2 + 30.0^2 + 2\times 20.0\times 30.0\cos 60.0^\circ} = 43.6 \text{ N}$

$\alpha = \tan^{-1}\left(\frac{F_1\sin\theta}{F_2 + F_1\cos\theta}\right) = \tan^{-1}\left(\frac{20.0\sin 60.0^\circ}{30.0 + 20.0\cos 60.0^\circ}\right) = 23.4^{\circ\,o}$ counterclockwise from the x axis, i.e. direction of the force of magnitude 30.0 N.

Activity 9.3

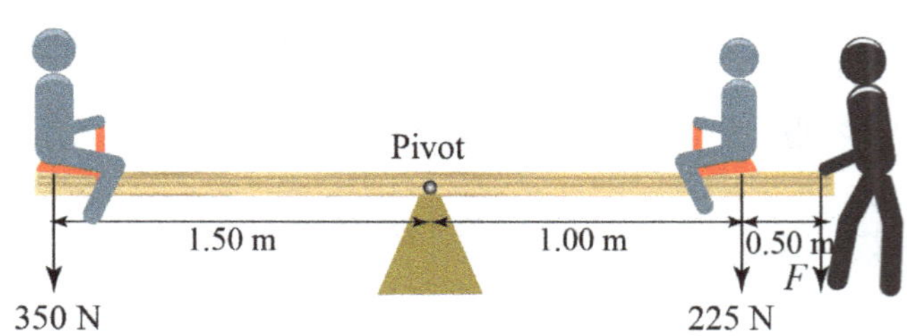

Since the system is in balanced state, the sum of the clockwise moments about any point is equal to the sum of the counterclockwise moments about the same point. Let take the moments about the pivot.

$$350 \times 1.50 = 225 \times 1.00 + F \times 1.50$$.

Therefore, $F = \dfrac{525 - 225}{1.5} = 200$ N : downwards direction.

Activity 9.4

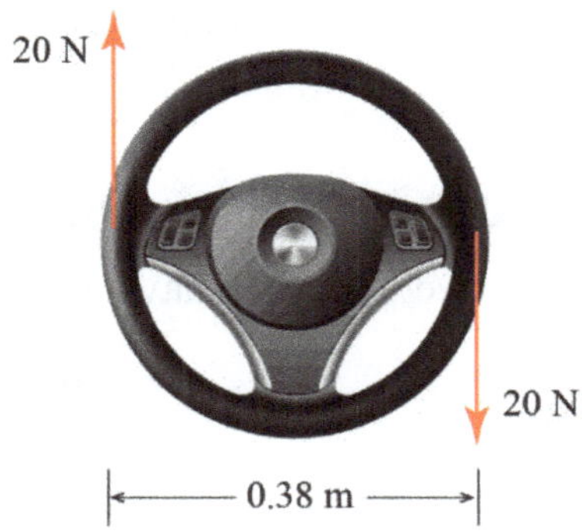

The magnitude of the moment or torque produced by a couple force is equal to the product of one of the forces and the perpendicular distance between the two forces. Thus,

$$\tau = 20 \times 0.38 = 7.6 \text{ N m}$$

Activity 9.5

Imagine that you are scooping up a shovelful of sand and swinging the shovel toward the side. Both the sand and the shovel are moving. However, during the process, suddenly the motion of the shovel stops, but the sand continues to move because of its inertia and it slides forward off the shovel, to be pulled down to the ground by gravity. The sand does not stop moving forward when the forward force applied to the shovel is removed. This procedure works best with fairly dry sand. However, wet sand, which might be sticky, tends to cling to the shovel.

Activity 9.6

Since the motion is along only the x axis, we can simplify each situation by writing the Newton's second law for x components only.

In figure (a) only one force is acting on the disc. Thus, if we apply the Newton's second law along the +ve x axis, we have $F_1 = ma_x$.

Hence, $4.0 = 0.160 \times a_x \qquad \Rightarrow \quad a_x = \dfrac{4.0}{0.160} = 25 \text{ m s}^{-2}$.

The positive answer for a_x indicates that the acceleration is in the positive direction of the x axis.

In figure (b) two forces F_1 and F_2 are acting on the disc. Thus, if we apply the Newton's second law along the +ve x axis we have $F_1 - F_2 = ma_x$.

Hence, $4.0 - 2.4 = 0.160 \times a_x \qquad \Rightarrow \quad a_x = \dfrac{1.6}{0.160} = 10 \text{ m s}^{-2}$.

The positive answer for a_x indicates that the acceleration is in the positive direction of the x axis.

In figure (c) three forces F_1, F_2 and F_3 are acting on the disc. Also the force F_3 is acting at an angle 60°. Thus, if we apply Newton's second law along the +ve x axis we have $F_1 + F_3 \cos 60^\circ - F_2 = ma_x$.

Hence, $4.0 + 1.6 \times 0.5 - 2.4 = 0.160 \times a_x \qquad \Rightarrow \quad a_x = \dfrac{2.4}{0.160} = 15 \text{ m s}^{-2}$.

The positive answer for a_x indicates that the acceleration is in the positive direction of the x axis.

Activity 9.7

The free-body diagrams

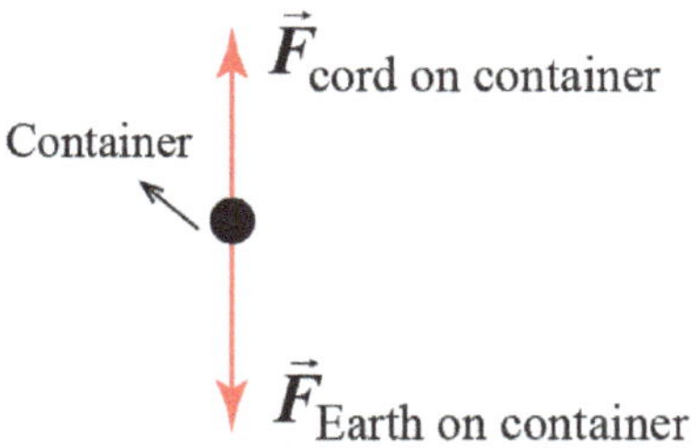

(i)

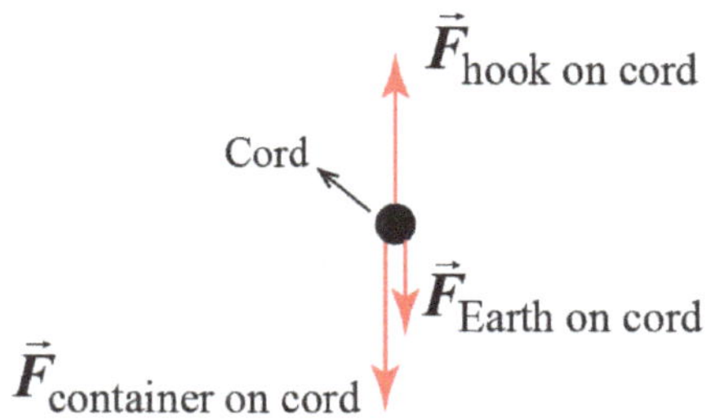

(ii)

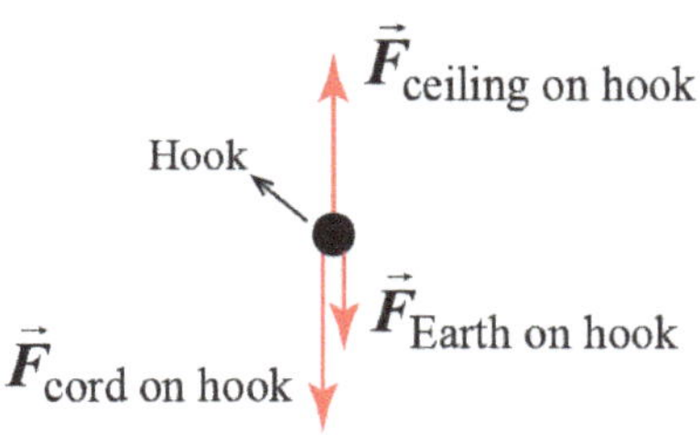

(iii)

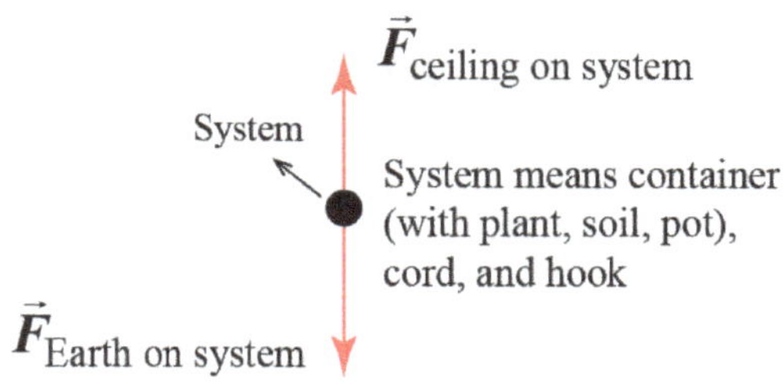

(iv)

Activity 9.8

(i) Person's weight on Earth is $mg = 70 \times 9.8 = 686\ \text{N}$.

(ii) Person's weight on Earth's Moon is $mg = 70 \times 1.6 = 112\ \text{N}$.

(iii) Person's weight on Venus is $mg = 70 \times 8.9 = 623\ \text{N}$.

(iv) Person's weight on Mars is $mg = 70 \times 3.7 = 259\ \text{N}$.

(v) Person's weight on Mars is $mg = 70 \times 3.7 = 259\ \text{N}$.

Activity 9.9

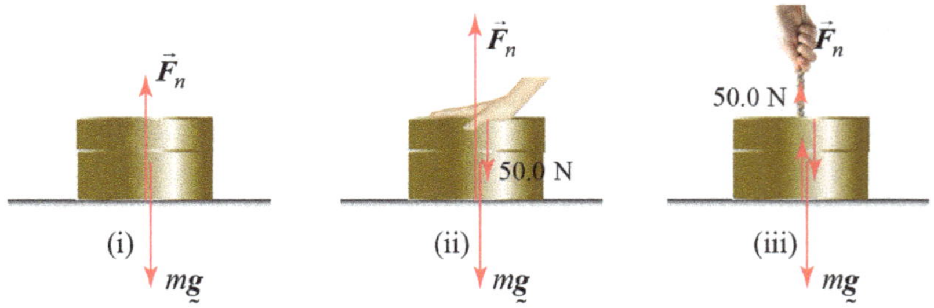

(i) The weight of the container $mg = 10.0 \times 10 = 100$ N and acts downwards direction. The only other force on the container is the normal force exerted by the table on the container and acts upward. Take the upward direction as the positive y axis. Then the net force $\sum F_y$ acting on the container would be $\sum F_y = F_n - mg$. Since the container is at rest, the acceleration $a = 0$, and the net force must be zero. Hence, Newton's second law gives

$$F_n - mg = 0.$$

Therefore, $F_n = 10.0 \times 10 = 100$ N. Hence, the magnitude of the normal force is equal to the weight of the container.

(ii) In this case, three forces act on the container. The pushing force in the downward direction, the weight, and the normal force. Then the net force $\sum F_y$ acting on the container would be $\sum F_y = F_n - mg - 50.0$. Since the container is at rest, the acceleration $a = 0$ and the net force must be zero. Hence, Newton's second law gives

$$F_n - mg - 50.0 = 0.$$

Therefore, $F_n = 10.0 \times 10 + 50.0 = 150$ N. Hence, the magnitude of the normal force is greater than that obtained in part (i). When pushing the container downward, the table pushes back with more force. Therefore, the normal force is not equal to the weight of the container.

(iii) In this case also, three forces act on the container. The pulling force in the upward direction, the weight, and the normal force. Then the net force $\sum F_y$ acting on the container would be $\sum F_y = F_n - mg + 50.0$. Since the container is at rest, the acceleration $a = 0$ and the net force must be zero. Hence, Newton's second law gives

$$F_n - mg + 50.0 = 0.$$

Therefore, $F_n = 10.0 \times 10 - 50.0 = 50.0$ N. Hence, the magnitude of the normal force is less than that obtained in parts (i) and (ii). When the container is pulled upward, the table does not push back against the full weight of the container. Therefore, the normal force is not equal to the weight of the container.

Activity 9.10

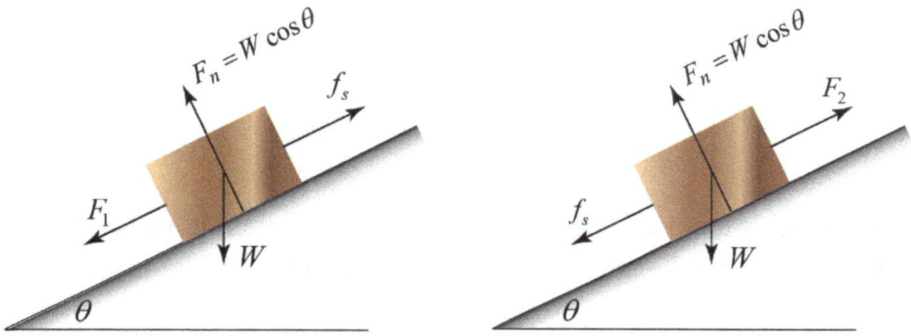

(a) (b)

Consider the figure (a) shown above. The normal reaction force exerted on the block by the plane is cos . The component of the weight W acting downwards along the plane is $W\sin\theta$. Let F_1 be the minimum force required to start the block moving down the plane. Then,

$$F_1 + W\sin\theta = f_s \quad \text{and} \quad f_s = \mu_s F_n = \mu_s W\cos\theta$$

Hence, $$F_1 + W\sin\theta = \mu_s W\cos\theta$$

where μ_s is the coefficient of static friction between the cube and plane.

Consider the figure (b) as shown above. Let F_2 be the minimum force required to start the block moving up the plane. Then,

$$F_2 - W\sin\theta = f_s \quad \text{and} \quad f_s = \mu_s F_n = \mu_s W\cos\theta$$

Hence, $$F_2 - W\sin\theta = \mu_s W\cos\theta$$

Therefore, the difference between the minimum force necessary to start the cube moving up the plane and moving down the plane is

$$F_2 - F_1 = 2W\sin\theta$$

Activity 9.11

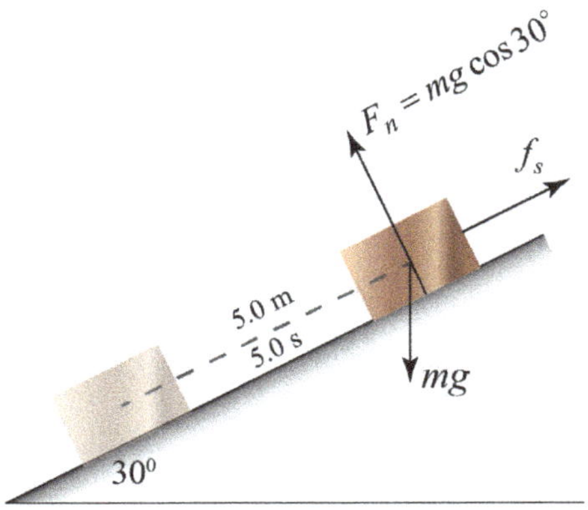

In the figure shown above, the component of the weight of the block acting downwards along the plank is $mg\sin 30^\circ$, where m is the mass of the block and g the acceleration due to gravity.

Since the block is not moving perpendicular to the plank, the magnitude of the normal reaction force F_n on the block is $mg\cos 30^\circ$. Thus, , if μ_s is the coefficient of static friction, then we have

$$mg\sin 30^\circ = f_s = \mu_s mg\cos 30^\circ$$

Hence, we have $\mu_s = \tan 30^\circ = 0.5774 \approx 0.58$.

The cube travels a distance of $5.0\ \text{m}$ from rest along the plank in $2.0\ \text{s}$. Assume its acceleration along the plank to be $\underset{\sim}{a}$. Using the equation of motion $\underset{\sim}{s} = \underset{\sim}{u}t + \frac{1}{2}\underset{\sim}{a}t^2$, we have

$$5.0 = 0 + \frac{1}{2}a(5.0)^2 \Rightarrow \quad a = \frac{2\times 5.0}{(2.0)^2} = 2.5\ \text{m s}^{-2}$$

If we apply the Newton's 2nd law to the block along the plank, then we have:

$$mg\sin 30^\circ - f_k = ma, \quad \text{but we have} \quad f_s = \mu_k mg\cos 30^\circ$$

Here, μ_k is the coefficient of kinetic friction.

Thus, $mg\sin 30^\circ - \mu_k mg\cos 30^\circ = m\times 2.5$

$$\text{Hence, } \mu_k = \frac{g\sin 30^\circ - 0.40}{g\cos 30^\circ} = \frac{10\times\sin 30^\circ - 2.5}{10\times\cos 30^\circ} = 0.2887 \approx 0.29$$

Activity 9.12

(i) The free-body diagram for the passenger. Take the $+x$ direction as the direction of the acceleration, opposite to the initial velocity of the car. Let $\underset{\sim}{v}$ be the velocity of the car, $\vec{F}$ be the force the air bag exerts and $\vec{F}_n$ be the normal force from the car seat.

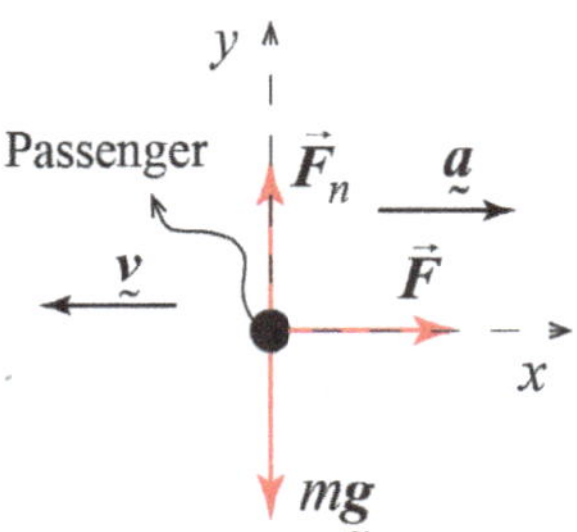

(ii) Apply Newton's second law in $+x$ direction , we have,

$$\sum F_x = ma_x \quad \Rightarrow \quad F = 70\times 60\times 10 = 4.2\times 10^4 \text{ N}$$

Activity 9.13

(i) The free-body diagrams for the bucket filled with mixed cement and the counter - weight are given below.

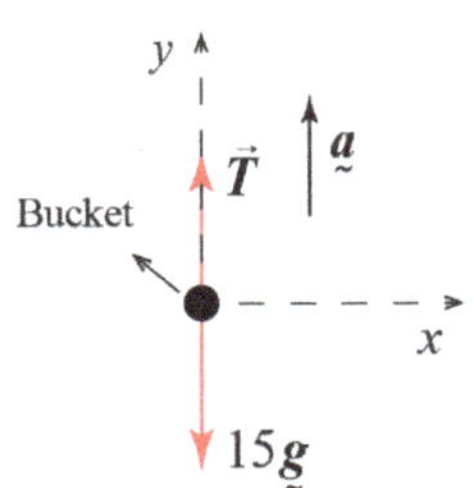

Bucket filled with mixed cement

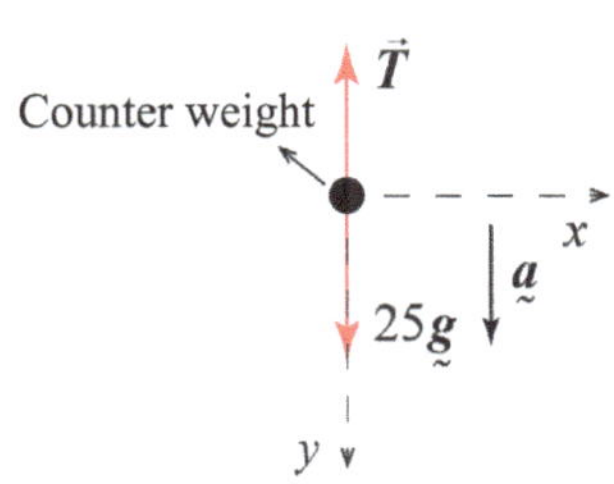

Counter weight

(ii) Apply $\sum F_y = ma_y$ to each object. For the bucket take $+y$ to be upward since $\underset{\sim}{a}$ for the bucket is upward. For the counter-weight, take $+y$ to be downward since $\underset{\sim}{a}$ is downward. The acceleration magnitude is the same for both objects. 25

For the bucket, $T - 15g = 15a$, and for the counter-weight, $25g - T = 25a$

When adding these two equations, we have

$$25g - 15g = (25 + 15)a \text{, thus } a = \frac{10 \times 10}{40} = 2.5 \text{ m s}^{-2}$$

(iii) If you take the equation $T - 15g = 15a$, we have,

$$T = 15(g + a) = 15 \times 12.5 = 187.5 \text{ N} \approx 188 \text{ N}$$

(iv) The tension is 1.25 times the weight of the bucket, and this causes the bucket to accelerate upward.

(v) The tension is 0.63 times the weight of the counter weight, and this causes the counter-weight to accelerate downward.

Activity 9.14

(i) If the weight is mg and the spring force is F_{spring}, then $mg + F_{spring} = 0$, thus

$$F_{spring} = -mg$$

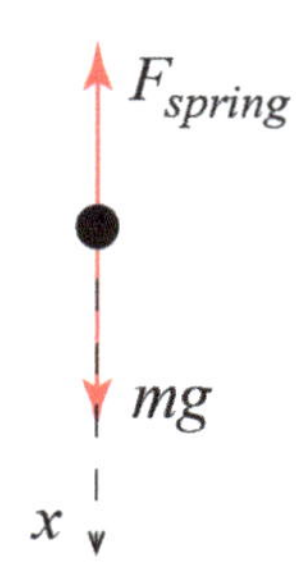

From Hooke's law, we have $F_{spring} = -kx$. Further, the force of magnitude 15.0 N is required to stretch the spring 1.25 cm on balance. Thus,

$$-15.0 = -k \times 1.25 \times 10^{-2} \quad \Rightarrow \quad k = 1200 \text{ N m}^{-1}$$

(ii) If 2.50 kg vegetable is placed, then $F_{spring} = -mg = -2.50 \times 10 = -25.0$ N Thus, $-25.0 = -1200 \times x \Rightarrow x = 2.08$ cm

Activity 9.15

The elevator and the student have the same acceleration. Let $+y$ be upward. The free-body diagrams for the student and for the composite object (elevator + student + scale) are given in the figures.

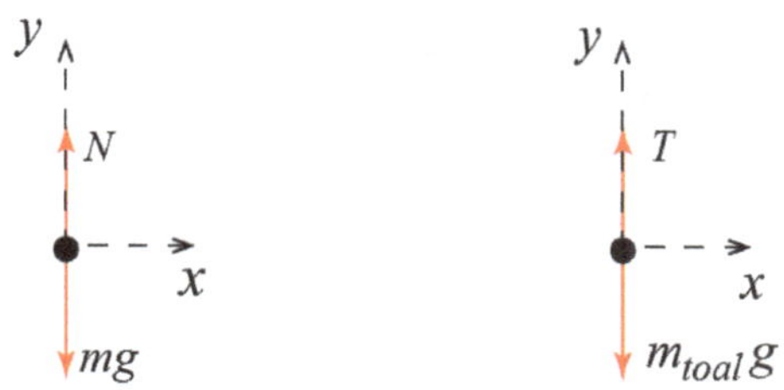

Free body diagram of student

Free body diagram of composite (elevator + student + scale)

(i) Apply Newton's second law to the student in the $+y$ direction.

$$\sum F_y = N - mg = ma_y, \text{ i.e. } a_y = \frac{N - mg}{m} = \frac{500 - 60.0 \times 10}{60.0} = -1.67 \text{ m s}^{-2}$$

Thus, the elevator has a downward acceleration of 1.67 m s^{-2}.

(ii) Apply Newton's second law to the student in the $+y$ direction.

$$\sum F_y = N - mg = ma_y, \text{ i.e. } a_y = \frac{N - mg}{m} = \frac{720 - 60.0 \times 10}{60.0} = 2.00 \text{ m s}^{-2}$$

Thus, the elevator has an upward acceleration of 2.00 m s^{-2}.

(iii) If the weighing scale reads zero, then $N = 0$. Thus, $a_y = -g$. Therefore, the composite object (elevator + student + scale) is under free fall. The student should worry.

(iv) Apply Newton's second law to the composite object (elevator + student + scale) in the $+y$ direction.

For part (a) $\sum F_y = T - m_{total} g = m_{total} a_y$

i.e. $T = m_{total}\left(a_y + g\right) = 860.0(-1.67+10) = 7\ 163.8\ \text{N} \approx 7\ 164\ \text{N}$

For part (b) $\sum F_y = T - m_{total}g = m_{total}a_y$

i.e. $T = m_{total}\left(a_y + g\right) = 860.0(2.00+10) = 10\ 320\ \text{N}$

This shows that, when the acceleration is upward the tension is greater than the weight of the composite object and when the acceleration is downward the tension is less than the weight of the composite object.

For part (c), $T = m_{total}\left(a_y + g\right)$ and $a_y = -g$. Therefore, $T = 0$.

Chapter 10

Activity 10.1

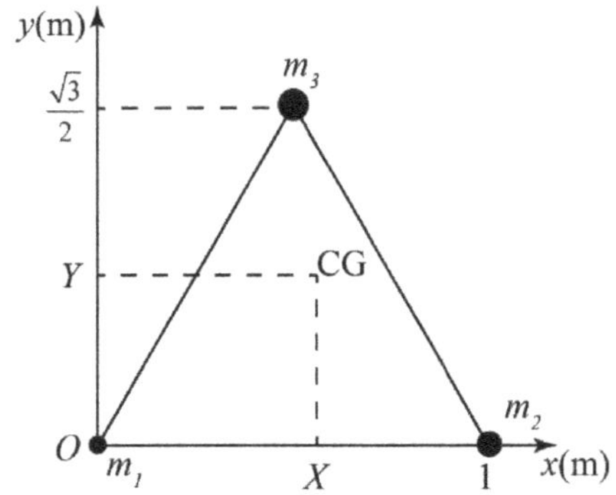

$$X = \frac{\sum m_i x_i}{\sum m_i} = \frac{1.0\times 0 + 2.0\times 1 + 3.0\times 0.5}{1+2+3} = \frac{3.5}{6}\ \text{m}$$

$$Y = \frac{\sum m_i y_i}{\sum m_i} = \frac{1.0\times 0 + 2.0\times 0 + 3.0\times \frac{\sqrt{3}}{2}}{1+2+3} = \frac{\sqrt{3}}{4}\ \text{m}$$

Activity 10.2

Figure shows the situation where the block just starts to topple over. At that position the line of action of the weight passes through the pivot.

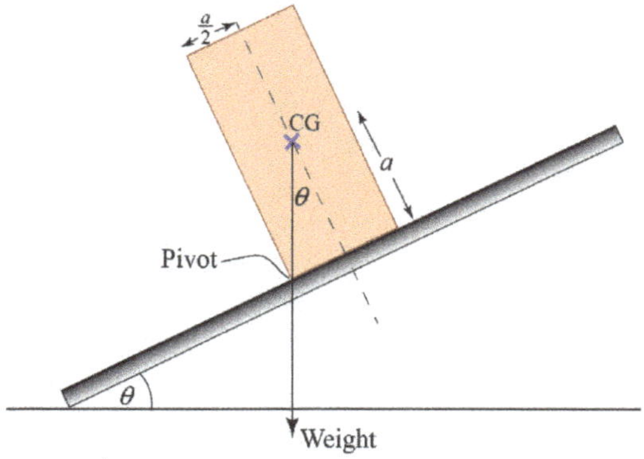

At this position $\tan\theta = \dfrac{a/2}{a} = 0.5$. Thus, $\quad = 26.6^\circ$

Activity 10.3

If an object is not moving and is balanced, then the total torque (or turning effect) acting on the object must be zero. Thus,

$$350 \times 150 \times 10^{-2} = 225 \times 100 \times 10^{-2} + F \times 150 \times 10^{-2}$$

$$\Rightarrow \quad F = \frac{350 \times 1.5 - 225 \times 1}{1.5} = 200 \text{ N}$$

Chapter 11

Activity 11.1

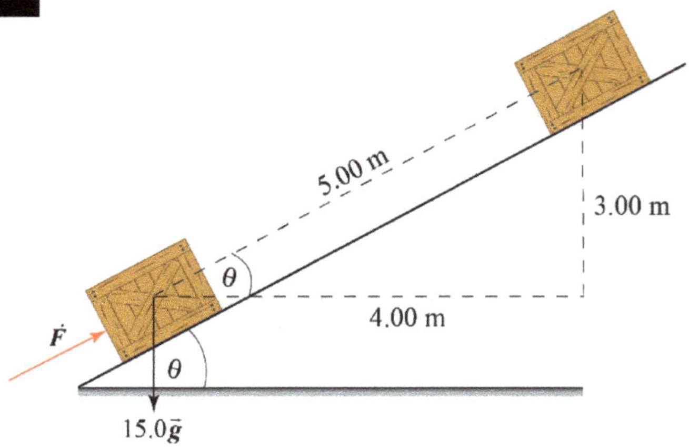

Since the crate is moving with constant speed, there is no resultant force acting on the crate parallel to the incline. Thus, $F - 15.0g\sin\theta = 0$, i.e. $F = 15.0\times10\times\dfrac{3.00}{5.00} = 90.0$ N.

(i) The work done by the force $\vec{F}$ is $W = F\times d = 90.0\times5.00 = 450$ J.

(ii) Since the crate is raised vertically up with the constant speed, there is no resultant force acting on the crate verticaly up. Thus, the magnitude of the force F is $F = 15.0g = 15.0\times10 = 150$ N.

The work done by the force $\vec{F}$ is $W = F\times d = 150\times3.00 = 450$ J.

Activity 11.2

The work done by the spring is negative. Thus, the work done to stretch the spring $W = \dfrac{1}{2}kx^2 = \dfrac{1}{2}\times 21.0\times 0.50^2 = 2.63$ J.

Activity 11.3

The network needed to accelerate the object from a speed zero to v is

$$W = \frac{1}{2}mv^2 - 0 = \frac{1}{2}mv^2$$

The work required to accelerate the object from speed v to speed $3v$ is

$$W' = \frac{1}{2}m(3v)^2 - \frac{1}{2}mv^2 = \frac{1}{2}mv^2(9-1) = 8\left(\frac{1}{2}mv^2\right) = 8W$$

Activity 11.4

(i) Initial kinetic energy $K_i = \frac{1}{2} \times 1750 \times 25.0^2 = 546\ 875$ J

(ii) Use the work-kinetic energy theorem: $W = K_f - K_i$

$W = F_{net} \times d = -f_k \times d$, but $f_k = \mu_k \times mg$

Therefore, $-\mu_k \times mg \times d = 0 - K_i$, since the car stops $K_f = 0$

Thus, $d = \dfrac{K_i}{\mu_k \times mg} = \dfrac{v_i^2}{2\mu_k g} = \dfrac{25.0^2}{2 \times 0.72 \times 10} = 43.4$ m

(iii) Since d is proportional to ^{2}, doubling v_i increases d by a factor of 4;

Therefore d becomes 173.6 m.

(iv) The initial kinetic energy will be converted into thermal and sound energy.

Activity 11.5

(i) The vertical displacement $y_f - y_i = 5.00 \sin 60°$.

(ii) Thus the increase in gravitational potential energy

$$\Delta U_{grav} = mg\left(y_f - y_i\right) = 50.0 \times 10 \times 5.00 \sin 60° = 2.17 \times 10^3 \text{ J}$$

$$\Delta U_{grav} = -W_{by\ gravity} \quad \Rightarrow \quad W_{by\ gravity} = -2.17 \times 10^3 \text{ J}$$

Activity 11.6

For both displacements (to the left and to the right), the direction of the friction force is opposite to the displacement

(i) $f = \mu_k R_n$, but $R_n = mg$. Therefore, $\mu_k = \dfrac{f}{R_n} = \dfrac{2.2}{750 \times 10^{-3} \times 10} = 0.29$.

(ii) When the block moves to the left, the friction force is to the right, and the work done on the block by the friction force is $-(2.2)\times 2.0 = -4.4$ J.

(iii) The friction force is now to the left, and the work done on the block by the friction force is $-(2.2)\times 2.0 = -4.4$ J.

(iv) The total work done on the block by friction force during the complete round trip is -8.8 J.

(v) From parts (b), (c) and (d), we see that the net work done by friction force during the entire round trip is not zero. Since the work done by the friction force on the block during the complete round-trip is not zero, the friction force is nonconservative.

Activity 11.7

Take the ground as the potential reference point. Initial speed $v_i = 200.0 \text{ m s}^{-1}$, initial vertical distance $h_i = 300.0$ m, final speed $v_f = ??$, final vertical distance $h_f = 0$. Let mass of the projectile be m. Thus, from the principle of conservation of mechanical energy, we have

$$\frac{1}{2}mv_i^2 + mgh_i = \frac{1}{2}mv_f^2 + mgh_f$$

$$\Rightarrow \quad v_f = \sqrt{v_i^2 + 2g(h_i - h_f)}$$

Therefore,

$$\Rightarrow \quad v_f = \sqrt{200.0^2 + 2\times 10(300.0 - 0)} = 214.5 \text{ m s}^{-1}$$

Activity 11.8

A slope of 1 in 20 means that the plane is inclined to the horizontal at an angle θ, where $\sin\theta = \frac{1}{20} = 0.05$. The weight of the vehicle $1700g$ acts vertically downwards. Its component along the incline is $1700g\sin\theta$. The vehicle has to do work against this component of force.

The velocity of the vehicle is $36 \text{ km h}^{-1} = \frac{36\times 1000}{3600} \text{ m s}^{-1} = 10 \text{ m s}^{-1}$.

This means that the point of application of the force $1700g \sin\theta$ is displaced through a distance of 10 m per second. Thus, the work done per second by the engine is

$$= 1700g \sin\theta \times 10 = 1700 \times 10 \times 0.05 \times 10 = 8500 \text{ W}$$

Thus, the power output of the engine is 8500 W.

Chapter 12

Activity 12.1

(i) The car's initial momentum $p_i = mv_i$, the direction is north.

$p_i = 1.2 \times 10^3 \times 15 = 18 \times 10^3 \text{ kg m s}^{-1}$ in northward direction

(ii) Th car's momentum $p_f = mv_f$ after turning the bend, the direction is east.

$p_f = 1.2 \times 10^3 \times 7.0 = 8.4 \times 10^3 \text{ kg m s}^{-1}$ in eastward direction

(iii) Change in momentum $\Delta\vec{p} = \vec{p}_f - \vec{p}_i$,

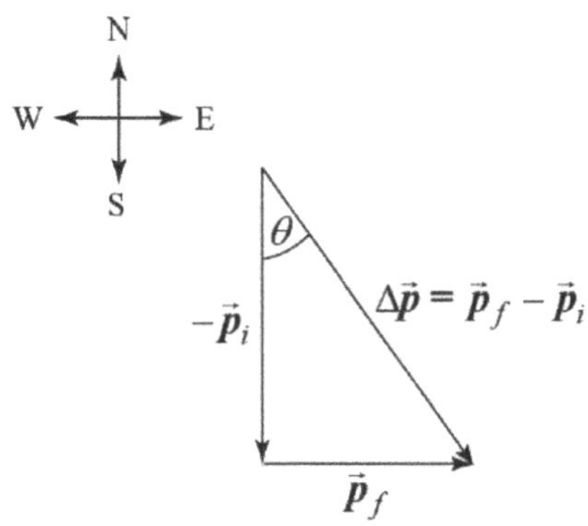

Therefore $|\Delta\vec{p}| = \sqrt{p_f^2 + p_i^2} = \sqrt{\left(18 \times 10^3\right)^2 + \left(8.4 \times 10^3\right)^2}$

Thus, $|\Delta\vec{p}| = 1.99 \times 10^4 \text{ kg m s}^{-1}$

The direction, $\tan\theta = \dfrac{|\vec{p}_f|}{|\vec{p}_i|} = \dfrac{8.4 \times 10^3}{18 \times 10^3} = 0.47$, thus, $\theta = 25.0°$

Therefore, the change in momentum of the car is 2.0×10^4 kg m s^{-1} and directed $\theta = 25^\circ$ east of south.

Activity 12.2

(i) There are no external forces in the horizontal direction, and the system is not moving. The total horizontal momentum of the system before the arrow is fired is zero. According to the conservation of linear momentum, the total horizontal momentum of the system after the arrow is fired must also be zero. Thus,

$$p_i = p_f \quad \Rightarrow \quad 0 = 70\times v + 300\times10^{-3}\times45$$

$$v = -\frac{300\times10^{-3}\times45}{70} = -0.19 \text{ m s}^{-1}$$

(ii) The negative sign of the velocity indicates that the archer is moving to the left after the arrow is fired, i.e. in the direction opposite to the direction of motion of the arrow, in accordance with Newton's third law.

Activity 12.3

Since the collision is elastic, the ball rebounds with a velocity equal but opposite to its incident velocity.

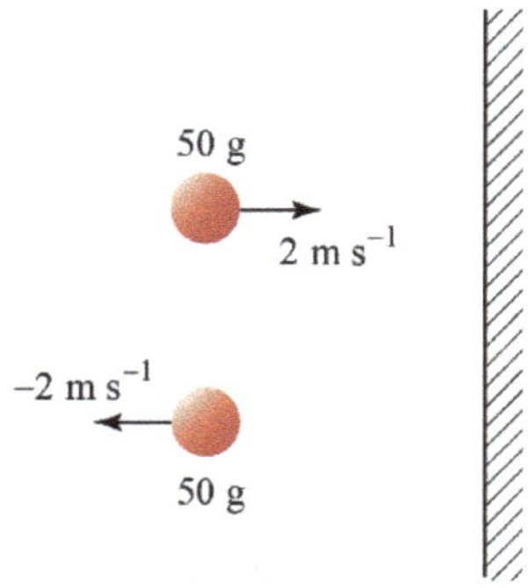

(i) Change of momentum at wall is

$$\Delta p = m\left(v_f - v_i\right) = 50\times10^{-3}\left(-2-2\right) = -0.2 \text{ kg m s}^{-1}$$

Thus the

(ii) The collision lasts for 0.05 s. The rate of change of momentum at the wall is the average force on the wall. Therefore, $F_{avg} = \frac{\Delta p}{\Delta t}$.

$$F_{avg} = \frac{\Delta p}{\Delta t} = \frac{0.2}{0.05} = 4 \text{ N} \text{ the average force on the wall.}$$

Activity 12.4

Let the mass of the ball *P* be $m_P = 150$ g and the mass of the ball *Q* be $m_Q = 300$ g. Thus, $m_Q = 2m_P$. Using conservation of linear momentum, we have

$$m_P v_{P,i} + 0 = m_P v_{P,f} + m_Q v_{Q,f} \Rightarrow \quad v_{P,i} = v_{P,f} + \frac{m_Q}{m_P} v_{Q,f} = v_{P,f} + 2v_{Q,f}$$

Since the collision is perfectly elastic, kinetic energy is conserved. Therefore,

$$\frac{1}{2} m_P v_{P,i}^2 + 0 = \frac{1}{2} m_P v_{P,f}^2 + \frac{1}{2} m_Q v_{Q,f}^2 \quad \Rightarrow \quad v_{P,i}^2 = v_{P,f}^2 + \frac{m_Q}{m_P} v_{Q,f}^2 = v_{P,f}^2 + 2v_{Q,f}^2$$

$$\Rightarrow \quad \left(v_{P,f} + 2v_{Q,f}\right)^2 = v_{P,f}^2 + 2v_{Q,f}^2 \quad \Rightarrow \quad v_{Q,f}^2 + 2v_{P,f} v_{Q,f} = 0,$$

i.e. $\Rightarrow \quad v_{Q,f} = -2v_{P,f}$

Hence, $\Rightarrow \quad v_{P,i} = v_{P,f} + 2\left(-2v_{P,f}\right) = -3v_{P,f}$.

Therefore, $v_{P,f} = -\frac{1}{3} \times 6.00 = -2.00 \text{ m s}^{-1}$ and $\Rightarrow \quad v_{Q,f} = 4.00 \text{ m s}^{-1}$.

Activity 12.5

Let the right hand side be the *x* - axis. Momentum is conserved along the *x* axis. Initial velocity of the clay along the *x*- axis is zero. Thus

$$\left(M_{car} + M_{clay}\right) v_f = M_{car} v_i$$

Therefore, $$v_f = \frac{150 \times 10^{-3}}{\left(150 \times 10^{-3} + 35 \times 10^{-3}\right)} \times 0.80 = 0.65 \text{ m s}^{-1}$$

Activity 12.6

Object A (proton) moving with velocity $v_{A,i}$ collides head-on elastically with a stationary object B (alpha particle), the final velocities are (see section 12.6)

$$v_{A_f} = \left(\frac{m_A - m_B}{m_A + m_B}\right) v_{A_i} \text{ and } v_{B_f} = \left(\frac{2m_A}{m_A + m_B}\right) v_{A_i}$$

Here, $m_A = 1.67 \times 10^{-27}$ kg and $m_B = 6.65 \times 10^{-27}$ kg

(i) $$v_{A_f} = \left(\frac{1.67 \times 10^{-27} - 6.65 \times 10^{-27}}{1.67 \times 10^{-27} + 6.65 \times 10^{-27}}\right) \times 300 \times 10^3 \approx -180 \text{ km s}^{-1} \sim$$

$$v_{B_f} = \left(\frac{2 \times 1.67 \times 10^{-27}}{1.67 \times 10^{-27} + 6.65 \times 10^{-27}}\right) \times 300 \times 10^3 \approx 120 \text{ km s}^{-1}$$

The proton recoils to the left at 180 km s^{-1} and the alpha particle travels to the right at 120 km s^{-1}.

(ii) The proton has initial kinetic energy

$$K_{A,i} = \frac{1}{2} m_A v_{A,i}^2 = \frac{1}{2} \times 1.67 \times 10^{-27} \times \left(300 \times 10^3\right)^2 = 7.52 \times 10^{-17} \text{ J}$$

and final kinetic energy

$$K_{A,f} = \frac{1}{2} m_A v_{A,f}^2 = \frac{1}{2} \times 1.67 \times 10^{-27} \times \left(180 \times 10^3\right)^2 = 2.71 \times 10^{-17} \text{ J}$$

Change in proton kinetic energy is

$$\Delta K_A = K_{A,f} - K_{A,i} = 2.71 \times 10^{-17} - 7.52 \times 10^{-17} = -4.8 \times 10^{-17} \text{ J}$$

Minus sign indicates it is a loss of kinetic energy.

(iii) Change in alpha particle kinetic energy is

$$\Delta K_B = K_{B,f} - K_{B,i} = \frac{1}{2} \times 6.65 \times 10^{-27} \times \left(120 \times 10^3\right)^2 - 0 = -4.8 \times 10^{-17} \text{ J}$$

The energy lost by the proton is gained by the alpha particle. The total kinetic energy of the system is constant and the collision is elastic.

Activity 12.7

M mass of the wooden block, m mass of the bullet and v_i its initial velocity before striking the block.

Let v_f be the speed of the system (wooden block + bullet) after the collision.

The initial momentum of the bullet is mv_i and the momentum of the system just after the collision is $(M+m)v_f$.

Total linear momentum is conserved. Thus, $mv_i = (M+m)v_f$

After the collision, the block and the bullet swing up to a height h. The mechanical energy of the system is conserved. Let's take the block's initial level as our reference level of zero gravitational potential energy.

Therefore, $0+\frac{1}{2}(M+m)v_f^2 = 0+(M+m)gh$

i.e. $\frac{1}{2}(M+m)v_f^2 = (M+m)gh$

Solving these two equations we have

$$v_i = \left[\frac{M+m}{m}\right]\sqrt{2gh}$$

Therefore, $v_i = \left[\frac{5.0+8.0\times10^{-3}}{8.0\times10^{-3}}\right]\sqrt{2\times10\times6.0\times10^{-2}} = 6.86\times10^2 \text{ m s}^{-1}$

Chapter 13

Activity 13.1

(i) $300 \text{ rev min}^{-1} = \frac{300}{60} \text{ rev s}^{-1} = 5.0 \text{ rev s}^{-1}$, i.e. $f = 5.0$ Hz

Thus, angular speed $\omega = 2\pi f = 2\pi\times5.0 = 10\pi \text{ rad s}^{-1}$

(ii) The linear speed $v = r\omega = 0.150\times10\pi = 1.50\pi \text{ m s}^{-1}$

(iii) Since $f = 5.0$ Hz, each revolution takes $\frac{1}{5.0} = 0.20$ s.

Therefore, $T = 0.20$ s

Activity 13.2

The frequency $f = \frac{400}{60}$ rev s^{-1}.

At a location 4.0 cm away from the CD's centre

The angular speed $\omega = 2\pi f = 2\pi \times \frac{400}{60} = 41.9$ rad s^{-1}

The linear speed $v = r\omega = 0.02 \times 2\pi \times \frac{400}{60} = 0.84$ m s^{-1}

Activity 13.3

(i) The frictional force of the road on the wheels.

(ii) Tension in the string supporting the pendulum.

(iii) The gravitational pull of the Earth on the Moon.

Activity 13.4

(i) The ball moves in a horizontal circle of radius r with a speed v, let the tension on the string be T.

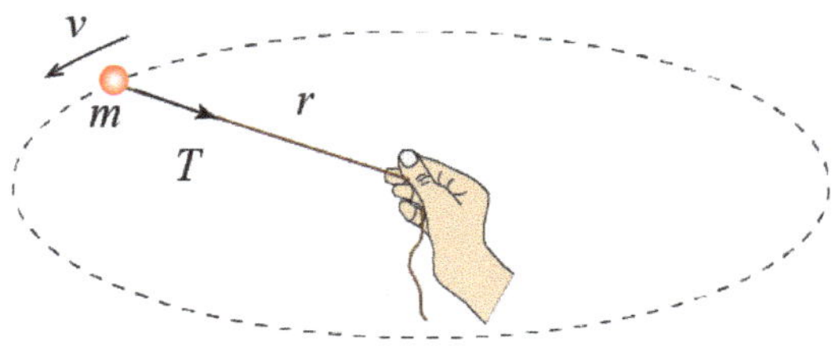

Incorporating the tension and the centripetal acceleration into Newton's second law, we have, $T = m\frac{v^2}{r}$. Therefore, $v = \sqrt{\frac{Tr}{m}}$.

The maximum speed of the ball corresponds to the maximum tension the string can withstand. Thus, $v_{max} = \sqrt{\frac{T_{max}r}{m}} = \sqrt{\frac{20.0 \times 1.00}{0.200}} = 10.0$ m s^{-1}

(ii) The larger radius means that the change in the direction of the velocity vector will be smaller in a given time interval. Therefore, the acceleration is smaller and the required tension in the string is smaller. As a result, the string is less likely to break when the ball travels in a circle of larger radius.

Activity 13.5

$F_g = \dfrac{Gmm_E}{r^2}$, incorporating this with the centripetal acceleration, we have

$$F_g = \frac{Gmm_E}{r^2} = \frac{mv^2}{r} \Rightarrow \quad v = \sqrt{\frac{Gm_E}{r}} = \sqrt{\frac{6.674\times10^{-11}\times5.98\times10^{24}}{8000\times10^3}} = 7.06\times10^3 \text{ m s}^{-1}$$

Activity 13.6

(i) The radial acceleration required for the recommended speed of 12 m s^{-1} is,

$$a_r = \frac{v^2}{r} = \frac{10^2}{25} = 4.0 \text{ m s}^{-2}$$

(ii) The only force with a horizontal component is the static frictional force f_s acting on the tyres due to the road. Therefore, $f_s = m\dfrac{v^2}{r}$. The car can go around the curve without skidding as long as $f_s \le \mu_s N$. Since $N = mg$, we have

$$f_s \le \mu_s mg \quad \Rightarrow \quad m\frac{v^2}{r} \le \mu_s mg \quad \Rightarrow \quad v \le \sqrt{\mu_s gr}$$

i.e. $v \le \sqrt{0.70\times10\times24} = 13 \text{ m s}^{-1}$

Since 10 m s^{-1} is less than the maximum safe speed of 13 m s^{-1}, the car can safely go around the curve.

(iii) At 17 m s^{-1}, the car moves at a speed higher than the maximum safe speed of 13 m s^{-1}. The frictional force cannot supply the radial acceleration needed for the car to go around the curve, thus the car will skid.

(iv) When $\mu_s = 0.04$, the maximum safe speed would be

$v_{\max} = \sqrt{\mu_s g r} = \sqrt{0.40 \times 10 \times 25} = 10 \text{ m s}^{-1}$; which is the same maximum speed recommended by the road sign.

(v) Assuming that friction is negligible, the horizontal component of the normal force is the only horizontal force. Let θ be the angle and the x-axis pointing toward the centre of curvature and the y-axis vertical, then the horizontal and vertical component of the normal force will be $N\sin\theta$ and respectively. Thus,

$$N\sin\theta = m\frac{v^2}{r} \quad \text{and} \quad N\cos\theta = mg$$

Therefore, $\dfrac{N\sin\theta}{N\cos\theta} = \dfrac{m\dfrac{v^2}{r}}{mg} = \dfrac{v^2}{gr}$,

i.e. $\theta = \tan^{-1}\left(\dfrac{v^2}{gr}\right) = \tan^{-1}\left(\dfrac{13^2}{10 \times 25}\right) = 34^\circ$

Chapter 14

Activity 14.1

In one revolution the angle swept out is 2π. Thus,

$$\omega_0 = \frac{120 \times 2\pi}{60} = 4\pi \text{ rad s}^{-1} \text{ and } \omega = \frac{300 \times 2\pi}{60} = 10\pi \text{ rad s}^{-1}$$

(i) Using $\omega = \omega_0 + \alpha t$, we get $\alpha = \dfrac{\omega - \omega_0}{t} = \dfrac{10\pi - 4\pi}{10} = 0.6\pi \text{ rad s}^{-2}$

(ii) Let θ be the angular displacement.

Using $\Delta\theta = \omega_0 t + \frac{1}{2}\alpha t^2$, we get

$$\Delta\theta = 4\pi \times 10 + \frac{1}{2}0.6\pi \times 10^2 = 70\pi \text{ rad}$$

Activity 14.2

In one revolution the angle swept out is 2π. Thus, $\omega_0 = \dfrac{90 \times 2\pi}{60} = 3\pi \text{ rad s}^{-1}$.

$\omega = 0$, and the angular displacement $\theta = 5.0 \times 2\pi = 10\pi$ rad.

Using $\omega^2 = \omega_0^2 + 2\alpha\Delta\theta$, we get

$$0^2 = (3\pi)^2 + 2\alpha \times 10\pi \quad \Rightarrow \quad \alpha = -\frac{9\pi^2}{20\pi} = -0.45\pi \text{ rad s}^{-2}$$

The negative sign inicates that the wheel is slowing down i.e. negative angular acceleration.

We have $I = \tau\alpha$, $\Rightarrow \quad \tau = 0.80 \times (-0.45\pi) = -0.36\pi$ N m.

The negative sign inicates that the torque is applied in opposite direction to that of the original rotation.

Activity 14.3

We have, $I = \tau\alpha \Rightarrow \quad \alpha = \dfrac{\tau}{I} = \dfrac{15.0}{2.5} = 6.0 \text{ rad s}^{-2}$

For constant angular acceleration, after 10.0 s the angular displacement will be

$$\Delta\theta = \omega_0 t + \frac{1}{2}\alpha t^2 = 0 + \frac{1}{2} \times 6.0 \times 10.0^2 = 300 \text{ rad}$$

(i) the work done by the motor in 10.0 s is $W = \tau\Delta\theta = 15.0 \times 300 = 4.50 \times 10^3$ J

(ii) the kinetic energy at the end of 10.0 s is $K = \dfrac{1}{2}I\omega^2$,

but $\omega = \alpha t = 6.0 \times 10.0 = 60 \text{ rad s}^{-1}$, therefore, $K = \dfrac{1}{2} \times 2.5 \times 60^2 = 4.50 \times 10^3$ J

(iii) the average power is the total work divided by the time interval:

i.e. $P_{avg} = \dfrac{W}{t} = \dfrac{4500}{10.0} = 450$ W

Activity 14.4

(i) The kinetic energy of translation is

$$K_{trans} = K_{cm} = \frac{1}{2}mv_{cm}^2 = \frac{1}{2}\times 15.0\times 12.0^2 = 1.08\times 10^3 \text{ J}$$

(ii) Let the radius of the cylinder be R. An observer at the centre of the cylinder sees the rough surface and the circumference of the cylinder moving at 12.0 m s^{-1}. Therefore the angular speed of the cylinder is $\omega = \frac{v_{cm}}{R}$.

The moment of inertia about the rolling axis through the centre of mass is $I = \frac{1}{2}mR^2$. Therefore, the rotational kinetic energy about the rolling axis through its centre of mass is

$$K_{rot} = \frac{1}{2}I\omega^2 = \frac{1}{2}\left(\frac{1}{2}mR^2\right)\left(\frac{v_{cm}}{R}\right)^2 = \frac{1}{4}mv_{cm}^2 = \frac{1}{4}\times 15.0\times 12.0^2 = 540 \text{ J}$$

(iii) The total energy: $K = K_{trans} + K_{rot} = 1.08\times 10^3 + 540 = 1.62\times 10^3 \text{ J}$.

Activity 14.5

The moment of inertia of a uniform solid disk is $I = \frac{1}{2}mr^2$.

The magnitude of the angular momentum of the rotating disk

$$L = I\omega = \left(\frac{1}{2}mr^2\right)\omega = \frac{1}{2}\times 3.00\times 0.100^2\times 1.00\times 2\pi = 9.42\times 10^{-2} \text{ kg m}^2 \text{ s}^{-1}$$

Activity 14.6

The rotational kinetic energy of the object $K_{rot} = \frac{1}{2}I\omega^2$

But we have $L = I\omega \Rightarrow \omega = \frac{L}{I}$. Thus, $K_{rot} = \frac{1}{2}I\left(\frac{L}{I}\right)^2 = \frac{L^2}{2I}$

Activity 14.7

(i) Using the conservation of angular momentum, we have $I\omega = I'\omega'$

$$\Rightarrow \quad \omega' = \left(\frac{I}{I'}\right)\omega \qquad \Rightarrow \quad \omega' = \left(\frac{2.35}{0.35}\right)\times 0.70 = 4.70 \text{ rev s}^{-1}$$

(ii) Rotational kinetic energy is given by $K_{rot} = \frac{1}{2} I\omega^2$

The initial rotational kinetic energy is

$$K_{rot-i} = \frac{1}{2}\times 2.35\times\left(0.70\times 2\pi\right)^2 = 22.7 \text{ J}$$

The final rotational kinetic energy is

$$K_{rot-f} = \frac{1}{2}\times 0.35\times\left(4.70\times 2\pi\right)^2 = 1.53\ \times 10^2 \text{ J}$$

Activity 14.8

The moment of inertia of the disk is

$$I = \frac{1}{2} mr^2 = \frac{1}{2}\times 400\times 10^{-3}\times\left(4.00\times 10^{-2}\right)^2 = 3.20\times 10^{-4} \text{ kg m}^2$$

The angular velocity of the disk is $\omega = 25.0\times 2\pi$ rad s^{-1}

The precessional angular velocity Ω is

$$\Omega = \frac{mgd}{I\omega} = \frac{400\times 10^{-3}\times 10\times 5.0\times 10^{-2}}{3.20\times 10^{-4}\times 25.0\times 2\pi} = 3.98 \text{ rad s}^{-1}$$

Answers

Chapter 2

1. (i) $\left(-2x^2y^4\right)^2 = (-2)^2\left(x^2\right)^2\left(y^4\right)^2 = 4x^4y^8$

(ii) $\left(\dfrac{6x^3}{3y^2}\right)^3 = \left(\dfrac{2x^3}{y^2}\right)^3 = \left(2x^3y^{-2}\right)^3 = 8x^9y^{-6}$

(iii) $\left(-\dfrac{x^2y^4}{xy^{-2}}\right)^2 = \left(-x^1y^6\right)^2 = x^2y^{12}$

2. $$\frac{x^3y^{-3}xy^{\frac{4}{3}}}{x^{-4}y^{\frac{1}{3}}\left(x^2\right)^3} = x^3y^{-3}x^1y^{\frac{4}{3}}x^4y^{-\frac{1}{3}}x^{-6} = x^{(3+1+4-6)}y^{\left(-3+\frac{4}{3}-\frac{1}{3}\right)} = x^2y^{-2}$$

when $x = 9$ and $y = 3$

$$\frac{x^3y^{-3}xy^{\frac{4}{3}}}{x^{-4}y^{\frac{1}{3}}\left(x^2\right)^3} = x^2y^{-2} = 9^2\times3^{-2} = \left(3^2\right)^2\times3^{-2} = 3^4\times3^{-2} = 3^2 = 9$$

(i) We raise each side of the equation to the $\frac{1}{4}$ power.

$\left(x^4\right)^{\frac{1}{4}} = (81)^{\frac{1}{4}} \Rightarrow x = (81)^{\frac{1}{4}} = \left(3^4\right)^{\frac{1}{4}}$ or $\left[(-3)^4\right]^{\frac{1}{4}}$.

Thus $x = +3$ or $x = -3$.

(ii) The quadratic formula: $x = \dfrac{-b\pm\sqrt{b^2-4ac}}{2a}$

For $4.9t^2 + 2t - 20 = 0$, $a = 4.9$, $b = 2$ and $c = -20$

$$x = \frac{-2\pm\sqrt{2^2-4\times4.9\times(-20)}}{2\times4.9} = \frac{-2\pm\sqrt{396}}{9.8} = -2.2 \text{ or } 1.8$$

4. (i) $T = 2\pi\sqrt{\frac{l}{g}}$, when squaring the equation on both sides, we get

$$T^2 = (2\pi)^2 \frac{l}{g} \quad \Rightarrow \quad \frac{l}{g} = \frac{T^2}{4\pi^2} \quad \Rightarrow \quad l = \frac{T^2 g}{4\pi^2}$$

(ii) $F = k\frac{m_1 m_2}{r^2} \quad \Rightarrow \quad r^2 = k\frac{m_1 m_2}{F} \quad \Rightarrow \quad r = \sqrt{k\frac{m_1 m_2}{F}}$

5. (i) $3x + y - 11 = 0$ and $2x + y - 8 = 0$

When subtracting the 2nd equation from the 1st equation, we get

$$(3x + y - 11) - (2x + y - 8) = 0$$
$$3x - 2x - 11 + 8 = 0$$
$$x = 3$$

Substituting this to the 1st equation, we get

$$3 \times 3 + y - 11 = 0$$
$$y - 11 + 9 = 0$$
$$y = 2$$

(ii) $3x + 2y = 17$ and $4x - y = 30$

Multiply the 2nd equation by 2 and then add it to the 1st equation, we get

$$(3x + 2y) + 2(4x - y) = 17 + 60$$
$$3x + 2y + 8x - 2y = 77$$
$$11x = 77$$
$$x = 7$$

Substituting this to the 1st equation, we get

$$3 \times 7 + 2y = 17$$
$$2y = 17 - 21$$
$$2y = -4$$
$$y = -2$$

6. $u = u_0 + at$ and $x = u_0 t + \frac{1}{2}at^2$

Multiply the 1st equation by $\frac{1}{2}t^2$ and then subtract the 2nd equation, we get

$$\frac{1}{2}ut - x = \frac{1}{2}t(u_0 + at) - \left(u_0 t + \frac{1}{2}at^2\right)$$

$$\frac{1}{2}ut - x = \frac{1}{2}tu_0 + \frac{1}{2}at^2 - u_0 t - \frac{1}{2}at^2$$

$$\frac{1}{2}ut - x = -\frac{1}{2}u_0 t$$

$$x = \frac{1}{2}ut + \frac{1}{2}tu_0$$

$$x = \left(\frac{u + u_0}{2}\right)t$$

7. Let T be the temperature of the gas (in kelvins) and V be its volume. Thus, we have $V \propto T$, when the pressure is fixed. i.e. $\frac{V_1}{T_1} = \frac{V_2}{T_2}$. When $T = 300$ K let the volume be V_2, therefore $\frac{4}{100} = \frac{V_2}{200}$, i.e. $V_2 = 8$ litres.

8. Let F be the force required to compress the spring by a distance x. Thus, $F \propto x$, and we have $\frac{F}{x} = \frac{80}{0.02}$. Therefore, when $x = 0.05$,

$$\frac{F}{0.05} = \frac{80}{0.02}, \text{ i.e. } F = \frac{80 \times 0.05}{0.02} = 200 \text{ N}$$

9. We have $a = 0.20$ m, $b = 0.8$ m and $c = 0.15$ m. The total surface area of the rectangular solid is

$$S = 2ab + 2bc + 2ca = 2 \times 0.20 \times 0.8 + 2 \times 0.8 \times 0.15 + 2 \times 0.15 \times 0.20 = 0.62 \text{ m}^2$$

Its volume is

$$V = abc = 0.20 \times 0.80 \times 0.15 = 0.02 \text{ m}^3 \text{ (rounding to the nearest hundredth)}$$

10. Assume that each step forms a right angle so that a line drawn between the corners of adjacent steps forms a right angled triangle with a base of 30 cm and a height of h. The angle between the hypotenuse and the base is 30°. Therefore,

$$\tan 30^\circ = \frac{\quad}{30} \quad \Rightarrow \quad = 30 \tan 30^\circ = 17.32 \text{ cm}$$

Chapter 3

1. Scientific notation eliminates the need to write many zeros in very large or small numbers. Also, the appropriate number of significant digits is unambiguous when written this way.

2. It is important to list the correct number of significant figures so that we can indicate how precisely a quantity is known and not mislead the reader by writing digits that are not at all known to be correct.

3. (i) Move the decimal point seven places to the left and multiply by 10^7. Thus,

$$23\ 000\ 000 \text{ people} = 2.3 \times 10^7 \text{ people}$$

(ii) Move the decimal point 15 places to the right and multiply by 10^{-15}. Thus,

$$0.000\ 000\ 000\ 000\ 003\ 8 \text{ m} = 3.8 \times 10^{-15} \text{ m}$$

4. (i) 6.28 m has 3 significant figures since all three digits are significant.

(ii) In 0.720 kg, the first zero is not significant since it is used only to place the decimal point. The digits 7, 2 and the final zero are significant. Therefore, 0.720 kg has 3 significant figures.

(iii) In 0.057 m, the first two zeros are not significant since they are used only to place the decimal point. The digits 5 and 7 are significant. Therefore, 0.057 m has 2 significant figures.

(iv) 9.85×10^2 g has 3 significant figures since all three digits are significant.

(v) In 3.80×10^5 m, the zero is significant since it comes after the decimal point. Also, the digits 3 and 8 are significant. Therefore, 3.80×10^5 m has 3 significant figures.

(vi) In 6.400×10^3 s, both 6 and 4 are significant. The two zeros are also significant since they come after the decimal point. Therefore, 6.400×10^3 s has 4 significant figures.

5. (i) Rewrite the numbers so that the power of 10 is the same for each. Then add and give the answer with the number of significant figures determined by the less precise of the two numbers. Thus,

$$2.723 \times 10^6 \text{ kg} + 2.35 \times 10^8 \text{ kg} = 0.02723 \times 10^8 \text{ kg} + 2.35 \times 10^8 \text{ kg} = 2.38 \times 10^8 \text{ kg}$$

(ii) Rewrite the numbers so that the power of 10 is the same for each. Then add and give the answer with the number of significant figures determined by the less precise of the two numbers. Thus,

$$1.83 \times 10^{-5} \text{ m} + 8.5 \times 10^{-7} \text{ m} = 1.83 \times 10^{-5} \text{ m} + 0.085 \times 10^{-5} \text{ m} = 1.92 \times 10^{-5} \text{ m}$$

(iii) Rewrite the numbers so that the power of 10 is the same for each. Then subtract and give the answer with the number of significant figures determined by the less precise of the two numbers. Thus,

$$9.65 \times 10^7 \text{ g} - 3.849 \times 10^5 \text{ g} = 9.65 \times 10^7 \text{ g} - 0.03849 \times 10^7 \text{ g} = 9.61 \times 10^7 \text{ g}$$

(iv) Significant figures determined by the number with the fewest significant figures.

$$(2.2 \text{ m}) \times (3.1 \times 10^{-3} \text{ m}) \times (2.3 \times 10^{-8} \text{ m}) = 1.6 \times 10^{-10} \text{ m}^3$$

(v) Significant figures determined by the number with the fewest significant figures.

$$\frac{4.563 \times 10^6 \text{ m}}{2.8 \times 10^{-2} \text{ s}} = 1.6 \times 10^8 \text{ m s}^{-1}$$

(vi) Here, the significant figures determined by the number with the fewest significant figures. In this case the quantity with the fewest number of significant figures is $27.2\ \text{m}$ and it has 3 significant figures. Thus,

$$\frac{27.2\ \text{m} \times 15.63\ \text{m}}{1.846\ \text{s}} = 230\ \text{m}^2\ \text{s}^{-1}$$

6. Significant figures determined by the number with the fewest significant figures.

$$\frac{7.33\ \text{m}}{8.10\ \text{ms}} = \frac{7.33\ \text{m}}{8.10\ \times 10^{-3}\ \text{s}} = 905\ \text{m s}^{-1}$$

7. Significant figures determined by the number with the fewest significant figures. The least number of significant figures among the dimensions is $1.8\ \text{cm}$. It contains two significant figures. Thus,

$$\left(1.02\ \times 10^1\ \text{cm}\right) \times \left(0.824 \times 10^1\ \text{cm}\right) \times \left(0.18 \times 10^1\ \text{cm}\right) = 0.15 \times 10^3\ \text{cm}^3$$

8. If the radius, r, of a circle is $5.15\ \text{cm}$ and has 3 significant figures, then the diameter d of the circle would be reported to three significant figures. Thus,

$$d = 2r = 2 \times 5.15\ \text{cm} = 10.3\ \text{cm}$$

Note: When a constant is present in a calculation involving measured values, the constant should be viewed as being infinitely precise. This means that a constant will be the most precise quantity in the calculation.

9. If d is the diameter, the circumference C of the circle is πd. Also we have $\pi = 3.142$ and $d = 10.3\ \text{cm}$. Hence, the least number of significant figures in the calculation would be 3. Therefore,

$$C = \pi d = 3.142 \times 10.3\ \text{cm} = 32.4\ \text{cm}$$

10. Let r_{in} be the inner radius and r_{out} outer radius of the hollow sphere. $r_{\text{in}} = 3.550$ cm and $r_{\text{out}} = 5.21$ cm. When calculating the intermediate results $\left(r_{\text{out}}^3 - r_{\text{in}}^3\right)$, use the rules for subtraction. Then the volume V of the sphere is

$$
\begin{aligned}
V &= \frac{4}{3}\pi\left(r_{\text{out}}^3 - r_{\text{in}}^3\right) \\
&= \frac{4}{3}\pi(141.420761 - 44.738875) \\
&= \frac{4}{3}\pi\left(1.41420761\times10^2 - 0.44738875\times10^2\right) \\
&= \frac{4}{3}\times3.142\times0.967\times10^2 \\
&= 4.05\times10^2\ \text{cm}^3
\end{aligned}
$$

Chapter 4

1. $$\left(7.86\ \frac{\cancel{\text{g}}}{\cancel{\text{cm}}^3}\right)\left(\frac{\text{kg}}{10^3\ \cancel{\text{g}}}\right)\left(\frac{10^2\ \cancel{\text{cm}}}{\text{m}}\right)^3 = \frac{7.86\times10^6}{10^3}\ \text{kg m}^{-3} = 7.86\times10^3\ \text{kg m}^{-3}$$

2. To obtain the volume V, we multiply the length, width, and height of the room, and use the conversion 1 m = 3.281 ft ,1 m = 3.281 ft.

$$V = 50\ \text{m}\times25\ \text{m}\times12\ \text{m} = \left(1.5\times10^4\ \text{m}^3\right)\left(\frac{3.281\ \text{ft}}{\text{m}}\right)^3 = 5.3\times10^5\ \text{ft}^3$$

3. $$0.425\ \text{liter} = \left(0.425\ \cancel{\text{liter}}\right)\left(\frac{1000\ \text{cm}^3}{1\ \cancel{\text{liter}}}\right) = \left(425\ \cancel{\text{cm}}^3\right)\left(\frac{1\ \text{in}}{2.54\ \cancel{\text{cm}}}\right)^3 = 25.9\ \text{in}^3$$

4. $$\text{Average speed} = \frac{\text{distance}}{\text{time}} = \frac{10\ \text{mi}}{20\ \text{min}} = 0.5\ \text{mi min}^{-1}$$

$$\text{Average speed} = \left(0.5\ \frac{\cancel{\text{mi}}}{\cancel{\text{min}}}\right)\left(\frac{1609\ \text{m}}{1\ \cancel{\text{mi}}}\right)\left(\frac{\cancel{\text{min}}}{60\ \text{s}}\right) = 13\ \text{m s}^{-1}$$

5. (i) and (ii)

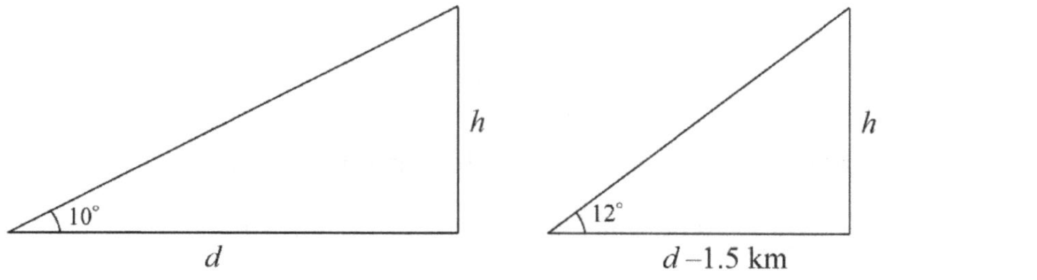

(iii) From the triangles, $\tan 10^\circ = \dfrac{h}{d}$

$$\tan 12^\circ = \frac{h}{d - 1.5\text{ km}} \quad \rightarrow \quad d = \frac{h}{\tan 12^\circ} + 1.5\text{ km}$$

(iv) Equating the two expressions for d, we get $\dfrac{h}{\tan 10^\circ} = \dfrac{h}{\tan 12^\circ} + 1.5\text{ km}$

Thus, $h = \dfrac{1.5\text{ km}}{\left(\dfrac{1}{\tan 10^\circ} - \dfrac{1}{\tan 12^\circ}\right)} = 1.55\text{ km}$

6. No lake is a perfect circle and cannot be expected to have flat bottom. Here we can only estimate the volume of water. To estmate we can make a simple model of the lake as a cylinder.

The volume V of the cylinder is $V = \left(\pi r^2\right)h$, where $r\left(= \frac{1}{2}\text{ km}\right)$ is the radius of the circular base and $h\left(=10\text{ m}\right)$ is the height. Thus,

$$V = \pi \times \left(0.5 \times 10^3\text{ m}\right)^2 \times \left(10\text{ m}\right) \approx 8 \times 10^6\text{ m}^3$$

Therefore the estimated volume of water will be in the order of 10^7 m^3.

7. We have $1\text{ in} = 2.54\text{ cm}$ and $1\text{ h} = 3600\text{ s}$.

$$1\text{ m s}^{-1} = \left(\frac{1\text{ m}}{1\text{ s}}\right)\left(\frac{100\text{ cm}}{1\text{ m}}\right)\left(\frac{1\text{ in}}{2.54\text{ cm}}\right)\left(\frac{1\text{ ft}}{12\text{ in}}\right)\left(\frac{1\text{ mi}}{5280\text{ ft}}\right)\left(\frac{3600\text{ mi}}{1\text{ h}}\right) = 2.24\text{ mi h}^{-1}$$

8. So, for a quick, approximate conversion, multiply by 2.

$$1\text{ y} = (365\text{ d})\left(\frac{24\text{ h}}{1\text{ d}}\right)\left(\frac{60\text{ min}}{1\text{ h}}\right) = 5.3\times10^5\text{ min}$$

Assume 10 breaths/min and there are four astronauts, thus

The volume is $4\times\dfrac{10\text{ breaths}}{\text{min}}\times\dfrac{450\times10^{-6}\text{ m}^3}{1\text{ breath}}\times\dfrac{5.3\times10^5\text{ min}}{1\text{ yr}} = 9.5\times10^3\text{ m}^3\text{ yr}^{-1}$

9. (i) $1\text{ y} = (365\text{ d})\left(\dfrac{24\text{ h}}{1\text{ d}}\right)\left(\dfrac{60\text{ min}}{1\text{ h}}\right)\left(\dfrac{60\text{ s}}{1\text{ min}}\right) = 3.154\times10^7\text{ s}$

$$1.00\text{ ly} = \left(\frac{2.998\times10^8\text{ m}}{1\text{ s}}\right)(3.154\times10^7\text{ s}) = 9.456\times10^{15}\text{ m}$$

(ii) $(1.00\text{ ly})\left(\dfrac{9.456\times10^{15}}{1.00\text{ ly}}\right)\left(\dfrac{1\text{ AU}}{1.50\times10^{11}\text{ m}}\right) = 6.30\times10^4\text{ AU}$

(iii) $\left(\dfrac{2.998\times10^8\text{ m}}{1\text{ s}}\right)\left(\dfrac{1\text{ AU}}{1.50\times10^{11}\text{ m}}\right)\left(\dfrac{3600\text{ s}}{1\text{ h}}\right) = 7.20\text{ AU h}^{-1}$

10. Write each quantity in U with its SI base units, Thus we have,

$U = mgh \quad \rightarrow \quad \text{J} = \text{kg m s}^{-2}\text{ m} = \text{kg m}^2\text{ s}^{-2}.$

(i) $v^2 = u^2 + 2as$ $\qquad [u^2] = (\text{LT}^{-1})^2 = \text{L}^2\text{T}^{-2}$

$$[v^2] = (\text{LT}^{-1})^2 = \text{L}^2\text{T}^{-2}$$

$$[as] = \text{LT}^{-2}\text{L} = \text{L}^2\text{T}^{-2}$$

Therefore, dimensions $[v^2] = [u^2] = [as]$. Hence, we have the same dimensions on both sides of the equation $v^2 = u^2 + 2as$ and it is dimensionally correct.

(ii) $s = ut + \frac{1}{2}at$

$[s] = \text{L}$ $\qquad [ut] = \text{LT}^{-1}\text{T} = \text{L}$ $\qquad [at] = \text{LT}^{-2}\text{T} = \text{LT}^{-1}$

Therefore, dimensions $[s]=[ut]\neq[at]$. Hence, dimensions are not equal on both sides of the equation $s=ut+\frac{1}{2}at$ and it is not dimensionally correct.

(iii) $v=\sin\left(\frac{at^2}{s}\right)$

$[v]=\mathrm{LT}^{-1}$ $\left[\frac{at^2}{s}\right]=\frac{\mathrm{LT}^{-2}\mathrm{T}^2}{\mathrm{L}}=1$

Therefore we can think of the quantity $\frac{at^2}{s}$ as an angle in radians, and we can take its sine. The cosine itself will be a pure number with no dimensions. Hence, dimensions are not equal on both sides of the equation $v=\sin\left(\frac{at^2}{s}\right)$ and it is not dimensionally correct.

12. $x=ka^{\alpha}t^{\beta}$

$[x]=\mathrm{L}$ $[a]=\mathrm{LT}^{-2}$ $[t]=\mathrm{T}$

The expression $x=ka^{\alpha}t^{\beta}$ is dimensionally consistent. Thus, we have

$$\mathrm{L}=\left(\mathrm{LT}^{-2}\right)^{\alpha}\mathrm{T}^{\beta}=\mathrm{L}^{\alpha}\mathrm{T}^{(\beta-2\alpha)}$$

The powers of L and T must be the same on each side of the equation. Therefore,

$$\alpha=1 \text{ and } \beta=2$$

The value of k, a dimensionless constant and cannot be obtained by dimensional analysis.

13. The equation $d=At^3+Bt$ is dimensionally consistent. Thus we have

$$\mathrm{L}=[A]\mathrm{T}^3+[B]\mathrm{T}$$

Since the equation is dimensionally consistent, it should have the same dimensions on both sides of the equation. Therefore,

$$[A]=\mathrm{LT}^3 \text{ and } [B]=\mathrm{LT}^{-1}$$

14. $T^2 = \dfrac{4\pi^2 r^3}{GM}$

$\left[T^2\right] = \mathrm{T}^2 \quad \left[r^3\right] = \mathrm{L}^3 \quad [G] = \mathrm{L}^3\mathrm{M}^{-1}\mathrm{T}^{-2}$ and $[M] = \mathrm{M}$

Therefore, $\left[\dfrac{r^3}{GM}\right] = \dfrac{\mathrm{L}^3}{\left(\mathrm{L}^3\mathrm{M}^{-1}\mathrm{T}^{-2}\right)\mathrm{M}} = \mathrm{T}^2$

Therefore, dimensions $\left[T^2\right] = \left[\dfrac{r^3}{GM}\right]$. Hence, we have the same dimensions on both sides of the equation $T^2 = \dfrac{4\pi^2 r^3}{GM}$ and it is dimensionally correct.

15. $[T] = \mathrm{T} \quad [L] = \mathrm{L} \quad [g] = \mathrm{LT}^{-2} \quad [m] = \mathrm{M}$

T depends on some, or all, of these properties: l, g and m. Thus,

$T \propto l^a g^b m^c \quad T = kl^a g^b m^c$, where k is a constant of proportionality.

Let $[T] = [l]^a [g]^b [m]^c$, then we have $\mathrm{T} = \mathrm{L}^a \left(\mathrm{LT}^{-2}\right)^b \mathrm{M}^c = \mathrm{L}^{a+b}\mathrm{T}^{-2b}\mathrm{M}^c$.

Hence, the powers of T, L and M must be the same on each side of the equation. Therefore, $a+b=0$, $-2b=1$, and $c=0$. Hence we get,

$a = \frac{1}{2}, \quad b = -\frac{1}{2}, \quad c = 0$

Therefore the equation should be $T = kl^{\frac{1}{2}} g^{-\frac{1}{2}} = k\sqrt{\dfrac{l}{g}}$

Chapter 5

1. The independent variable is the variable that you change during an experiment. The dependent variable is the variable that you measure during an experiment.
2. If you have a set of repeated measurements of a quantity, calculate the difference between the maximum and minimum values in the measurements and divide it by two. This will be the uncertainty of the mean.
3. If the results are repeatable, you can get the same or very similar results by repeating the same experiment. Whereas if the results are reproducible, another researcher is able to recreate the experiment using a different set of equipment and/or methods and get the same or similar results.
4. Automatic measurements taken at each set time interval will remove the random error caused by the student by manually reading the temperature measurements at each time interval looking at a stopwatch.
5. The percentage uncertainty in the length of each side of this square tile is given by:

$$\text{Percentage uncertainty} = \frac{0.5}{74} \times 100\% = 0.68\% = 0.7\%$$

6. The average of the three pairs of perpendicular measurements is 31.9 mm. The difference between the largest and smallest measurements is 0.5 mm. Hence, the uncertainty is $\pm\frac{0.5 \text{ mm}}{2} = \pm 0.25 \text{ mm}$.
   Thus, the cylinder has a diameter of $31.9 \pm 0.3 \text{ mm}$.
7. $$19 \text{ MSD} = 20 \text{ VSD}.$$

   This means,
   $$1 \text{ VSD} = \frac{19}{20} \text{ MSD}$$

   Since one MSD is equal to 0.5 mm, we have

   $$1 \text{ VSD} = \frac{19}{20} \times 1 \text{ mm}$$

The least count (LC) is defined as the difference between the length of one MSD and the length of one VSD. i.e.,

$$1\ \text{LC} = 1\ \text{MSD} - 1\ \text{VSD}$$

Therefore, the least count of the measuring instrument is

$$\text{LC} = \left(1\ - \frac{19}{20} \times 1\right)\ \text{mm} = \left(1 - \frac{19}{20}\right)\ \text{mm} = 0.05\ \text{mm}$$

8. Measured value = MSR + VSR × LC −zero error

   Final measurement $= 55\ \text{mm} + 4 \times 0.05\ \text{mm} = 55.2\ \text{mm} = 5.52\ \text{cm}$

9. Measured value = MSR + VSR × LC −zero error

   Final measurement $= 20\ \text{mm} + 22 \times 0.01\ \text{mm} - (-0.01)\ = 20.23\ \text{mm}$

10. (i) Type of wood, the thickness of wood, the width of the wood, length of wood from where it was fixed, l, the position of load and mass of the load.

    (ii) (a)

Length (l) in cm	Deflection (d) in cm	l^2 $\times 10^2$ cm^2	l^3 $\times 10^3$ cm^3	$\frac{l^2}{d}$ $\times 10^2$ cm	$\frac{l^3}{d}$ $\times 10^4$ cm^2
30.0	0.4	9.00	27.0	22.5	6.75
40.0	1.0	16.0	64.0	16.0	6.40
50.0	1.9	25.0	125	13.2	6.58
60.0	3.3	36.0	216	10.9	6.56
70.0	5.4	49.0	343	9.07	6.35
80.0	7.9	64.0	512	8.10	6.48
90.0	11.1	81.0	729	7.30	6.57
95.0	13.08	90.3	857	6.90	6.55

(b) The values $\frac{l^2}{d}$ are getting smaller for longer lengths, whereas the values $\frac{l^3}{d}$ are all very similar and may lead to a content value. From this, we could predict that the deflection d could be proportional to the length cubed, i.e. l^3.

(c)

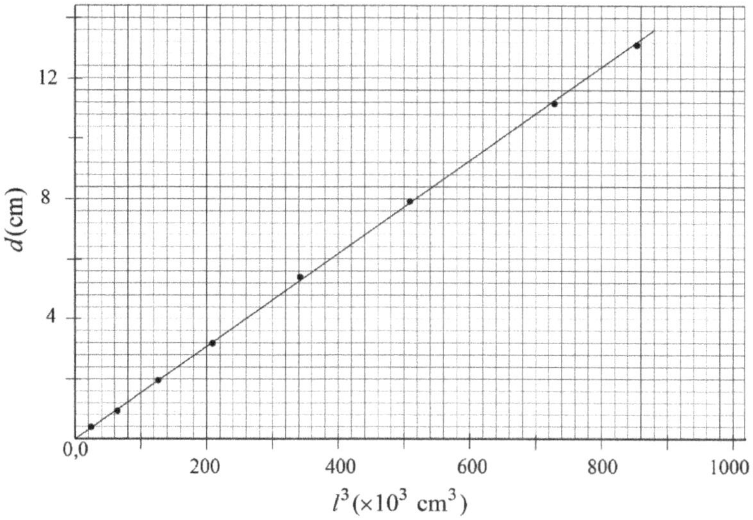

A graph of d against l^3 produces a straight line through the origin. Thus the deflection d is proportional to the length cubed, i.e. l^3.

(d) To improve the graph, the student could have used lengths to spread the graph's points along the line evenly.

11. (i)

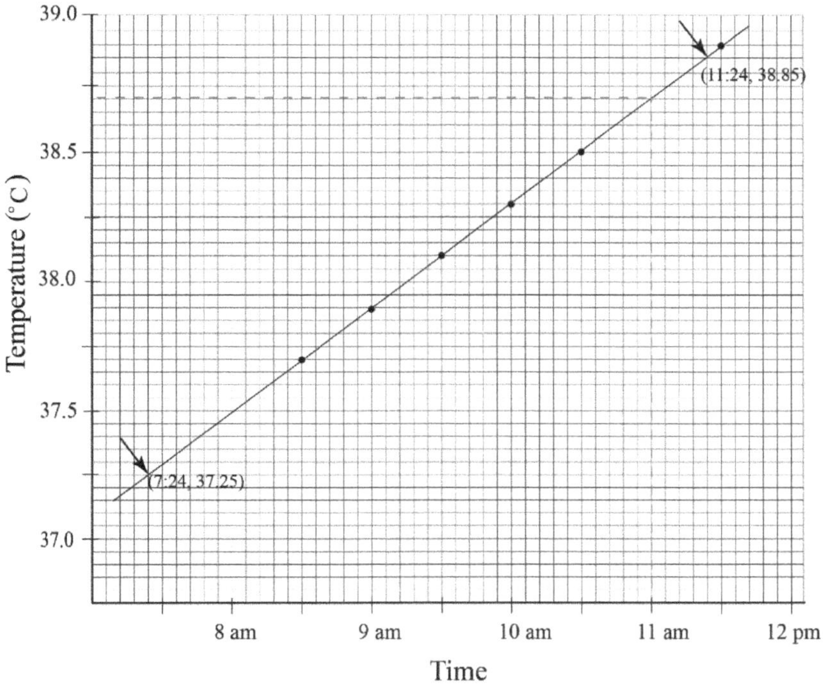

(ii) From the graph, the temperature at 11:00 am should have been 38.7°C.

(iii) The slope of the straight line $m = \dfrac{(38.85-37.25)^\circ\text{C}}{(11:24-7.24)\text{h}} = \dfrac{1.6^\circ\text{C}}{4\text{ h}} = 0.4^\circ\text{C h}^{-1}$.

(iv) The trend of the graph is linear, i.e., the patient's temperature increases linearly. If this trend continues for the next twelve hours, the temperature will be

$$T = (0.4\times12+38.9)^\circ\text{C} = 43.7^\circ\text{C}$$

The maximum body temperature a human can survive is 42.3°C. Thus, the patient would be dead before the temperature reached this level. So, the answer is no.

12. (i) Independent variable – Weight or Applied force F,
Dependent variable – measured length L

(ii)
$$F = k\Delta L$$
$$F = k(L - L_0)$$
$$\Rightarrow\ L = \left(\frac{1}{k}\right)F + L_0$$

(iii)

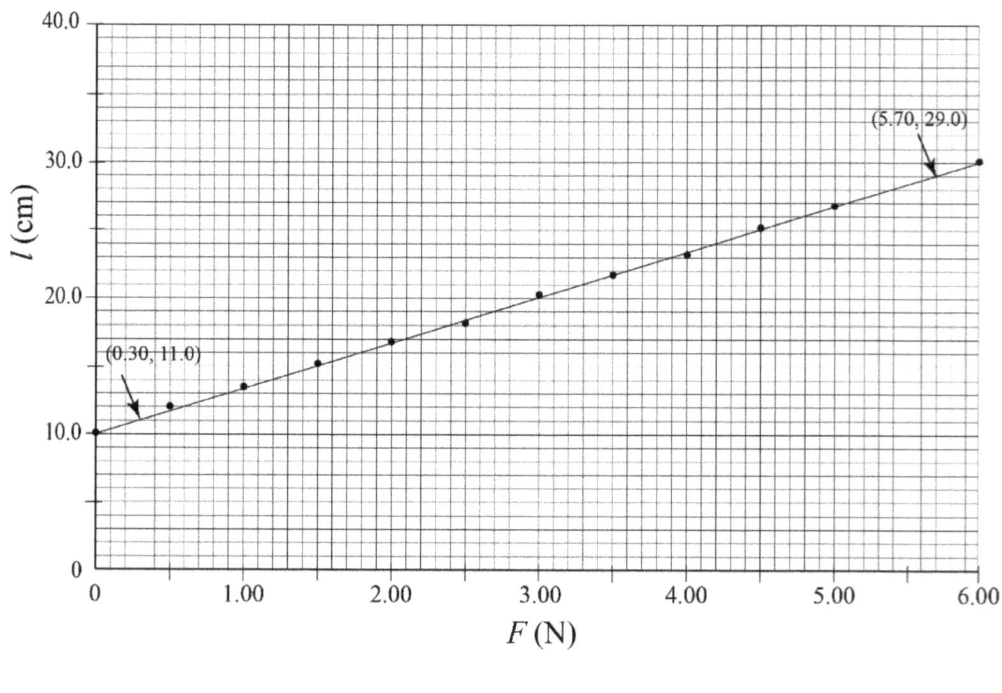

(iv) $(0.30\text{ N}, 11.0\text{ cm})$ and $(5.70\text{ N}, 29.0\text{ cm})$

(v) The slope (or gradient) of the straight-line

$$m = \frac{(29.0 - 11.0)}{(5.70 - 0.30)} = \frac{18.0}{5.40} = 3.33\text{ cm N}^{-1}$$

(vi) The value for the spring constant k of the spring is

$$k = \frac{1}{m} = \frac{5.40}{18.0} = 0.300\text{ N cm}^{-1} = 30.0\text{ N m}^{-1}$$

Chapter 6

1. We have $x = r\cos\theta$ and $y = r\sin\theta$

   (i) For the polar coordinates $\left(2, \frac{\pi}{4}\right)$

   We get $x = 2\cos\left(\frac{\pi}{4}\right) = 2 \times \frac{1}{\sqrt{2}} = \sqrt{2}$ and $y = 2\sin\left(\frac{\pi}{4}\right) = 2 \times \frac{1}{\sqrt{2}} = \sqrt{2}$

   Thus, the Cartesian coordinates $\left(\sqrt{2}, \sqrt{2}\right)$

   (ii) For the polar coordinates $\left(3, -\frac{\pi}{6}\right)$,

   We get $x = 3\cos\left(-\frac{\pi}{6}\right) = 3 \times \frac{\sqrt{3}}{2} = \frac{3\sqrt{3}}{2}$ and $y = 3\sin\left(-\frac{\pi}{6}\right) = 3 \times \left(-\frac{1}{2}\right) = -\frac{3}{2}$

   Thus, the Cartesian coordinates $\left(\frac{3\sqrt{3}}{2}, -\frac{3}{2}\right)$

2. We have $r = \sqrt{x^2 + y^2}$ and $\tan\theta = \frac{y}{x}$

   (i) For the Cartesian coordinates $\left(1, -\sqrt{3}\right)$

   We get $r = \sqrt{1^2 + \left(-\sqrt{3}\right)^2} = 2$ and $\tan\theta = \frac{-\sqrt{3}}{1} = -\sqrt{3}$

   This point is in the fourth quadrant, we have $\theta = -\frac{\pi}{3}$

Therefore the polar coordinates for the point $\left(1,-\sqrt{3}\right)$ is $\left(2,-\frac{\pi}{3}\right)$ or $\left(2,\frac{5\pi}{3}\right)$

(ii) For the Cartesian coordinates $\left(-1,\sqrt{3}\right)$

We get $r=\sqrt{(-1)^2+\left(\sqrt{3}\right)^2}=2$ and $\tan\theta=\frac{\sqrt{3}}{-1}=-\sqrt{3}$.

This point is in the second quadrant,, we have $\theta=\frac{2\pi}{3}$.

Therefore the polar coordinates for the point $\left(-1,\sqrt{3}\right)$ is $\left(2,\frac{2\pi}{3}\right)$.

3. We have $x=r\cos\theta$ and $y=r\sin\theta$

$$x=r\cos\theta \quad\Rightarrow\quad 3=r\cos 60^\circ=\frac{r}{2} \quad\Rightarrow\quad r=6$$

$$y=r\sin\theta \quad\Rightarrow\quad y=6\sin 60^\circ \quad\Rightarrow\quad y=6\times\frac{\sqrt{3}}{2}=3\sqrt{3}$$

4. Use the vector trianglar method to determine the resultant force. Let $\vec{F}_1$ be 5 N in a direction 60° above the positive horizontal axis, and $\vec{F}_2$ be 4 N along the positive vertical axis. Thus resultant force will be $\vec{R}=\vec{F}_1+\vec{F}_2$. Hence, first draw the force vector $\vec{F}_2$ on a paper, with its magnitude represented by a convenient length scale, and then draw the force vector $\vec{F}_1$ to the same scale, with its tail starting from the tip of $\vec{F}_2$ and complete the triangle as shown below. Measure the length of the resultant vector (using the ruler) and determine the magnitude by converting the scale to force. Measure the angle (above the horizontal axis) using the protractor to determine the direction of the resultant force.

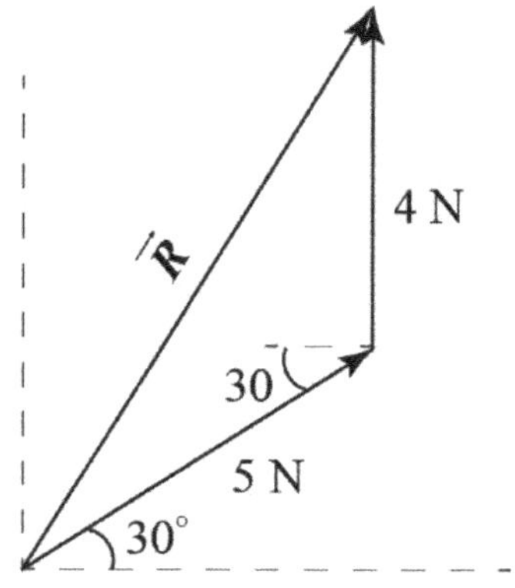

The resultant force $\vec{R}$ could be calculated using the cosine rule. Thus,

$$R = \sqrt{4^2 + 5^2 - 2 \times 4 \times 5 \cos 120^\circ} = 7.8 \text{ N}$$

Using the sine rule, $\dfrac{\sin(\theta - 30^\circ)}{4} = \dfrac{\sin 120^\circ}{|\vec{R}|}$

$$\sin(\theta - 30^\circ) = \frac{4 \sin 120^\circ}{|\vec{R}|} = \frac{4 \sin 120^\circ}{7.81} = 0.44$$

Thus, $\theta = 30^\circ + 26.3^\circ = 56.3^\circ$ above the horizontal axis.

5. When vector $\vec{P}$ and vector $\vec{Q}$ are added together, graphically the vector $\vec{Q}$ is positioned with its tail at the tip of the vector $\vec{P}$ as shown in figure below.

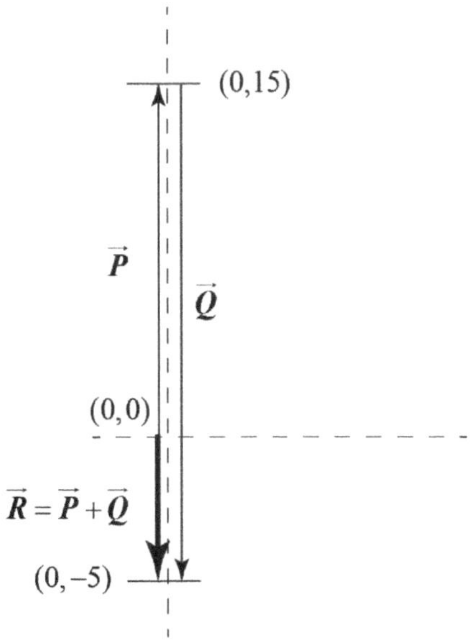

The resultant vector $\vec{R}$ then runs from the tail of the vector $\vec{P}$ to the tip of the vector $\vec{Q}$. In this case, vector $\vec{P}$ will be positioned with its tail at the origin and its tip at the point $(0,15)$. The resultant is then drawn, starting at the origin (tail of vector $\vec{P}$) and going 5 units in the negative y direction to the point $(0,-5)$.

The vector $\vec{\boldsymbol{Q}}$ must have started from the tip of $\vec{\boldsymbol{P}}$ at point $(0,15)$ and end on the tip of resultant vector $\vec{\boldsymbol{R}}$ at point $(0,-5)$ as shown in figure above. From this the vector $\vec{\boldsymbol{Q}}$ has the magnitude of 20 units and points in the negative y direction.

6. Draw the vectors to scale on graph paper, using the vector triangular method.

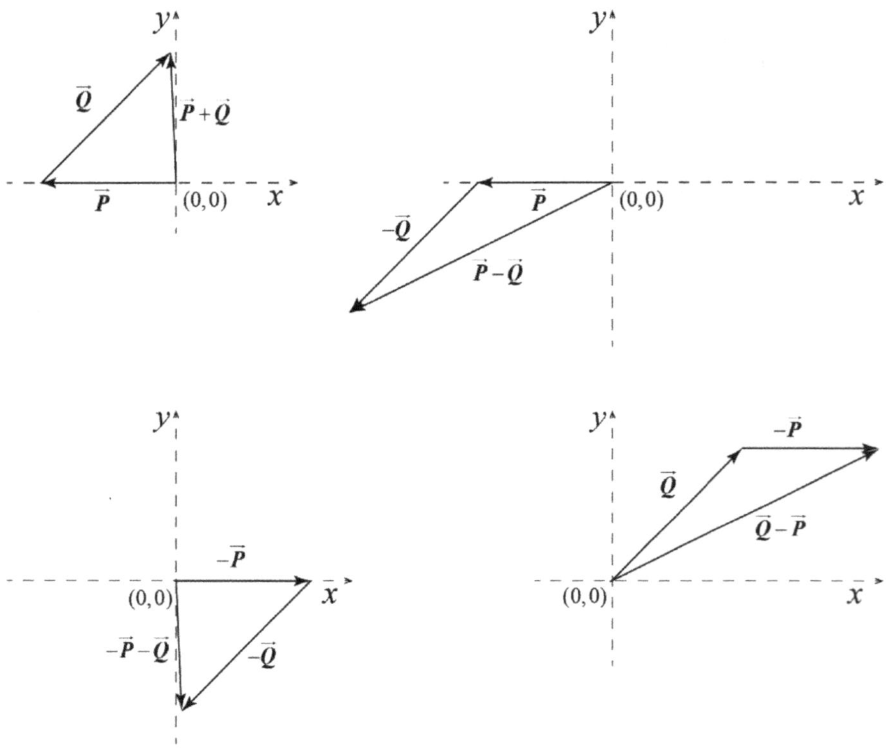

7. The vertical component of the eagle velocity v_y is

$$v_y = 30.0\sin 60.0^\circ \text{ m s}^{-1} = 26.0 \text{ m s}^{-1}$$

The horizontal component of the eagle velocity v_x is

$$v_x = 30.0\cos 60.0^\circ \text{ m s}^{-1} = 15.0 \text{ m s}^{-1}$$

8. Let the weight of the wooden box be W, let the weight along the slope downwards be W_x and the weight perperndicular to the slope be W_y.

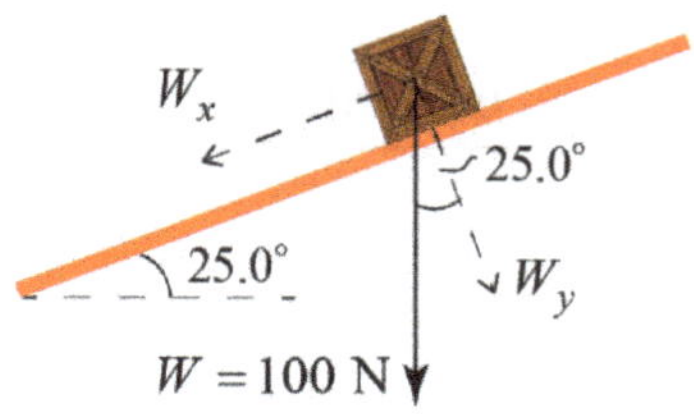

The weight along the slope downwards direction is $W_x = 100 \sin 25.0^0 = 42.3$ N.
The weight perpendicular to the slope is $W_y = 100 \cos 25.0^0 = 90.6$ N.

9. The vector diagram for the milkman's travel is

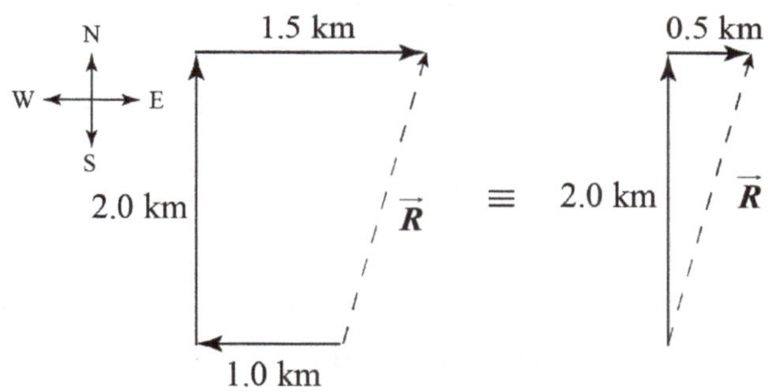

(i) The milkman's resultant displacement R is

$$R = \sqrt{(0.5)^2 + (2.0)^2} = 2.1 \text{ km}$$

(ii) The total distance (S) that the milkman travelled is

$$S = (1 + 2 + 1.5) \text{ km} = 4.5 \text{ km}$$

10. The horizontal component of the force F_x is

$$F_x = F\cos\theta = 8.0\cos 25.5^\circ \text{ kN} = 7.2 \text{ kN}$$

The vertical component of the force F_y is

$$F_y = F\sin\theta = 8.0\sin 25.5^\circ \text{ kN} = 3.4 \text{ kN}$$

11. (i) The suitable vector diagram for showing the travel route

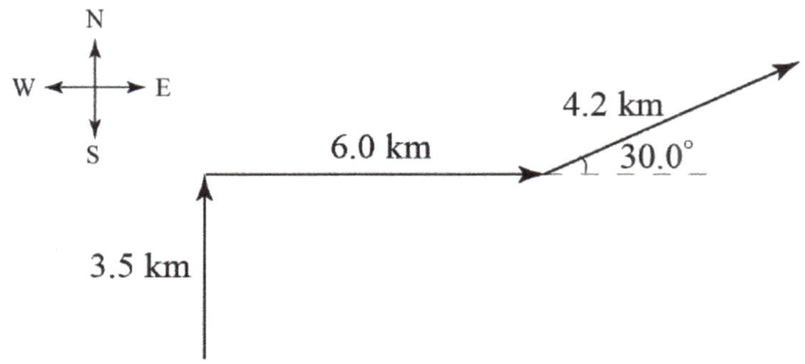

(ii) The north ward components of the travel is $d_N = \left(3.5 + 4.2\sin 30.0^\circ\right)$ km $= 5.6$ km

The eastward components of the travel is $d_E = \left(6.0 + 4.2\cos 30.0^\circ\right)$ km $= 9.6$ km

(iii) The magnitude and direction of the truck's resultant displacement

$$d = \sqrt{(5.6)^2 + (9.6)^2}\ \text{km} = 11.1\ \text{km}$$

$$\tan\theta = \frac{5.6}{9.6} \quad \Rightarrow \quad \theta = 30.3^\circ$$

12. Let the positive x axis be the east and the positive y axis be north. Let the $\vec{C}$ representing the third segment of the journey. The vector addition diagram for the journey

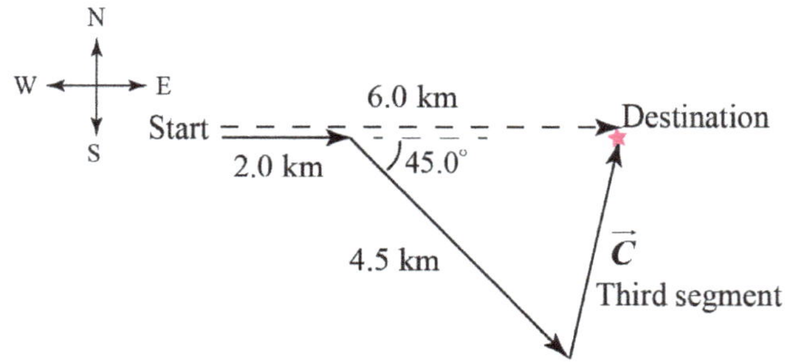

$$C_x = \left(6.0 - 2.0 - 4.5\cos 45.0^\circ\right)\ \text{km} = 0.82\ \text{km},$$

$$C_y = \left\{0 - 0 - \left(-4.5\sin 45.0^\circ\right)\right\}\ \text{km} = 3.18\ \text{km}$$

Therefore, $C = \sqrt{(0.82)^2 + (3.18)^2}$ km $= 3.3$ km °

$\tan\theta = \dfrac{3.18}{0.82}$, $\Rightarrow$ $\theta = 75.5^\circ$ north of east.

Chapter 7

1. The distance travelled is a scalar quantity equal to the total length of the path taken in moving from one point to another. Displacement is a vector quantity directed from the initial point towards the final point with a magnitude equal to the straight line distance between the initial and the final points.

2. The velocity of an object is a vector quantity equal to the displacement of the object per unit time interval. The speed of an object is a scalar quantity equal to the distance travelled by the object per unit time interval.

3. A car speedometer measures only speed. It does not give any information about the direction. Does not measure velocity.

4. If an object has a constant velocity, then both the object's speed and its direction of motion are constant. Therefore the object CANNOT have a varying speed if its velocity is constant.

5. The average velocity of an object is zero means, the net displacement must be zero. Thus the object could have moved away from its starting point and back again to the same point, i.e., it is at its initial position again at the end of the time interval.

6. Let east be the $+x$ direction.

   (i) The motor car's displacement from his starting point is $\Delta x = x_f - x_i$, $x_f = -800$ m i.e. 800 m west, and $x_i = 60$ m i.e. 60 m east.

   Therefore, $\Delta x = x_f - x_i = (-800 - 60)$ m $= -860$ m, i.e. 860 m west.

   (ii) Since the motor car is located 800 m west of the land mark, the motor car's displacement from the land mark is 800 m west, or -800 m.

(iii) The motor car's displacement from his starting point is $\Delta x = x_f - x_i$.

$x_f = 860$ m i.e. 800 m east, and $x_i = 60$ m i.e. 60 m east.

Therefore, $\Delta x = x_f - x_i = (860 - 60)$ m $= 800$ m, i.e. 800 m east.

(iv) The motor car has travelled a distance $60 \text{ m} + 800 \text{ m} = 860 \text{ m}$ during the 1st move.

the mortor car has travelled a distance $800 \text{ m} + 860 \text{ m} = 1660 \text{ m}$ during the 2nd move.

The total distance travelled is $860 \text{ m} + 1660 \text{ m} = 2520 \text{ m}$.

7. The time for the first part of the trip is calculated from the initial speed and the first distance. i.e. $\Delta t_1 = \frac{d_1}{v_1} = \frac{160}{80.0} = 2 \text{ hours} = 7200 \text{ s}$

The time for the second part of the trip is $\Delta t_2 = \Delta t_{total} - \Delta t_1 = 3 - 2 = 1 \text{ hour} = 3600 \text{ s}$.

The distance for the second part of the trip is $d_2 = v_2 \times \Delta t_2 = 60.0 \times 1 = 60.0 \text{ km}$.

(i) The total distance is $d_{total} = d_1 + d_2 = 160.0 + 60.0 = 220.0 \text{ km}$.

(ii) The average speed is NOT the average of the two speeds. Use the definition of average speed. Thus, average speed $= \frac{d_{total}}{\Delta t_{total}} = \frac{220.0}{3} = 73.3 \text{ km h}^{-1}$

8. (i) The average velocities in the first half of the swim

$$v_{1,avg,x} = \frac{\text{displacement}}{\text{time interval}} = \frac{l - 0}{t_1} = \frac{l}{t_1}$$

(ii) The average velocities in the second half of the swim

$$v_{2,avg,x} = \frac{\text{displacement}}{\text{time interval}} = \frac{0 - l}{t_2} = -\frac{l}{t_2}$$

(iii) The average velocities of the round trip is zero, since the displacement is zero.

(iv) The swimmer's average speed for the round trip

$$v_{avg} = \frac{\text{total distance}}{\text{time interval}} = \frac{2l}{t_1 + t_2}$$

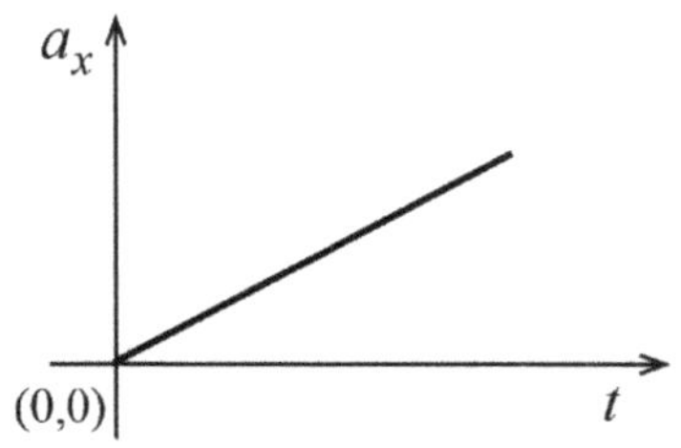

(a)

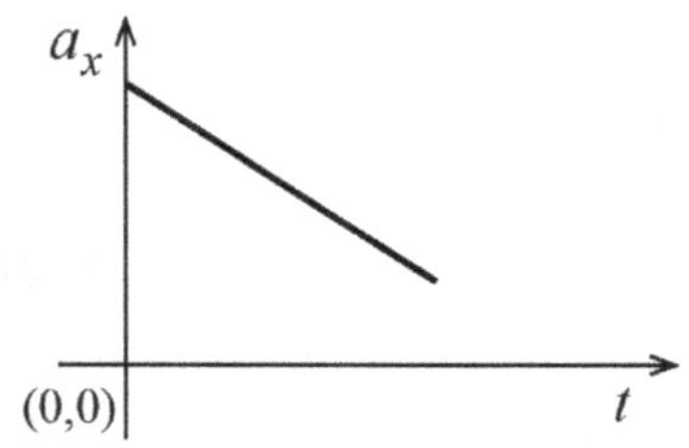

(b)

(a) Acceleration is positive and increasing.

(b) Acceleration is positive and decreasing.

10. (i) For average velocity, we find the slope joining the points $(11 \text{ s}, 25.0 \text{ m})$ and $(17 \text{ s}, 11.0 \text{ m})$, thus

$$\text{The average velocity } = \frac{(11.0 - 25.0) \text{ m}}{(17 - 11) \text{ s}} = -\frac{14}{6} \text{ m s}^{-1} = -2.3 \text{ m s}^{-1}$$

(ii) The instantaneous velocity at point P ($t = 12$ s) is the slope of the tangent line at point P.

$$\text{The instantaneous velocity at point } P = \frac{(7.0 - 41.0) \text{ m}}{(16 - 6) \text{ s}} = -\frac{34}{10} \text{ m s}^{-1} = -3.4 \text{ m s}^{-1}$$

The negative sign shows that the direction of the velocity is along the negative x direction.

11. Let the point starting from rest be the origin, where $x = 0$. The initial velocity $u = 0$ and after 5.0 s its velocity $v = \dfrac{72.0 \times 1000}{3600} \text{ m s}^{-1} = 20.0 \text{ m s}^{-1}$.

(i) Use the equation $v = u + at$.

$$\Rightarrow \quad 20.0 = 0 + a \times 2.5 \qquad \Rightarrow \quad a = \frac{20.0}{2.5} = 8.0 \text{ m s}^{-2}$$

(ii) Use the equation $s = ut + \frac{1}{2}at^2$

$$\Rightarrow \quad s = 0\times 2.5 + \frac{1}{2}\times 8.0\times(2.5)^2 \Rightarrow s = \frac{1}{2}\times 8.0\times(2.5)^2 = 25 \text{ m}$$

12. The car's velocity $= \frac{54\times 1000}{3600} \text{ m s}^{-1} = 15 \text{ m s}^{-1}$

(i) Let us choose the position of the lamp post as the origin and set t 0 when the traffic police start to move. At that moment, since the car is moving with a constant velocity, the car would already have travelled a distance of 15 m from the lamp post. Thus, the initial position of the car at $t = 0$ is 15 m.

Say, at the time $t = t$, the traffic police have just overtaken the car. Thus, the distance travelled by the traffic police.

$$s = 0\times t + \frac{1}{2}\times 2.0\times t^2 = t^2$$

This distance should be equal to the distance covered by the car from the origin. Therefore,

$$t^2 = 15 + 15t \quad \Rightarrow \quad t^2 - 15t - 15 = 0$$

Thus, $t = \frac{15 \pm \sqrt{(-15)^2 - 4\times(-15)}}{2} = \left(\frac{15 \pm 16.9}{2}\right) \text{ s} \approx 16 \text{ s}$ Since time can take only +ve value, the time taken for the traffic police to overtake the car is 16 s.

(ii) Let the speed v of the traffic police when $t = 16$ s be v, thus

$$v = u + at \quad \Rightarrow \quad v = 0 + 2.0\times 16 = 32 \text{ m s}^{-1}$$

(iii) $s = 0\times t + \frac{1}{2}\times 2.0\times t^2 = t^2 = 16^2 \text{ m} = 256 \text{ m}$

(iv) If the traffic police had a more powerful motorcycle with higher acceleration, the police would have caught up to the car sooner. This means that the rate of change in velocity would be greater, and the motorcycle will move faster and will be able to catch up the car much less than the time obtained in part (a).

13. (i)

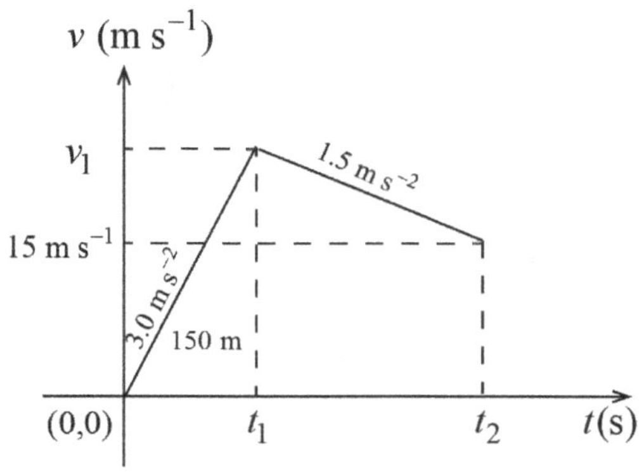

(ii) Acceleration is $3.0\ \text{m s}^{-2}$. Let the velocity be v_1 when it reaches the distance $150\ \text{m}$. Thus, $3.0 = \dfrac{v_1}{t_1}$ and $150 = \dfrac{1}{2} \times t_1 \times v_1$.

Therefore, $150 = \dfrac{1}{2} \times \dfrac{v_1}{3.0} \times v_1 \quad \Rightarrow \quad v_1^2 = 900$, hence $v_1 = 30\ \text{m s}^{-1}$.

(iii) Acceleration is $1.5\ \text{m s}^{-2}$. Therefore, $-1.5\ \text{m s}^{-2} = \dfrac{15-30}{t_2 - t_1}$.

But we have, $3.0 = \dfrac{v_1}{t_1} \quad \Rightarrow \quad t_1 = \dfrac{30}{3.0} = 10\ \text{s}$

Hence, $t_2 = \dfrac{15}{1.5} + 10 = 20\ \text{s}$

The total time required by the boat to reach the buoy is $20\ \text{s}$

(iv) The distance (d) between the dock and the buoy

$$d = 150 + \frac{1}{2} \times (30+15) \times (20-10) = 150 + 225 = 375\ \text{m}$$

14.

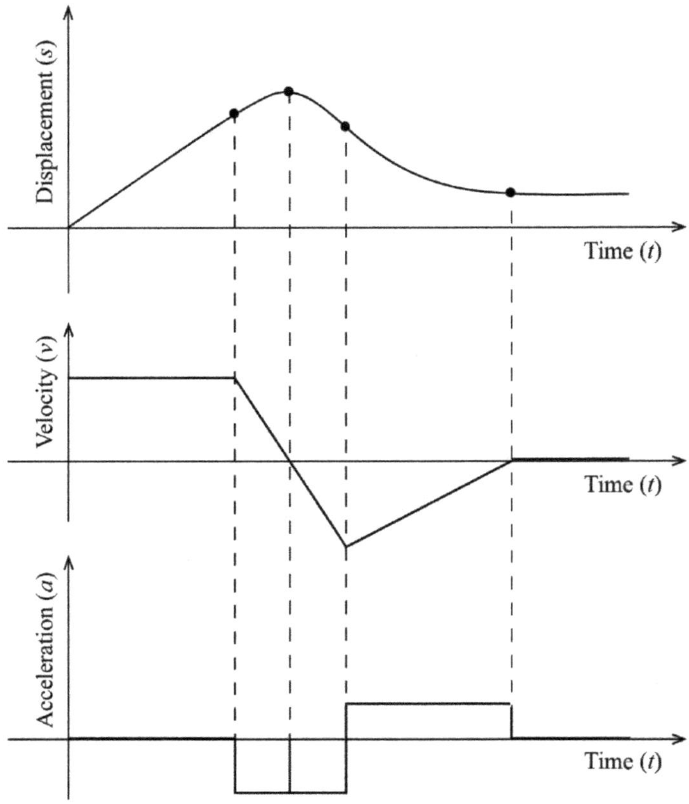
Displacement (s)
Time (t)
Velocity (v)
Time (t)
Acceleration (a)
Time (t)

15. (i)

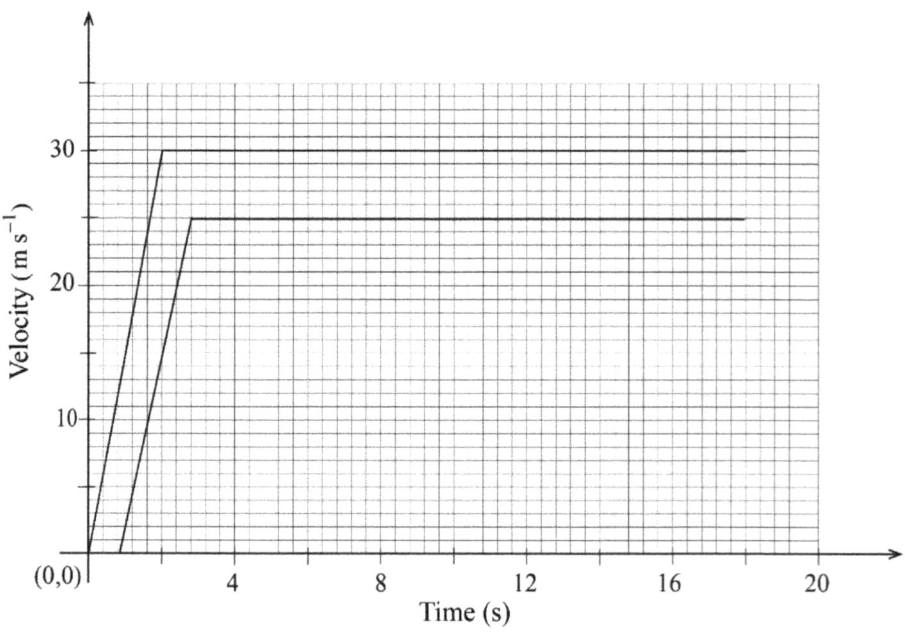
30
20
10
(0,0)
4
8
12
16
20
Velocity (m s⁻¹)
Time (s)

(ii) Cheetah's acceleration $a = \dfrac{30}{2} = 15\ \text{m s}^{-2}$.

(iii) The distance travelled by the cheetah while accelerating

$$= \frac{1}{2} \times 2 \times 30 = 30\ \text{m}$$

(iv) The distance travelled by the cheetah while it is moving at a constant velocity $= 30 \times 16 = 480\ \text{m}$.

(v) The distance covered by the antelope $18\ \text{s}$ after the cheetah started to run

$$= \frac{1}{2}\{(18 - 0.8) + (16 - 0.8)\} \times 25 = 405\ \text{m}$$

(vi) The distance apart between the cheetah and the antelope $18\ \text{s}$ after the cheetah starts to run $= (480 + 30) - (100 + 405) = 5\ \text{m}$

16. Let the vertical y axis upward direction be positive. Take the origin O as shown in the figure. The magnitude of the initial velocity $u = 8.0\ \text{m s}^{-1}$, this is in upward direction. Motion is under the acceleration of gravity and is downward direction.

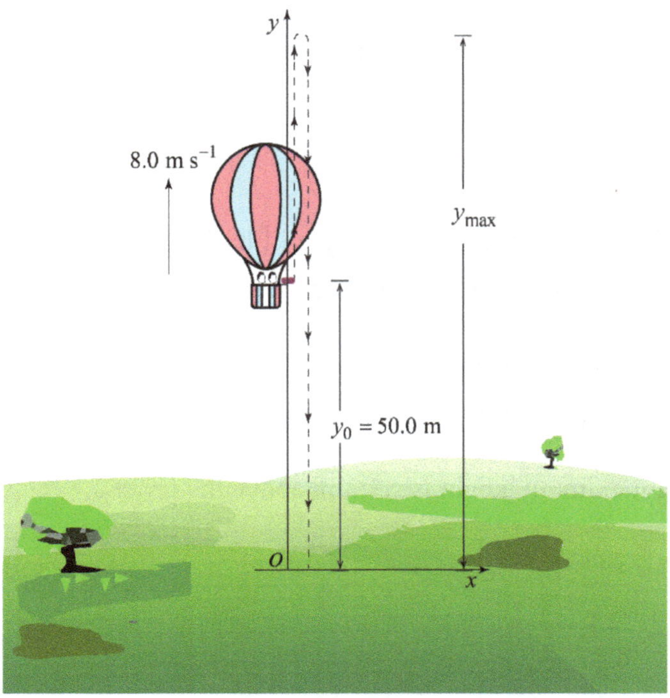

(i) Let s_1 be the distance above the ground at $t = 0.5$ s. Use the equation $s = ut + \frac{1}{2}at^2$. Thus, $s_1 - 50.0 = 8.0 \times 0.5 + \frac{1}{2}(-9.8) \times (0.5)^2 = 2.775$ m.

Therefore, the sand bag is 52.8 m above the ground at $t = 0.5$ s

Let s_2 be the distance above the ground at $t = 1.5$ s. Use the equation $s = ut + \frac{1}{2}at^2$.Thus, $s_2 - 50.0 = 8.0 \times 1.5 + \frac{1}{2}(-9.8) \times (1.5)^2 = 0.975$ m.

Therefore, the sand bag is 51.0 m above the ground at $t = 1.5$ s.

(ii) Let t_1 be the time taken to strike the ground. Use the equation $s = ut + \frac{1}{2}at^2$. Thus, $-50.0 = 8.0 \times t_1 + \frac{1}{2}(-9.8) \times t_1^2$

$\Rightarrow \quad 4.9t_1^2 - 8.0t_1 - 50.0 = 0$. This is a quadratic equation in time t_1.

Therefore, $t_1 = \dfrac{8.0 \pm \sqrt{(-8.0)^2 - 4 \times 4.9 \times (-50.0)}}{2 \times 4.9} = 4.1$ s.

Time must be positive, hence $t_1 = 4.1$ s

(iii) Let v_1 be the velocity of the sand bag upon reaching the ground. Use the equation $v^2 = u^2 + 2as$. Thus,

$v_1^2 = (8.0)^2 + 2 \times (-9.8) \times (-50)$

$v_1 = \pm 32.3$ m s^{-1}, since the velocity will be in the doward direction. The velocity of the sand bag upon reaching the ground $v_1 = -32.3$ m s^{-1}.

(iv) At the higest point the velocity is zero. . Use the equation $v^2 = u^2 + 2as$. Thus, $0 = (8.0)^2 + 2 \times (-9.8)(y_{max} - 50.0)$, therefore, $y_{max} = 53.3$ m.

(v) Hence, the maximum height is 53.3 m above the ground.

(vi)

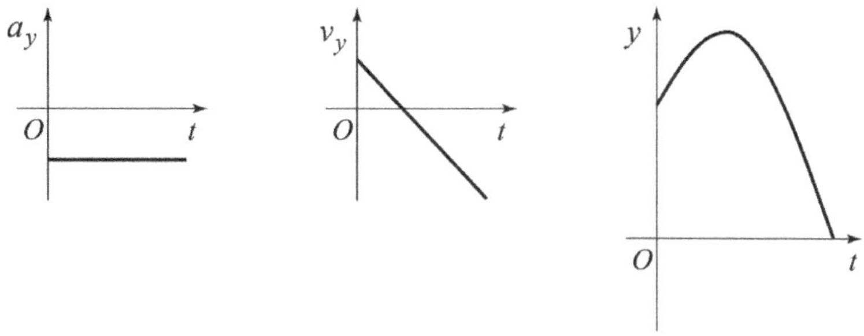

17. Let the vertical downward direction be positive.
    - (i) Ball travels upwards on bouncing, i.e. reverses its direction on bouncing.
    - (ii) Yes. In both cases, the ball is accelerating due to gravity only.
    - (iii) Initial height of the ball above the ground.
    - (iv) No. Ball does not bounce as high as initial position.
    - (v) (a) Use the equation $v^2 = u^2 + 2as$ in the downward direction.

      Thus, $v^2 = 0 + 2(9.8)\times(1.5)$. Therefore, $v = \pm 5.42 \text{ m s}^{-1}$. Hence, the speed of the ball just before hitting the ground is 5.4 m s^{-1}.

      (b) Use the equation $v^2 = u^2 + 2as$, in the downward direction.
      Thus, $0 = u^2 + 2(9.8)\times(-1.0)$. Therefore, $u = \pm 4.43 \text{ m s}^{-1}$. Hence the speed of the ball immediately after hitting the ground is 4.4 m s^{-1}.

      (c) Use the equation $v = u + at$, in the downward direction.

      Thus, $-4.4 = 5.4 + a\times(0.2)$. Therefore, $a = -49 \text{ m s}^{-2}$. Hence, the magnitude of the acceleration of the ball during contact with the ground is 49 m s^{-2} and the acceleration will be in the upwards direction.

18. Let the northward direction be positive.

    (i) Use the equation $v^2 = u^2 + 2as$ in the north direction during the skidding of 150 m. Thus, $0 = u^2 + 2\times(-2.0)\times 144$. Therefore, $u = \pm 24 \text{ m s}^{-1}$. Hence, the initial speed of the car before the brakes are applied is 24 m s^{-1}.

    (ii) Initially, the car is travelling at a constant velocity, thus acceleration is zero. Thus, $50 = 24\times t$. Therefore, $t = 2.1 \text{ s}$. The reaction time is approximately 0.3 s, hence the driver was alert.

    (iii) The speed limit is 100 km h^{-1}, i.e. $\frac{100\times 1000}{3600} \text{ m s}^{-1} \approx 27.8 \text{ m s}^{-1}$. Thus, the driver was not speeding, as the speed of 24 m s^{-1} is less than the speed limit.

19. Let *A* be the car, *B* be the ambulance and *E* be the Earth. The +ve x direction is north. Therefore, $\underset{\sim}{v}_{A,E} = 70 \text{ km h}^{-1} \uparrow$ and $\underset{\sim}{v}_{B,E} = 100 \text{ km h}^{-1} \uparrow$.

    The velocity of the ambulance *(B)* relative to the car *A*, $v_{B,A}$ is

    $$\underset{\sim}{v}_{B,A} = \underset{\sim}{v}_{B,E} + \underset{\sim}{v}_{E,A} = \underset{\sim}{v}_{B,E} - \underset{\sim}{v}_{A,E}$$

    Thus, $\underset{\sim}{v}_{B,A} = \underset{\sim}{v}_{B,E} - \underset{\sim}{v}_{A,E} = \left(100 \text{ km h}^{-1} \uparrow\right) - \left(70 \text{ km h}^{-1} \uparrow\right) = 30 \text{ km h}^{-1} \uparrow$

    i.e. the velocity of the ambulance relative to the car *A*, $v_{B,A}$ is 30 km h^{-} towards north direction.

20. Let *T* be the train, *M* be the motorcycle rider and *E* be the Earth. The +ve x direction is east. Therefore, $\underset{\sim}{v}_{T,E} = 75 \text{ km h}^{-1} \rightarrow$ and $\underset{\sim}{v}_{M,E} = 85 \text{ km h}^{-1} \rightarrow$.

    The velocity of the motorcycle rider (M) relative to the train T, $v_{M,T}$ is

    $$\underset{\sim}{v}_{M,T} = \underset{\sim}{v}_{M,E} + \underset{\sim}{v}_{E,T} = \underset{\sim}{v}_{M,E} - \underset{\sim}{v}_{T,E}$$

    Thus, $\underset{\sim}{v}_{M,T} = \underset{\sim}{v}_{M,E} - \underset{\sim}{v}_{T,E} = \left(85 \text{ km h}^{-1} \rightarrow\right) - \left(75 \text{ km h}^{-1} \rightarrow\right) = 10 \text{ km h}^{-1} \rightarrow$

Let the time taken by the motorcycle rider to cross the entire length of the train be t. Thus, $t = \frac{\text{displacement}}{v_{M,T}} = \frac{175\ \text{m}}{10\ \text{km h}^{-1}} = \frac{175\ \text{m} \times 3600\ \text{s}}{10 \times 1000\ \text{m}} = 63\ \text{s}$.

Chapter 8

1. (i) Yes, the projectile is in free fall.

   (ii) Its vertical component of acceleration is the downward acceleration of gravity.

   (iii) Its horizontal component of acceleration is zero.

2. Yes, it is possible for two different projectiles with identical initial speeds but different angles of elevation to land in the same spot. Horizontal range of a projectile is proportional to its horizontal velocity and also its time of flight.

3.

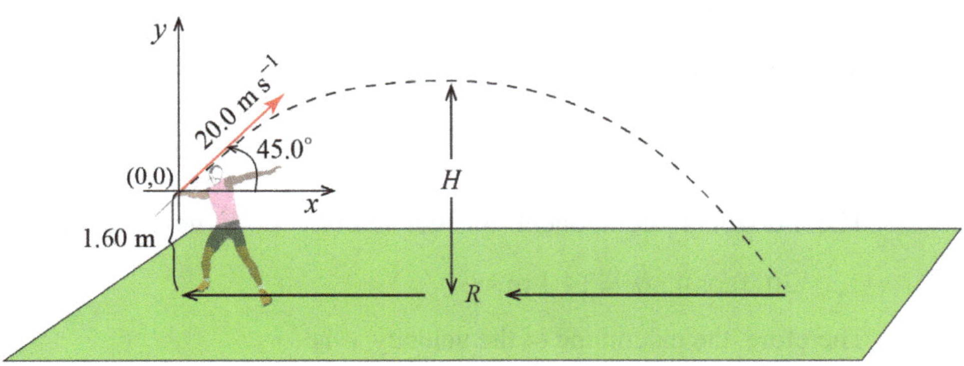

   (i) Use the equation $v^2 = u^2 + 2as$ for the projectile motion in the vertical direction from the initial point to the highest point.

$$0 = \left(20.0 \sin 45.0^\circ\right)^2 + 2(-10)(H - 1.60)$$

$$\Rightarrow\ H = \frac{\left(20.0 \sin 45.0^\circ\right)^2}{2 \times 10} + 1.60 = 11.6\ \text{m}$$

   Therefore, the maximum height that the stone reaches above the ground is $11.6\ \text{m}$.

(ii) Use the equation $s = ut + \frac{1}{2}at^2$ for the projectile motion in the vertical direction from the initial point to the point where the javelin lands on the ground.

$$-1.60 = \left(20.0\sin 45.0^\circ\right)t_f + \frac{1}{2}(-10)t_f^2$$

$$\Rightarrow \; 5t_f^2 - 14.1t_f - 1.60 = 0 \quad \Rightarrow \; t_f = \frac{14.1 \pm \sqrt{(-14.1)^2 - 4\times 5\times(-1.60)}}{2\times 5}$$

$$\Rightarrow \; t_f = 2.93 \text{ s}$$

(iii) Use the equation $s = ut + \frac{1}{2}at^2$ for the projectile motion in the horizontal direction from the initial point to the point where the javelin lands on the ground.

$$R = \left(20.0\cos 45.0^\circ\right)\times 2.93 + 0 = 41.4 \text{ m}.$$

(iv) Use the equation $v = u + at$ for the projectile motion in the vertical direction from the initial point to the point where the javelin lands on the ground.

$$v_y = 20.0\sin 45.0^\circ + (-10)\times 2.93 = -15.2 \text{ m s}^{-1},$$

The horizontal component of the velocity does not change. Thus $v_x = 20.0\cos 45.0^\circ = 14.1 \text{ m s}^{-1}$.

Therefore, the magnitude of the velocity v is

$$|v| = \sqrt{v_x^2 + v_y^2} = \sqrt{14.1^2 + 15.2^2} = 20.7 \text{ m s}^{-1}.$$

The direction with the horizontal ground

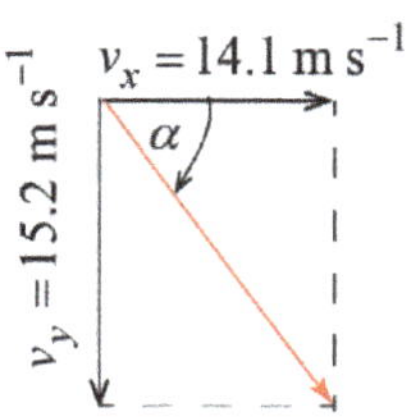

$$\tan\alpha = \frac{15.2}{14.1} \quad \Rightarrow \quad \alpha = \tan^{-1}\left(\frac{15.2}{14.1}\right) = 47.1^\circ \text{ Clockwise from the ground.}$$

4.

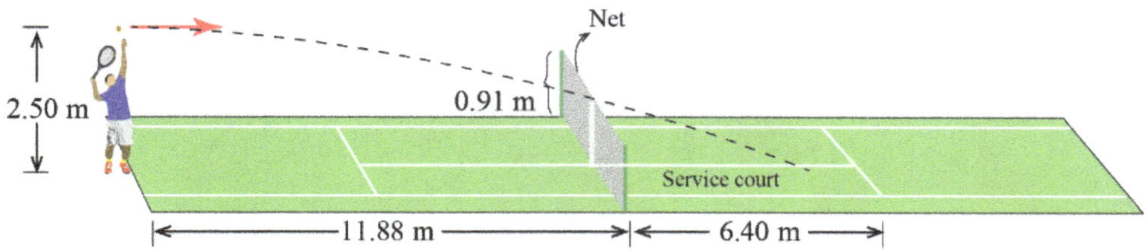

Let the origin be the point on the ground directly below the ball's initial position, and define the upward direction as the positive y axis.

(i) Use the equation $s = ut + \frac{1}{2}at^2$ for the projectile motion in the vertical direction from the initial point of the ball to the point where the ball to clear the 0.91 m height net.

$$0.91 - 2.50 = 0 + \frac{1}{2}(-10)t^2 \qquad \Rightarrow \qquad t = 0.564 \text{ s}$$

Use the equation $s = ut + \frac{1}{2}at^2$ for the projectile motion in the horizontal direction from the initial point of the ball to the point where the ball to clear the 0.91 m height net.

$$11.88 = v_x \times 0.564 \text{ s} + 0 \qquad \Rightarrow \qquad v_x = 21.1 \text{ m s}^{-1}$$

(ii) Use the equation $s = ut + \frac{1}{2}at^2$ for the projectile motion in the vertical direction from the initial point of the ball to the point where the ball hits the ground.

$$0 - 2.50 = 0 + \frac{1}{2}(-10)t^2 \qquad \Rightarrow \qquad t = 0.707 \text{ s}$$

(iii) Use the equation $s = ut + \frac{1}{2}at^2$ for the projectile motion in the horizontal direction from the initial point of the ball to the point where the ball hits the ground. Let R be the horizontal displacement when the ball hits the ground. Thus,

$$R = 21.1 \times 0.707 \text{ s} + 0 \quad \Rightarrow \quad R = 14.9 \text{ m}$$

Since the horizontal displacement (R) of the ball when it hits the ground is between 11.88 m and 18.28 m, the ball will land on the service court with the minimum velocity of $v_x = 21.1 \text{ m s}^{-1}$.

5.

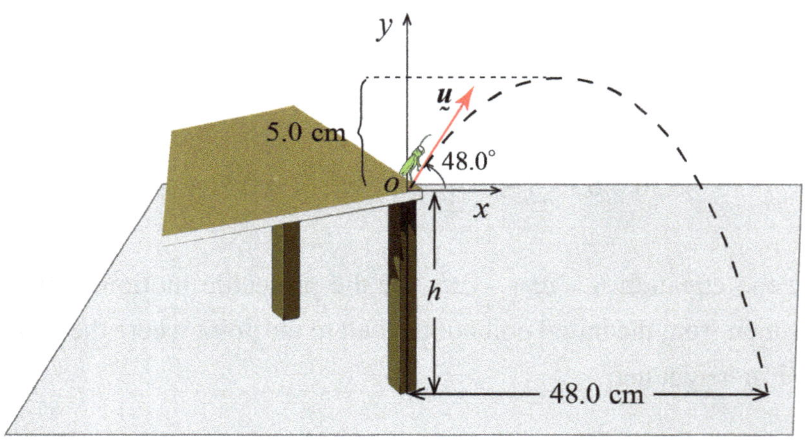

(i) Use the equation $v^2 = u^2 + 2as$ for the projectile motion in the vertical direction from the initial point of the highest point.

$$0 = \left(u \sin 48.0^\circ\right)^2 + 2(-10) \times \left(5.0 \times 10^{-2}\right)$$

$$\Rightarrow \quad u = \frac{\sqrt{2 \times 10 \times 5.0 \times 10^{-2}}}{\sin 48.0^\circ} = 1.35 \text{ m s}^{-1}$$

Hence, the initial velocity of the grasshopper is 1.35 m s^{-1}.

(ii) Use the equation $s = ut + \frac{1}{2}at^2$ for the projectile motion in the horizontal direction from the initial point to the point where the grasshopper lands on the ground.

$$50.0 \times 10^{-2} = 1.35 \times \cos 48.0^\circ t + 0 \quad \Rightarrow \quad t = \frac{48.0 \times 10^{-2}}{1.35 \times \cos 48.0^\circ} = 0.53 \text{ s}$$

Hence, the time taken for the grasshopper to land on the ground is 0.53 s.

(iii) Use the equation $s = ut + \frac{1}{2}at^2$ for the projectile motion in the vertical direction from the initial point to the point where the grasshopper lands on the ground. Let h be the height of the table. Thus,

$$-h = 1.35 \times \sin 48.0^\circ \times 0.53 + \frac{1}{2} \times (-10) \times (0.53)^2 \Rightarrow \quad h = 0.873 \text{ m} = 87.3 \text{ cm}$$

Hence, the height of the table is 87.3 cm.

6. (i) Use the equation $s = ut + \frac{1}{2}at^2$ for the projectile motion of the soccer ball in the vertical direction from the initial point to the point where the ball hits the ground.

$0 = 20.0 \sin 45.0^\circ t + \frac{1}{2}(-10)t^2 \quad \Rightarrow \quad t\left(5t - 20.0 \sin 45.0^\circ\right) = 0$, thus $t = 2.83 \text{ s}$.

The time that the soccer ball remains in the air before it hits the ground is 2.83 s.

(ii) Use the equation $s = ut + \frac{1}{2}at^2$ for the projectile motion of the soccer ball in the horizontal direction from the initial point to the point where the ball hits the ground.

$R = 20.0 \cos 45.0^\circ \times 2.83 + 0 = 40.0 \text{ m}$. Hence, the horizontal distance travelled by the soccer ball from the initial kicking point is 40.0 m.

(iii) The average velocity of the running player to meet the ball just before it lands on the ground is $v_{avg} = \dfrac{\text{displavement}}{\text{time}} = \dfrac{40.0-50.0}{2.83} = -\dfrac{10.0}{2.83} = -3.53 \text{ m s}^{-1}$.

The average speed of the running player is 3.53 m s^{-1} and the direction of the running player will be opposite to the direction of the kicking (horizontal) direction.

7. Take the vertical upward direction as $+y$. Use the equation $s = ut + \frac{1}{2}at^2$ for the projectile motion in the vertical direction from the initial point where the water leave from the cannon to the water hit the blaze.

The vertical displacement $= (12-1.5) \text{ m} = 10.5 \text{ m}$, The vertical component velocity, $u_y = 30\sin 45.0° \text{ m s}^{-1}$. Thus,

$10.5 = 30\sin 45.0° t + \frac{1}{2}(-10)t^2 \quad \Rightarrow \quad 5t^2 - 21.21t + 10.5 = 0$. Hence,

$$t = \frac{21.21 \pm \sqrt{21.21^2 - 4\times 5\times 10.5}}{2\times 5} = \frac{21.21 \pm 15.49}{10}, \quad t = 3.67 \text{ s} \text{ or } t = 0.57 \text{ s}$$

Both answers for the time are acceptable. Now use the equation $s = ut + \frac{1}{2}at^2$ for the projectile motion in the horizontal direction from the initial point where the water leaves from the cannon to hit the blaze.

For $t = 0.57$ s $\quad R = 30\cos 45.0° \times 0.57 + 0 = 12.1$ m

For $t = 3.67$ s $\quad R = 30\cos 45.0° \times 3.67 + 0 = 77.9$ m

This shows that when the cannon is 12.1 m from the building the water hits the blaze on its way up to its maximum height. When the cannon is 77.9 m from the building the water hits the blaze after it has passed through its maximum height.

8. Let $\underset{\sim}{v}_{A,E}$ be the velocity of the air relative to Earth, $\underset{\sim}{v}_{P,E}$ be the velocity of the aeroplane relative to Earth, and $\underset{\sim}{v}_{P,A}$ be the velocity of the aeroplane relative to air. Since the pilot notices a westward heading on the compass, the aeroplane's velocity relative to the air must be pointing westward. Thus the vector diagram will be

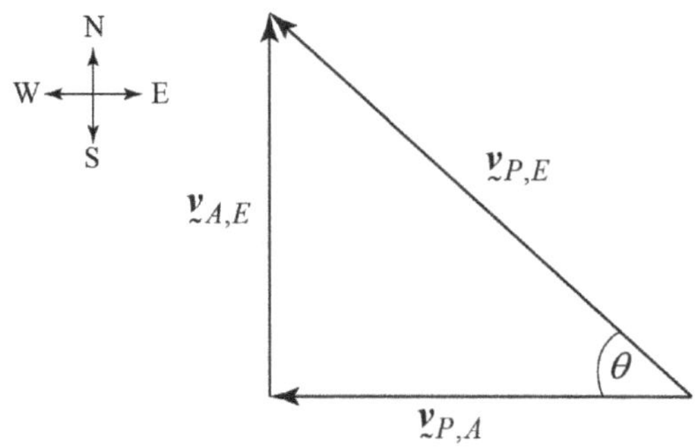

Therefore, using Pythagorean theorem, the magnitide of the velocity of airplane relative to the Earth will be $\left|\underset{\sim}{v}_{P,E}\right| = v_{P,E} = \sqrt{v_{P,A}^2 + v_{A,E}^2}$. Hence,

$$v_{P,E} = \sqrt{200^2 + 35.0^2} = 203 \text{ km h}^{-1}$$

$$\theta = \tan^{-1}\left(\frac{35.0}{200}\right) = 9.9^\circ \text{ north of west.}$$

9. Let $\underset{\sim}{v}_{W,E}$ be the velocity of the wind relative to Earth, $\underset{\sim}{v}_{B,E}$ be the velocity of the bird relative to Earth, and $\underset{\sim}{v}_{B,W}$ be the velocity of the bird relative to wind. We have $\underset{\sim}{v}_{B,W} = 115 \text{ km h}^{-1}$ in the north-south direction and $\underset{\sim}{v}_{W,E} = 44 \text{ km h}^{-1}$ from west to east. Thus, vector diagram would be

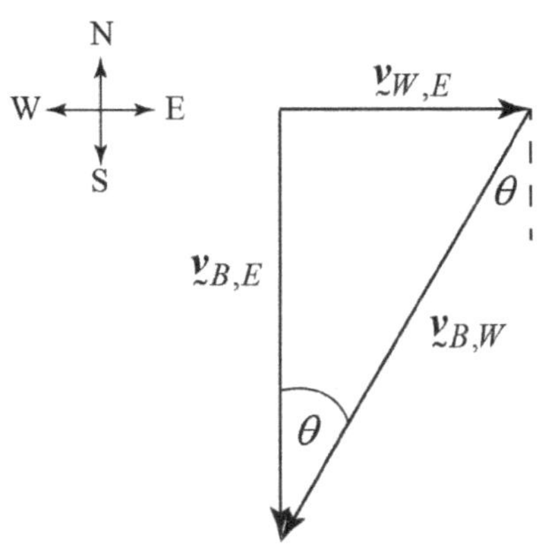

The relative velocity addition will be $\underset{\sim}{v}_{B,E} = \underset{\sim}{v}_{B,W} + \underset{\sim}{v}_{W,E}$

(i) $\sin\theta = \dfrac{|\underset{\sim}{v}_{W,E}|}{|\underset{\sim}{v}_{B,W}|} = \dfrac{44}{115} \quad \Rightarrow \quad \theta = 22.5^\circ$ west of south.

(ii) Using Pythagorean theorem, the magnitide of the velocity of bird relative to the Earth : $v_{B,E} = \sqrt{v_{B,W}^2 - v_{W,E}^2}$. Thus,

$$v_{B,E} = \sqrt{115^2 - 44^2} = 106 \text{ km h}^{-1}$$

The speed of the bird relative to the Earth is less than its speed relative to the wind. This is due to its velocity relative to the air is directed to oppose the effect of the wind.

(iii) Time taken by the bird to cover a ground distance of 500 km from north to south is $t = \dfrac{500 \text{ km}}{106 \text{ km h}^{-1}} = 4.7 \text{ h}$.

10. The cutter must move with the moving alumium sheet to cut perpendicularly to the direction of motion of the conveyor belt. Thus, the cutter must be set at some angle with respect to the width of the belt and toward the direction of motion of the belt. Let d be the distance that the alumium sheet travels in time t that it takes the cutter to cut it. Thus,

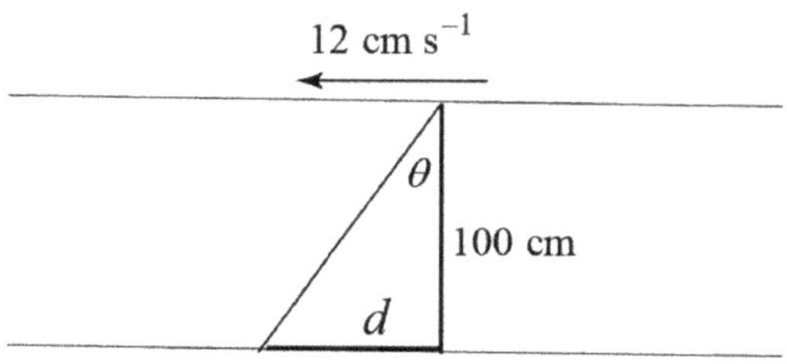

The speed of the convey belt is 12 cm s^{-1}. Therefore, $d = 12 \times t$ cm . Further the cutter moves across the width of 100 cm at a speed of $20 \text{ cm s}^{\ 1}$. Thus, the time t taken by the cutter to cut the sheet would be $t = \dfrac{100}{20} = 5.0$ s . Therefore, $d = 12 \times 5 = 60$ cm . Hence, $\tan\theta = \dfrac{d}{100} = \dfrac{60}{100} \quad \Rightarrow \quad \theta = \tan^{-1}(0.6) = 31^\circ$.

Chapter 9

1. In a car accident, the force caused by the collision changes the motion of the car, but the driver and passengers continue to move according to Newton's first Law. Seat belts supply the force necessary to change their motion and slow them down. For example, without seat belts, the driver and passengers would collide with the steering wheel, wind shield, or other objects inside the car and would be at increased risk of injury.
2. When the dog shakes, its wet fur changes velocity back and forth. Water will only remain in the fur if it has the same velocity as the fur, so there should be a sufficiently large force holding it while the dog shakes. When the force is not sufficient, drops of water lose contact with the fur and experience no more force from the dog's motion. From the principle of inertia (Newton's first Law), these drops resist changes in velocity, so they fly off the dog's body with whatever velocity they had when they lost contact. The drops in the air will then fall to the ground due to the force of gravity. As the dog continues to shake its body, more drops are shaken and fall to the ground.
3. If an object is influenced exclusively by a single non-zero force, it is impossible to have zero net force acting on it. Since the object is in free fall, it should be influenced solely by gravity and must have a non-zero net force. For an object to be in equilibrium, the net external force acting on it must equal zero. Therefore, an object in free fall cannot be in a state of equilibrium.
4. According to the laws of physics, an object cannot exert a force on itself. This is because the concept of force involves an interaction between two separate entities. An object cannot push or pull itself. It requires an external force or interaction with its surroundings to cause any change in its motion or acceleration. So, the idea of an object exerting a force on itself is not feasible within the framework of Newtonian mechanics.

5. Resultant force due to horizontal components

$$R_x = 4500\cos 30^\circ + 4500\cos 30^\circ = 7794 \text{ N}$$

Resultant force due to vertical components

$$R_y = 4500\sin 30^\circ - 4500\sin 30^\circ = 0$$

Thus the resultant force $R = 7794$ N ; along the horizontal forward direction

6. The force diagram

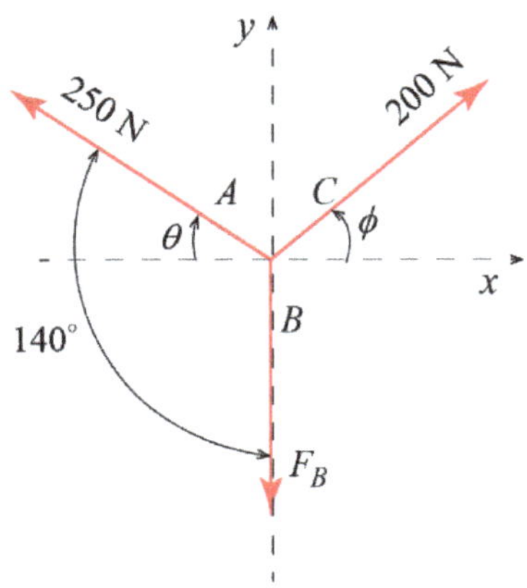

Despite the three pulls, the tyre remains stationary. Thus, the resultant component along x and y directions would be zero. Therefore,

$$R_x = 200\cos\phi - 250\cos\theta = 0 \quad \Rightarrow \quad \cos\phi = \frac{250\cos\theta}{200}$$

But we have, $90^\circ + \theta = 140^\circ \quad \Rightarrow \quad \theta = 140^\circ - 90^\circ = 50^\circ$,

hence $\cos\phi = \dfrac{250\cos 50^\circ}{200} \quad \Rightarrow \quad \phi = 36.5^\circ$

Also we have, $R_y = 200\sin 36.5^\circ + 250\sin 50^\circ - F_B = 0$. Therefore,

$F_B = 200\sin 36.5^\circ + 250\sin 50^\circ = 310.5 \text{ N} \approx 311 \text{ N}$.

7. The force diagram

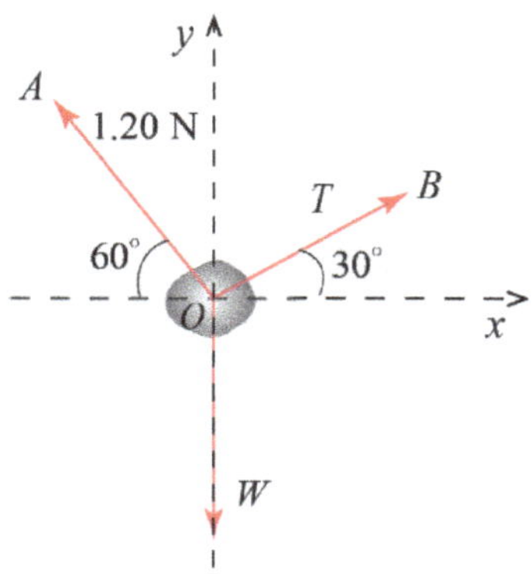

Stone suspended by two strings, *A* and *B*, at rest. Thus, the resultant component along x and y directions would be zero. Therefore,

(i) $R_x = T\cos 30^\circ - 1.20\cos 60^\circ = 0 \quad \Rightarrow \quad T = \dfrac{1.20\cos 60^\circ}{\cos 30^\circ} = 0.69 \text{ N}$

(ii) $R_y = 0.69\sin 30^\circ + 1.20\sin 60^\circ - W = 0$

$\Rightarrow \quad W = 0.69\sin 30^\circ + 1.20\sin 60^\circ = 1.38 \text{ N}$

8. (i) The free-body diagram for the crate

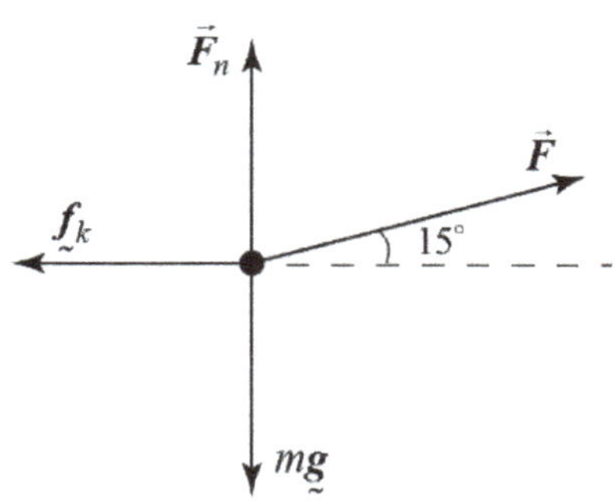

(ii) Motion is in the horizontal direction, so the net horizontal force is causing the crate to accelerate. Let the direction of the motion be the positive horizontal direction. Since there is no motion in the vertical direction, the vertical acceleration is zero. If the acceleration is zero, the net vertical force must be zero. Thus, $F\sin 15^\circ + F_n - mg = 0$

Hence, $F_n = 50 \times 10 - 120 \sin 15^\circ = 468.9$ N

(iii) Kinetic friction applies to the crate during the motion. Apply Newton's Second Law to the net horizontal force acting on the crate. Thus,

$F \cos 15^\circ - f_k = ma$ and $f_k = \mu_k F_n$

Hence, $a = \dfrac{120 \cos 15^\circ - 0.2 \times 468.9}{50} = 0.44 \text{ m s}^{-}$

(iv) Use the kinematic equation that relates acceleration, initial velocity, final velocity, and time. Thus,

$$v = u + a\Delta t$$

Hence,
$$\Delta t = \frac{v-u}{a} = \frac{1.5-0}{0.44} = 3.4 \text{ s}$$

9. Let $F_{B \text{ on } A}$ and $F_{A \text{ on } B}$ be the magnitude of the the force of contact between the two blocks A and B respectively, when force $\vec{F}$ acts on block A. According to Newton's Third Law, we have $F_{B \text{ on } A}$ $F_{A \text{ on } B}$.

(i) (a)

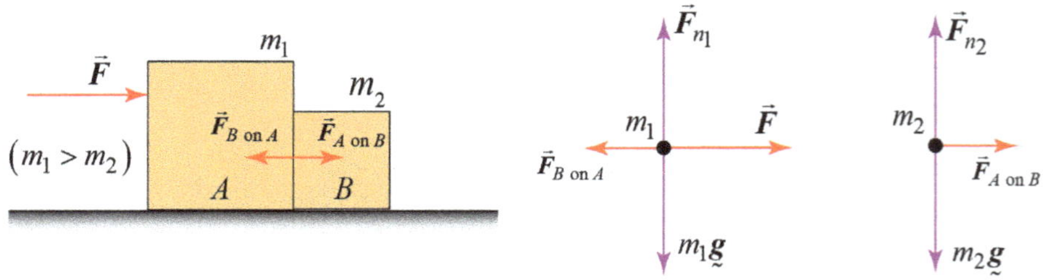

Free-body diagram for block A and B

(b) Let a be the magnitude of the acceleration along the direction of the applied force $\vec{F}$. Applying Newton's Second Law for block A, we have,

$$F - F_{B \text{ on } A} = m_1 a .$$

Since the blocks move together they have the same acceleration. Thus, applying Newton's Second Law for block B, we have,

$$F_{A \text{ on } B} = m_2 a .$$

But we have, $F_{B \text{ on } A} = F_{A \text{ on } B}$. Hence,

$$F - F_{B \text{ on } A} = m_1 a \qquad \Rightarrow \quad F - m_2 a = m_1 a$$

Therefore, $a = \dfrac{F}{m_1 + m_2}$

Thus, the force of contact between the two blocks A and B would be

$$F_{B \text{ on } A} = F_{A \text{ on } B} = \frac{m_2}{m_1 + m_2}$$

(ii) Let $F'_{B \text{ on } A}$ and $F'_{A \text{ on } B}$ be the magnitude of the force of contact between the two blocks A and B, respectively when force $\vec{F}$ acts on block B. According to Newton's Third Law, we have $F'_{B \text{ on } A} = F'_{A \text{ on } B}$.

(a)

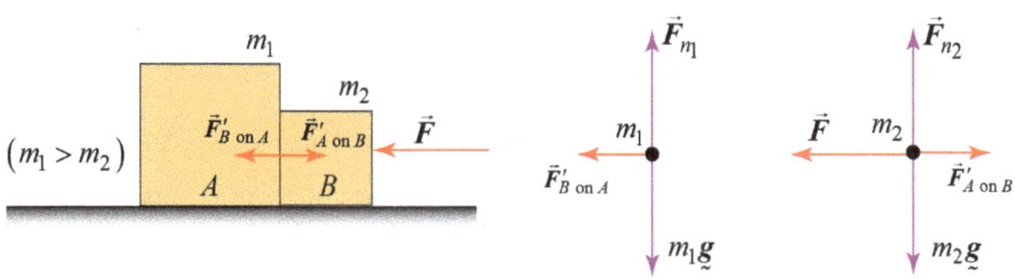

Free-body diagram for block A and B

(b) Let a be the magnitude of the acceleration along the direction of the applied force $\vec{F}$. Applying Newton's Second Law for block B we have,

$F - F'_{A \text{ on } B} = m_2 a$.

Since the blocks move together, they have the same acceleration. Thus, applying Newton's Second Law for block B we have,

$F'_{B \text{ on } A} = m_1 a$.

But we have, $F'_{B \text{ on } A} = F'_{A \text{ on } B}$. Hence,

$$F - F'_{A \text{ on } B} = m_2 a \qquad \Rightarrow \quad F'_{B \text{ on } A} = m_1 a$$

Therefore, $a = \frac{F}{m_1 + m_2}$.

Thus, the force of contact between the two blocks A and B would be

$$F'_{B \text{ on } A} = F'_{A \text{ on } B} = \frac{m_1}{m_1 + m_2}.$$

Since, $m_1 > m_2$, the $F'_{B \text{ on } A} > F_{B \text{ on } A}$.

This means that the contact force between blocks obtained in part (ii) must be greater than that obtained in part (i).

The magnitude of the acceleration is the same in both cases. In part (i), the contact force $F_{A \text{ on } B} \left(= F_{B \text{ on } A}\right)$ is the only horizontal force on block B (where its mass is m_2), whereas, in part (ii), the contact force $F'_{B \text{ on } A} \left(= F'_{A \text{ on } B}\right)$ is the only horizontal force on block A (where its mass is m_1).

10. (i) FBD for bricks, pulley P, and pulley Q.

(ii) Consider the FBD for bricks. Two forces act on the brick load, the

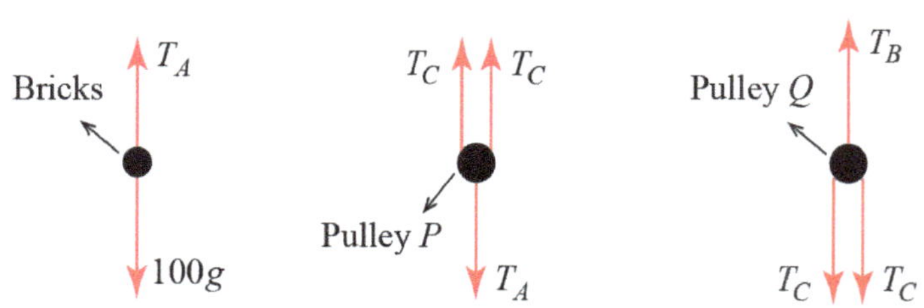

gravitational force, which is 1000 N downward and the upward pull (the tension T_A) of rope A. The load of bricks is moving upward at a constant speed. Thus, the acceleration is zero. Hence, the net force is zero. Therefore, $T_A - 1000 = 0$, and $T_A = 1000$ N.

Let us consider the FBD for the pulley P. Rope A pulls down with a force of magnitude T_A, and rope C pulls upward on each side. Since the rope is massless, the rope will have the same tension throughout, so all forces labelled T_C in the FBD for pulley P have the same magnitude. Since it is moving at a constant speed, the acceleration is zero. Thus, the net force is zero. Hence, $2T_C - T_A = 0$. Therefore, $T_C = \frac{1}{2}T_A = \frac{1}{2}\times 1000 = 500$ N.

Let us consider the FBD for the pulley *Q*. Rope *B* pulls upward with a force of magnitude T_B. On each side of the pulley *Q*, rope *C* pulls downward with a force of magnitude T_C. Since it is moving at a constant speed, the acceleration is zero. Thus, the net force is zero. Hence, $T_B - 2T_C = 0$. Therefore, $T_B = 2T_C = 1000$ N.

(iii) In this case, rope *C* is pulled on the load of bricks using the pulley *P* with the total magnitude of force $2T_C$ (rope *C* pulling upward on each side). Therefore, the worker pulling the rope only needs to exert a force equal to half the weight of the brick load.

11. The crow is resting on the clothes line. The system is at rest; thus, the acceleration is zero. Drawing a diagram.

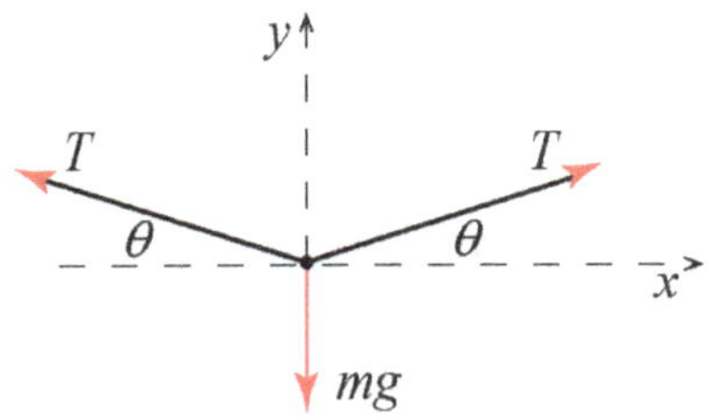

Applying Newton's Second Law in y direction, we have.

$$\sum F_y = T\sin\theta + T\sin\theta - mg = 0, \text{ hence}$$

$$T = \frac{mg}{2\sin\theta}$$

12. (i) Since a rope connects the crates, both have the same acceleration 2 m s^{-2}.

(ii)

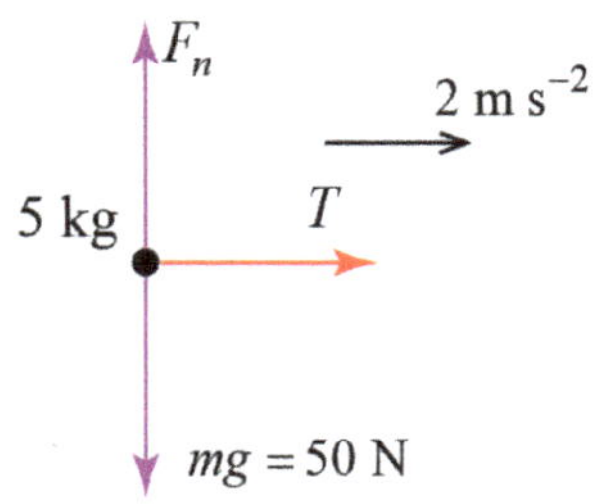

Free-body diagram for box A

(iii) Applying Newton's Second Law in the horizontal direction, we get

$$\sum F_x = T = 5.0 \times 2.0 = 10.0 \text{ N}, \text{ thus}$$

$$T = 10.0 \text{ N}$$

(iv)

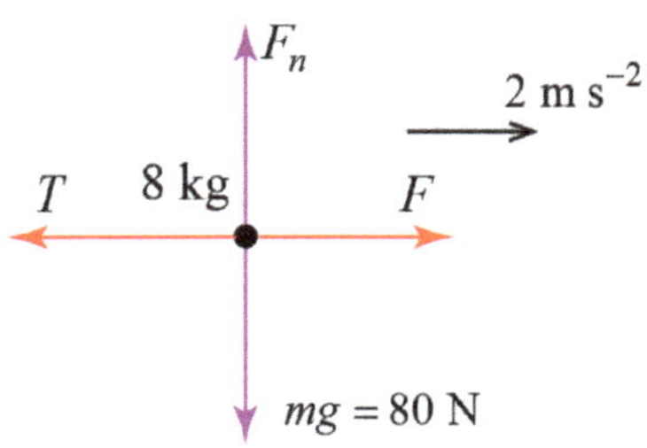

Free-body diagram for box B

(v) The box accelerates to the right, so the net force is to the right. Hence, F must be larger than T.

(vi) Applying Newton's Second Law in the horizontal direction, we get

$$\sum F_x = F - T = 8.0 \times 2.0 = 16 \text{ N}, \text{ thus}$$

$$F = T + 16.0 = 10 + 16 = 26 \text{ N}$$

13. (i) The free-body diagrams for the person. Consider the person as a point object.

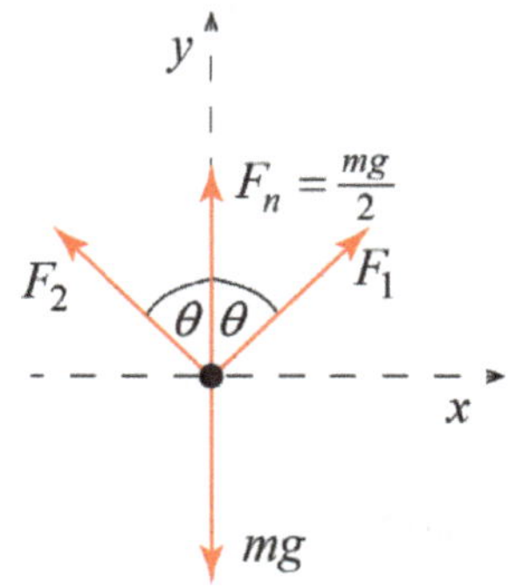

(ii) Let F_1 and F_2 be the magnitude of the compression force on the right and left crutches respectively. Applying Newton's Second Law along the +ve x direction, the person is not moving along the +ve x direction, thus the net force along that direction must be zero. $\sum F_x = F_1 \sin\theta - F_2 \sin\theta = 0$.

Therefore, $F_1 = F_2 = F$. Hence, both crutches will have compression force of F.

Applying Newton's Second Law along the +ve y (vertically upward) direction, the person is not moving in vertically upward direction, thus the net force along that direction must be zero. $\sum F_y = F\cos\theta + F\cos\theta + \frac{mg}{2} - mg = 0$.

Therefore, $2F\cos\theta = \frac{mg}{2}$. Hence $F = \frac{mg}{4\cos\theta}$.

(iii) The free-body diagrams of crutch tip. Let f be the friction force and $F_{n_{tip}}$ be the normal force.

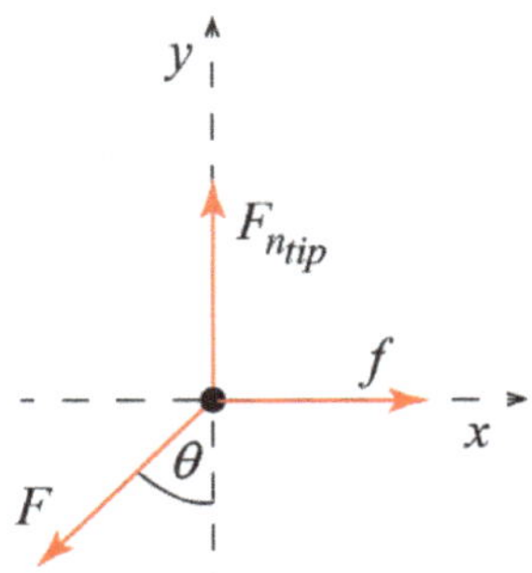

Applying Newton's Second Law along the +ve x direction, the person is not moving along the +ve x direction, thus the net force along that direction must be zero.

$$\sum F_x = f - F\sin\theta = 0$$

Therefore, $f = \dfrac{mg}{4\cos\theta}\sin\theta = \dfrac{1}{4}mg\tan\theta$

Applying Newton's Second Law along the +ve y (vertically upward) direction, the person is not moving vertically upward, thus the net force along that direction must be zero. i.e. $\sum F_y = F_{n_{tip}} - F\cos\theta = 0$.

Therefore, $F_{n_{tip}} = \dfrac{mg}{4\cos\theta}\cos\theta = \dfrac{mg}{4}$

For minimum coefficient of friction, the crutch tip will be on the verge of slipping. Hence, $f = (f_s)_{\text{max}} = \mu_s F_{n_{tip}}$

Therefore, $\mu_s = \dfrac{(f_s)_{\text{max}}}{F_{n_{tip}}} \Rightarrow \mu_s = \dfrac{\left(\dfrac{mg}{4}\tan\theta\right)}{\left(\dfrac{mg}{4}\right)} = \tan\theta \Rightarrow \mu_s = \tan\theta$

14. $\left|F_{spring}\right| = mg = kx$.

(i) $k = \dfrac{mg}{x} = \dfrac{900\times10^{-3}\times10}{(18.0-15.0)\times10^{-2}} = 300\text{ N m}^{-1}$

(ii) $m = \dfrac{kx}{g} = \dfrac{300\times(22.0-15.0)\times10^{-2}}{10} = 2.1\text{ kg}$

15. (i) Free-body diagrams of the two iron balls°

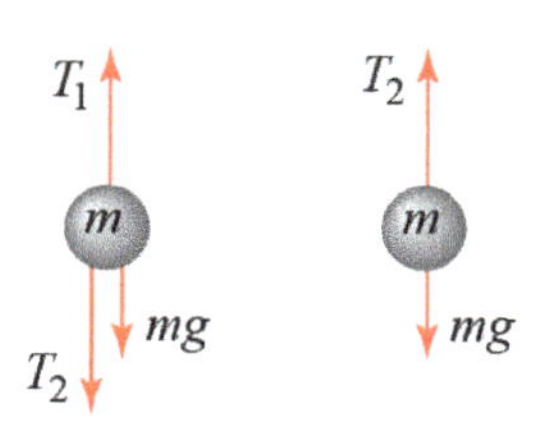

(ii) Applying Newton's Second Law to the lower iron ball in the $+y$ direction

$$T_2 - mg = ma_y \quad \Rightarrow \quad T_2 = m\left(g + a_y\right) = 3.0\left(10 + 1.50\right) = 34.5 \text{ N}$$

Hence, $T_2 = 34.5$ N.

Applying Newton's Second Law to the upper iron ball in the $+y$ direction

$$T_1 - T_2 - mg = ma_y \quad \Rightarrow \quad T_1 = T_2 + m\left(g + a_y\right) = 34.5 + 3.0\left(10 + 1.50\right) = 69.0 \text{ N}$$

Hence, $T_1 = 69.0$ N.

(iii) Note that the tension is greater in the upper string, and this string will break first as the acceleration of the system increases. Now, we need to find the value of a_y when $T_1 = 75.0$ N. Using the equation $T_1 = T_2 + m\left(g + a_y\right)$, we have

$$T_1 = m\left(g + a_y\right) + m\left(g + a_y\right) = 2mg + 2ma_y.$$

Therefore, $a_y = \dfrac{T_1 - 2mg}{2m} = \dfrac{75.0 - 2 \times 3.0 \times 10}{2 \times 3.0} = 2.5 \text{ m s}^{-2}$

Chapter 10

1. (i) A couple is a pair of forces of equal magnitude and opposite direction that act on an object but do not produce any translational motion. Instead, they create a rotational effect or moment on the object. The two forces in a couple have different lines of action, but they are parallel to each other.

   (ii) Torque is a measure of the turning effect produced by force around an axis or pivot point. It is also known as the moment of force. Torque is calculated as the product of the force and the perpendicular distance of the line of action of the force to the pivot point.

2. We apply a larger torque when we try to open a swinging door with force applied near the edge. However, to open the same swinging door, if the force is applied in the middle, we require a torque that is half what is needed when applied near the edge. Therefore, to open the swinging door in the middle, we require twice as much force as when the force is applied near the edge.

3. Consider the figure below. Distance of centre of mass from m_1 is $X_{CM} - x_1$. Distance of centre of mass from m_2 is $x_2 - X_{CM}$.

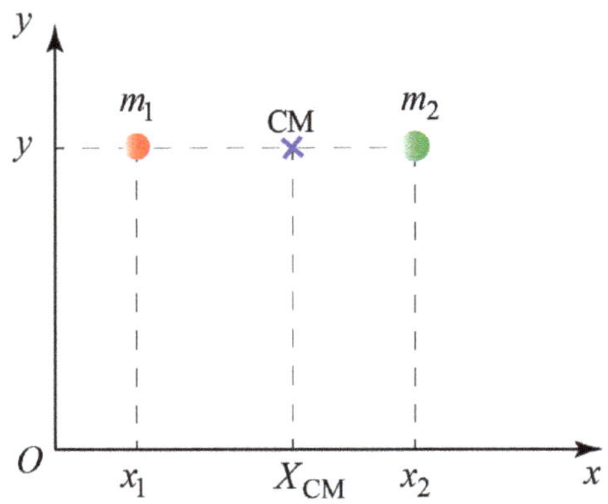

We have
$$X = \frac{m_1 x_1 + m_2 x_2}{m_1 + m_2}$$

Therefore the distance of centre of mass from m_1 is

$$X - x_1 = \frac{m_1 x_1 + m_2 x_2}{m_1 + m_2} - x_1 = \frac{m_2 (x_2 - x_1)}{m_1 + m_2}$$

The distance of the centre of mass from m_2 is

$$x_2 - X = x_2 - \frac{m_1 x_1 + m_2 x_2}{m_1 + m_2} = \frac{m_1 (x_2 - x_1)}{m_1 + m_2}$$

Thus,
$$\frac{\text{the distance of the CM from } m_1}{\text{the distance of the CM from } m_2} = \frac{m_2}{m_1}$$

4. The three masses are located as shown in the figure.

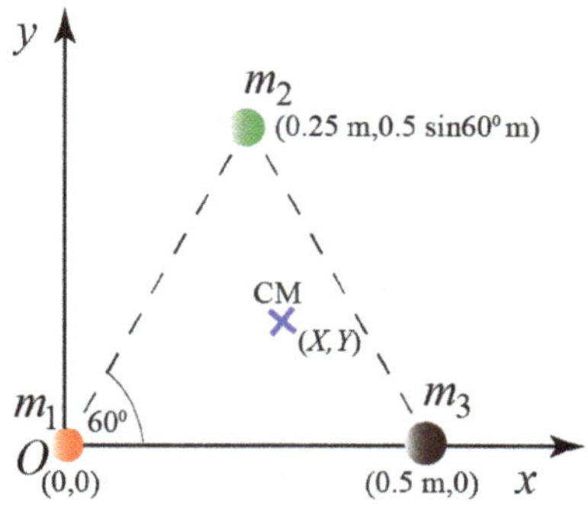

The x component of the CM

$$X = \frac{\sum m_i x_i}{\sum m_i} = \frac{2.0\times 0 + 4.0\times 0.5 + 6.0\times 0.25}{2.0+4.0+6.0} = 0.3 \text{ m}$$

The y component of the CM

$$Y = \frac{\sum m_i y_i}{\sum m_i} = \frac{2.0\times 0 + 4.0\times 0 + 6.0\times 0.5\sin 60^\circ}{2.0+4.0+6.0} = 0.2 \text{ m}$$

5. (i) The resultant centre of gravity (CG) will be in between CG_1 and CG_2 and on the horizontal line joining CG_1 and CG_2 as shown.

The ratio is

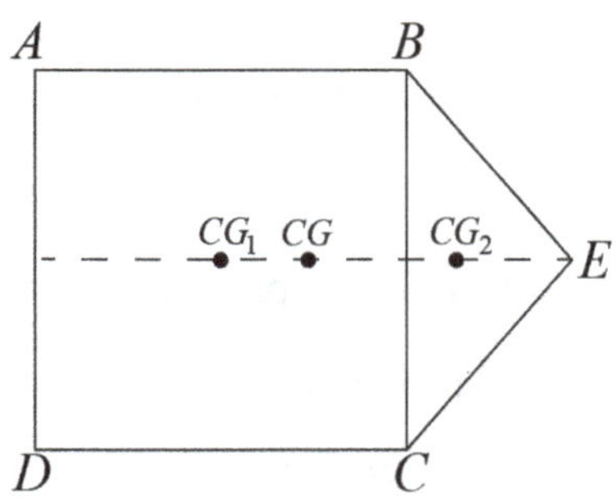

$$\frac{\text{the distance of the CG from } CG_1}{\text{the distance of the CG from } CG_2} = \frac{\text{mass of } BCE}{\text{mass of } ABCD} = \frac{\left(\frac{1}{2}\times 32.0\times 12.0\right)d\rho}{(32.0\times 32.0)\,d\rho} = \frac{3}{16}$$

where, d is the thickness and ρ is the density of the lamina.

(ii) At equlibrium the net moment of the system is zero.

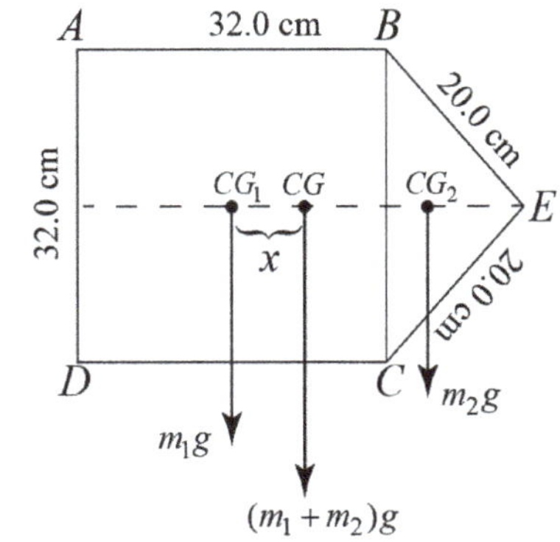

Thus, taking the moment about CG ↷, we have

$$m_2 g \times (20.0 - x) - m_1 x = (m_1 + m_2) g \times 0$$

$$\left\{\left(\frac{1}{2}\times 32.0\times 12.0\right)\times 10^{-4} d\rho g\right\}\times\left(20.0\times 10^{-2} - x\right) - \left\{(32.0\times 32.0)\times 10^{-4} d\rho g\right\} x = 0$$

Therefore, $x = \dfrac{120.0\times 10^{-2}}{38.0} = 3.16\times 10^{-2}$ m ≈ 3.2 cm

6. Diagram with components of the force.

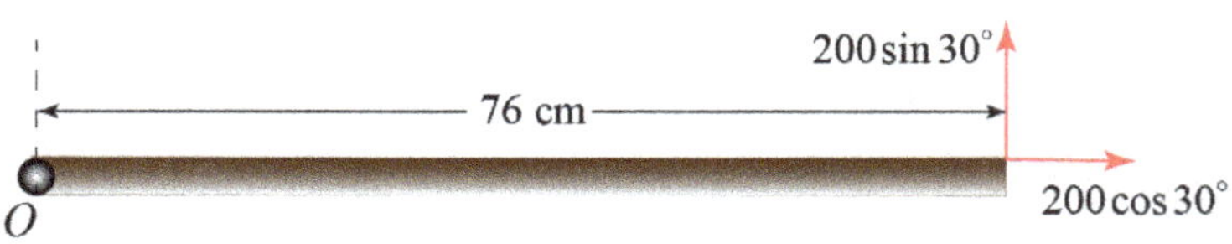

The force component $200\cos 30^\circ$ is passing through the hinge *O*, therefore this will not contribute to the moment. Thus, the moment of the force would be $\text{Moment}(\tau) = \left(200\sin 30^\circ\right)\times 0.76 = 76$ N m .

7. Moment = magnitude of the force × perpendicular distance between the forces.

   Thus, the moment applied to the handle $= 30 \times 0.70 = 21 \text{ N m}$.

8. (i) Rotation will be in the clockwise direction and the horizontal component of the force exerted by the road on the wheel will be in the forward direction.

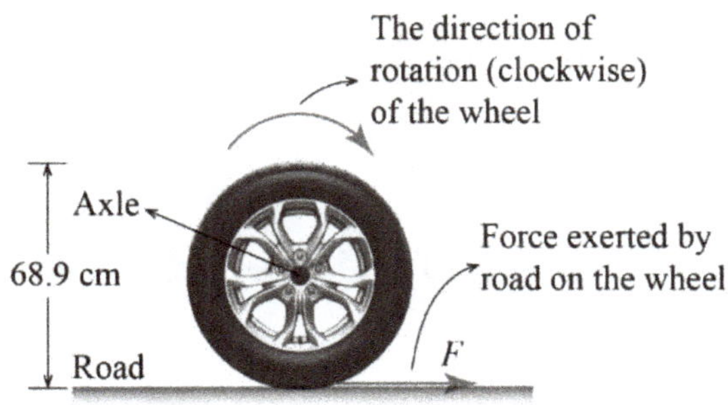

   (ii) The resultant torque acting on the wheel is zero. Because, the car moves at a constant velocity in the forward direction, the net torque acting on the wheel must be zero. Therefore, the torque produced by the engine must be equal and opposite to the torque produced by the horizontal component of the force exerted by the road on the wheel.

   (iii) $\tau = F \times \text{radius of the wheel}$,

$$F = \frac{250}{\left(\frac{0.689}{2}\right)} = \frac{2 \times 250}{0.689} = 725.7 \text{ N} \approx 726 \text{ N}$$

9. There are four forces acting on the system.

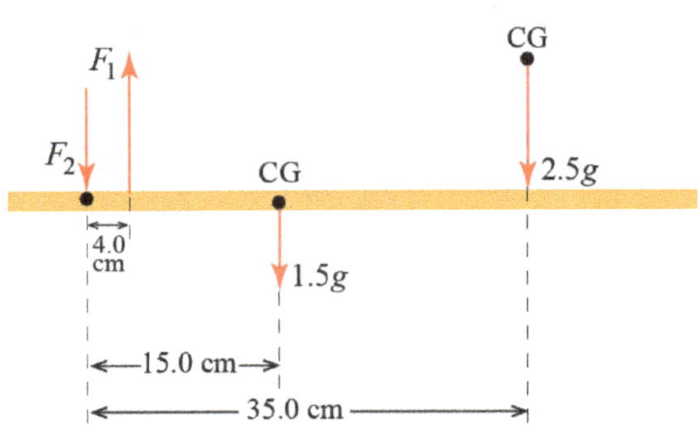

(i) The torques created by the weights of the block and the fore arm are clockwise relative to the elbow, while the torque created by the biceps is counter-clockwise relative to the elbow. At equilibrium, net torque is zero, i.e.,

$$\tau_{net} = 0$$

Thus, if we choose the pivot to be at the elbow, we have

$$2.5 \times 10 \times 35.0 \times 10^{-2} + 1.5 \times 10 \times 15.0 \times 10^{-2} - F_1 \times 4.0 \times 10^{-2} = 0$$

i.e. $F_1 \times 4.0 = 2.5 \times 10 \times 35.0 + 1.5 \times 10 \times 15.0 = 1100$

Therefore, $F_1 = 275$ N.

(ii) Free body diagram for the system

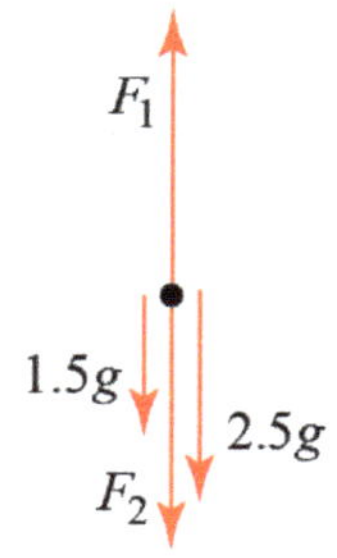

At equilibrium $F_2 + 1.5 \times 10 + 2.5 \times 10 = F_1$

i.e. $F_2 = 275 - 40 = 235$ N

(iii) The combined weight W of the arm and block is

$W = 1.5 \times 10 + 2.5 \times 10 = 40$ N

Hence, the ratio of the force exerted by the biceps to the total weight is

$$\frac{F_1}{W} = \frac{275}{40} = 6.875 \approx 6.9$$

This means that the biceps muscle exerts a force 6.9 times the combined weight of the arm and block.

The ratio of the force exerted at the elbow joint to the total weight is

$$\frac{F_2}{W} = \frac{235}{40} = 5.875 \approx 5.9$$

This means that the force exerted at the elbow joint is 5.9 times the combined weight of the arm and block.

Therefore, very large forces are created in the biceps muscle and the elbow joint. Since the biceps exerts its force on the arm at a point close to the elbow, it is much less effective in producing a rotation than forces exerted farther from the elbow, such as the weights of the fore arm and the block. To compensate for this, the biceps must exert a larger force. As a result, the force exerted on the elbow joint will also be very large.

10. The conditions for equilibrium are that the net external force equal zero and the net external torque equal zero:

$$\sum \vec{F}_{ext} = 0 \qquad \text{and} \qquad \sum \vec{\tau}_{ext} = 0$$

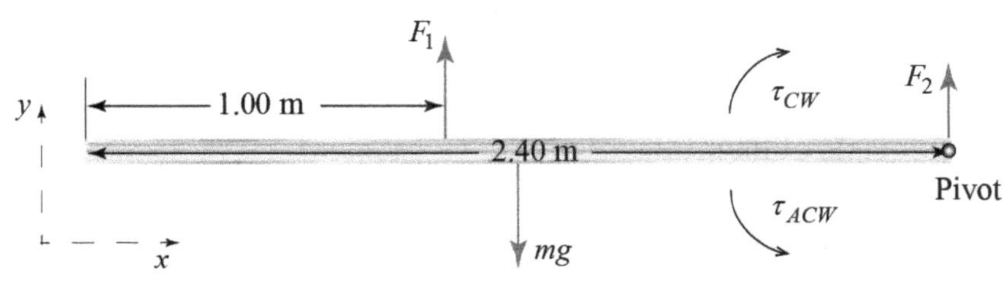

Let choose a rotation axis (pivot) perpendicular to the xy–plane and passing through the point of application of F_2. The distance of lever arm for F_1 from pivot is $(2.40-1.00)\,\text{m} = 1.40\ \text{m}$. Since the beam is uniform, its center of gravity is at its midpoint. Torque of F_1 is in clockwise (CW) direction and the torque of wooden beam is in anticlockwise (ACW) direction about the pivot. One of the conditions for equilibrium is that the net torque be zero. Thus,

$$\sum \vec{\tau}_{ext} = 0 \quad \Rightarrow \quad F_1 \times 1.40 - 50.0 \times 10 \times 1.20 = 0$$

Therefore, $F_1 = \dfrac{49.0 \times 10.0 \times 1.20}{1.40} = 420 \text{ N}$.

The other condition for equilibrium is that the net force be zero. Thus,

$$\sum F_y = 0 \quad \Rightarrow \quad F_1 + F_2 - 49.0 \times 10 = 0$$

Therefore, $F_2 = (490 - 420)\,\text{N} = 70 \text{ N}$

11. The weight of the rod 12 N acts at the centre of the rod.

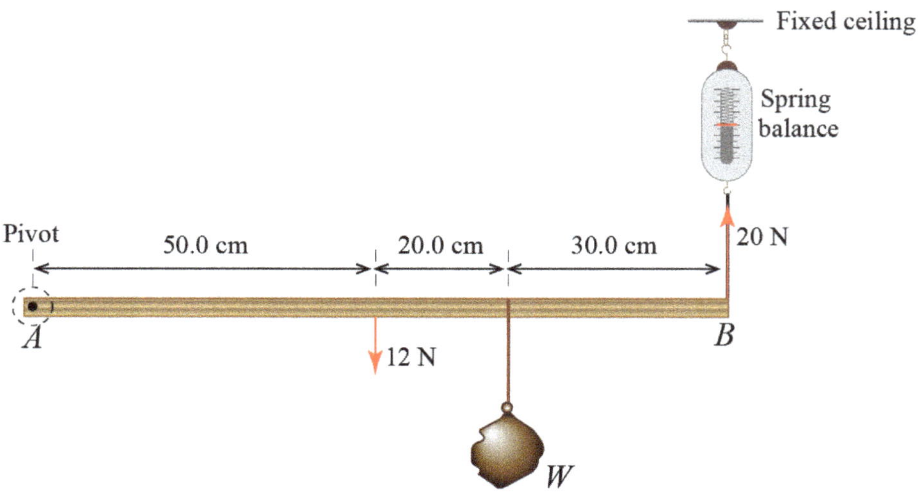

By the principle of moments

Sum of the clockwise moments about A = Sum of the anticlockwise moments about A

Thus, $W \times 70.0 \times 10^{-2} + 12 \times 50.0 \times 10^{-2} = 20 \times 100.0 \times 10^{-2}$

$$\Rightarrow \quad W = \frac{2000 - 600}{70.0} = 20 \text{ N}$$

Therefore, the mass of the irregular solid object is 2 kg.

12. (i) Choose the axis of rotation (pivot) at the point of contact between the plank and the right saw horse.

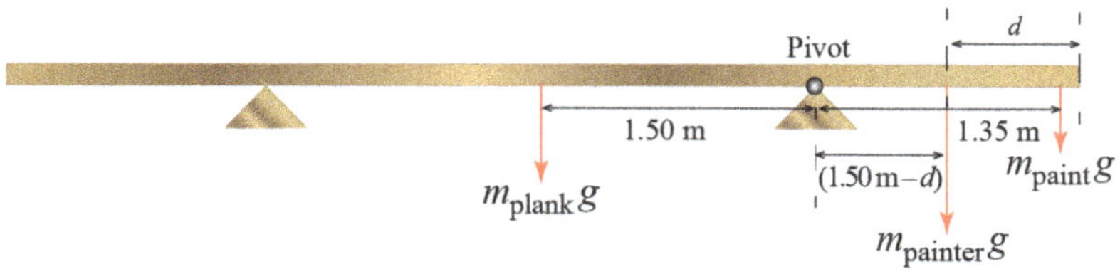

Let d be the distance of painter from the right-hand edge of the plank. At equilibrium $\sum \tau_{ext} = 0$, thus

$$20.0\times10\times1.50-70.0\times10\times(1.50-d)-4.0\times10\times1.35=0$$

i.e. $d=1.50-\dfrac{20.0\times10\times1.50-4.0\times10\times1.35}{60.0\times10}=1.50-0.41=1.09\text{ m}$

(ii) Choose the axis of rotation (pivot) at the point of contact between the plank and the left saw horse.

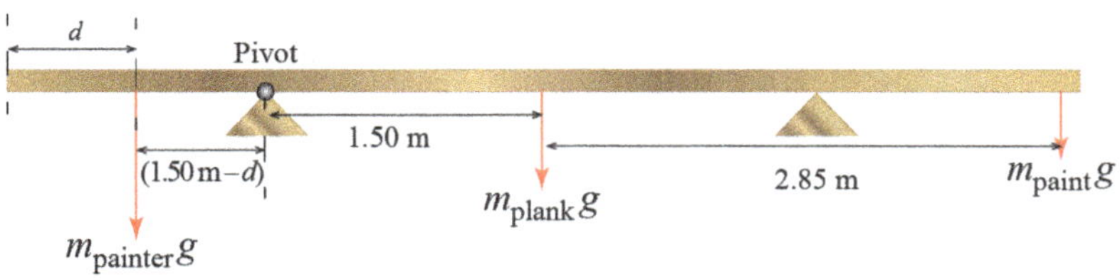

Let d be the distance of painter from the right-hand edge of the plank.

At equilibrium $\sum \tau_{ext} = 0$, thus

$$60.0\times10\times(1.50-d)-20.0\times10\times1.5-4.0\times10\times(1.5+2.85)=0$$

i.e. $d=1.50-\dfrac{20.0\times10\times1.5+4.0\times10\times(1.5+2.85)}{60.0\times10}=1.50-0.79=0.71\text{ m}$

Chapter 11

1. The student is engaged in reading a book. This may seem like the student is doing 'hard work', but no force is involved, so the student does not do any work in terms of physics.

2. During the fall, the force of gravity on the mango is parallel to the mango's displacement, so it does positive work.

3. Yes, static friction can do work. The force of static friction on the book is directed upward along the surface of the belt and has a component that is parallel to the book's displacement. Thus, the force of static friction does positive work on the book. At the same time, the work done on the book by gravity is negative, and the total work done on the book is zero.

4. (i) The gravitational potential energy is energy due to position in a gravitational field. When an object is raised above the Earth's surface, its gravitational potential energy increases. The elastic potential energy is the energy obtained when an object is stretched or compressed.

   (ii) Work done by the hand to draw back the rubber cord is stored as elastic potential energy in the catapult, and on releasing the pellet, it will be transferred as the kinetic energy of the pellet.

5. (i) In case (a), the force and the displacement are in the same direction. Thus the work done is positive, and the value is $W = Fx$.

   In case (b), the horizontal component of the force and the displacement are in the same direction. The vertical component of the force is perpendicular to the direction of the displacement. Therefore, only the horizontal component of the force will do the work, and its value is $W = \left(F\cos 60^\circ\right)x = \frac{1}{2}Fx$. Thus the work done is positive.

In case (c), the force and the displacement are in the opposite direction to each other. Thus, the work done is positive, and the value is $W = -Fx$.

In case (d), the force and the displacement are perpendicular to each other. Thus, the work done is zero.

(ii) The order of the work done by the force on the object, from most positive to most negative, is (a), (b), (d), and (c).

6.

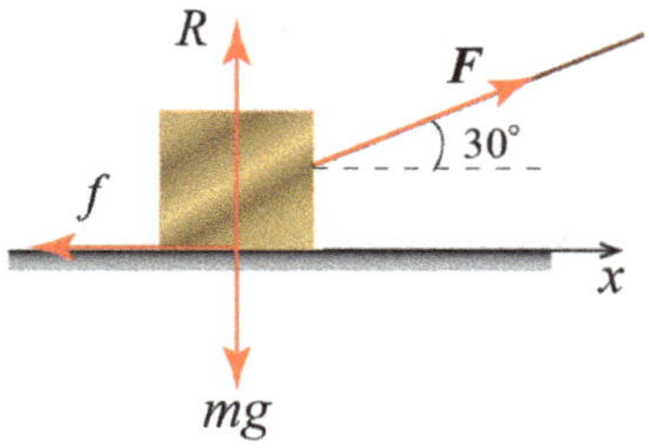

(i) Since the block is moving at a constant velocity, according to Newton's First Law of Motion, there cannot be any resultant force acting upon it. Hence the total work done on the block is zero.

(ii) The force exerted by the string on the block parallel to its displacement is $7.50\cos 30^\circ$ N, and the displacement is 4.00 m. Therefore, the work done by the string on the block is

$$= 7.50\cos 30^\circ \times 4.00 = 25.98 \text{ J}$$

(iii) No resultant force acts on the block horizontally. Therefore, the force of friction must be $7.50\cos 30^\circ$ N and acts opposite to the displacement. Since the displacement is 4 m, the work done by friction is

$$= \left(-7.50\cos 30^\circ\right) \times 4.00 = -25.98 \text{ J}$$

(iv) No resultant force acts on the block vertically.

Therefore, $R + 7.50\sin 30^\circ = 3.50 \times 10$. Thus, $R = 31.25$ N.

The force of friction on the block is $= 31.25 \times \mu_k$ N, where μ_k is the coefficient of kinetic friction. Therefore, $31.25 \times \mu_k = 7.5\cos 30^\circ$.

Thus, $\mu_k = \dfrac{7.5\cos 30^\circ}{31.25} = 0.21$.

7. The work done by the hammer on the nail is represented by the area between the curve and the x-axis. Thus, $W = \frac{1}{2}(20+50)\times 0.02 + 100\times 0.04 = 4.7$ J.

8. The work needed to stop the car is equal to the change in the car's kinetic energy. The work done on the car is by the force of friction. Let f be the friction force, m be the mass of the car including the driver and passengers and μ_k be the coefficient of kinetic friction. Thus, $f = \mu_k \times mg$.

   The friction force acts opposite to the distance moved. Therefore, the work done on the car by the force of friction will be $W = -f\times d = -\mu_k \times mg \times d$.

   According to the work-kinetic energy theorem: $W = K_f - K_i$

   Therefore, $-\mu_k \times mg\times d = 0 - K_i$, since the car stops $K_f = 0$.

   Thus, $v_i^2 = 2\mu_k \times g\times d = 2\times 0.400\times 10\times 90.0 \Rightarrow v_i = 26.8 \text{ m s}^{-1}$.

9. (i)

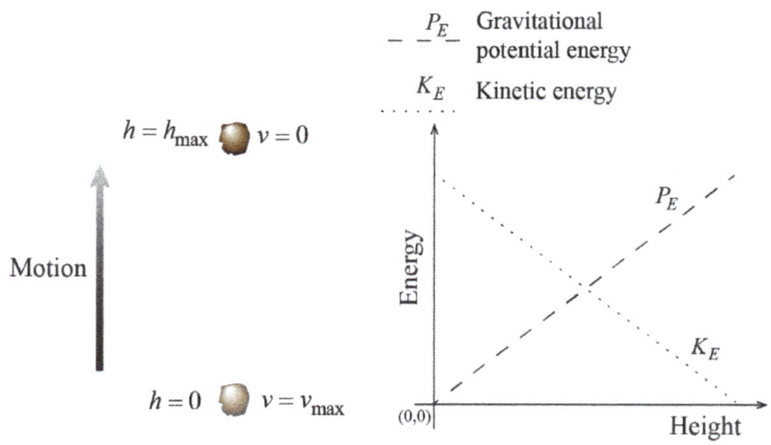

(a)

(ii)

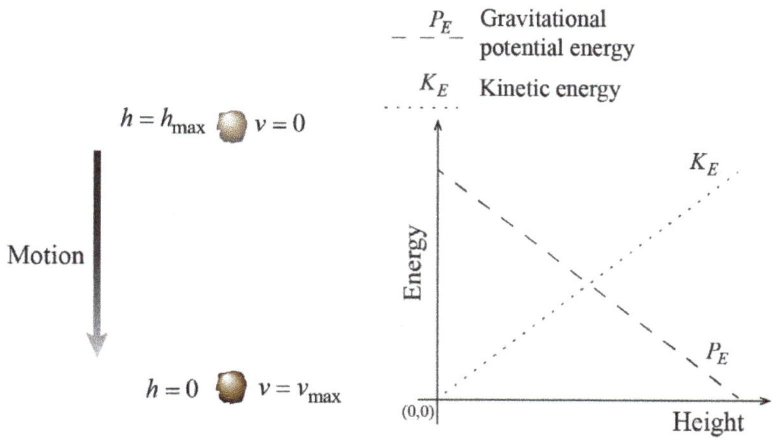

(b)

10. Potential energy $U = \frac{1}{2}kx^2$

(i) $k = \frac{2U}{x^2} = \frac{2\times 14.4}{\left(1.20\times 10^{-2}\right)^2} = 2.00\times 10^5\ \text{N m}^{-1}$

(ii) $x = \sqrt{\frac{2U}{k}} = \sqrt{\frac{2\times 900}{2.00\times 10^5}} = 9.49\ \text{cm}$

The spring is only $8.70\ \text{cm}$ long and thus cannot be compressed $9.49\ \text{cm}$ to achieve the desired $900\ \text{J}$ stored energy.

11. (i) The gravitational potential energy that the falcon loses is

$$U_{PE} = mg\Delta h = 750\times 10^{-3}\times 10\times 120 = 900\ \text{J}$$

(ii) Let v be the velocity at the end of the dive.

$$0.90\times 900 = \frac{1}{2}750\times 10^{-3}v^2$$

Thus $$v = \sqrt{\frac{2\times 0.90\times 900}{750\times 10^{-3}}} = 46.5\ \text{m s}^{-1}$$

(iii) Let h be the height that the falcon climbs before it flaps its wings

$$mgh = 0.75 \times \frac{1}{2} mv^2$$

$$h = \frac{0.75 \times 46.5^2}{2 \times 10} = 81.0 \text{ m}$$

12. (i) The forces acting on the objects are the Earth's gravitational force and the elastic force exerted by the spring.

(ii) Both forces are conservative forces.

(iii) Let v_1 be the velocity of the object when it touches the upper end of the vertical spring. Using the conservation of mechanical energy, we have

$$0 + 3.0 \times 10 \times 45.0 \times 10^{-2} = \frac{1}{2} \times 3.0 \times v_1^2 + 0$$

Thus, $v_1 = \sqrt{2 \times 10 \times 45.0 \times 10^{-2}} = 3 \text{ m s}^{-1}$

(iv) When the ball is fully compressed, the object comes to a resting position momentarily. Since both forces acting on the object are conservative forces, we can use the conservation of mechanical energy, considering kinetic energy and both types of potential energy. Let k be the spring constant of the coiled spring. Thus,

$$\frac{1}{2} \times 3.0 \times (3.0)^2 + 0 + 0 = 0 + 3.0 \times 10 \times \left(-15.0 \times 10^{-2}\right) + \frac{1}{2} \times k \times \left(15.0 \times 10^{-2}\right)^2$$

$$k = \frac{2}{\left(15.0 \times 10^{-2}\right)^2} \left[\frac{1}{2} \times 3.0 \times (3.0)^2 + 3.0 \times 10 \times 15.0 \times 10^{-2}\right] = 1600 \text{ N m}^{-1}$$

13. During sliding, the friction force on the tool box does a negative job and removes mechanical energy from the system. Take the edge of the roof as the potential reference point. Initial kinetic energy is zero, and final potential energy is zero. Let m the mass of the tool box, h the vertical height moved, f friction force, d distance moved by the tool box while sliding and v_f be the final velocity of the tool box at the roof edge. Thus, when applying the principle of conservation of energy to the system, we have

$$0 + mg \times h - f \times d = \frac{1}{2} mv_f^2 + 0$$

$$\text{Therefore, } v_f = \sqrt{\frac{2\left(6.0\times10\times5.0\sin30^\circ - 16.0\times5.0\right)}{6.0}} = 5.9 \text{ m s}^{-1}$$

14. (i) Once the box is on the moving belt, the speed of the box of mass 100 kg increases from 0 to 1.40 m s^{-1} relative to the factory ground. Therefore, the kinetic energy supplied to it is $K = \frac{1}{2}mv^2 = \frac{1}{2}\times100\times1.40^2 = 98.0$ J

(ii) The magnitude of the kinetic frictional force is

$f = \mu R_n = \mu mg = 0.400\times100\times10 = 400$ N

(iii) Suppose the distance the box moved relative to the conveyor belt before it stops slipping is d, then the energy change ΔE during slipping would be $\Delta E = fd$. Using the equation of motion we have $v^2 = 0 + 2ad$, but $f = ma$.

Therefore, $v^2 = 2\left(\frac{f}{m}\right)d \Rightarrow fd = \frac{1}{2}mv^2$. Hence, $\Delta E = fd = \frac{1}{2}mv^2 = K = 98.0$ J.

Thus, the total energy that must be supplied by the motor is

$$E = K + \Delta E = 2K = 2\times98.0 = 196 \text{ J}$$

(iv) The energy supplied by the motor (i.e. E) is the work done on the system and must be greater than the kinetic energy gained by the box computed in part (ii) because the part of the energy supplied by the motor is being used to compensate for the energy dissipated ΔE while it was slipping.

15. (i) The jumper's gravitational potential energy above the river just before he falls from the bridge is $U_{i,grav} = mgh_i = 70.0\times10\times60.0 = 42$ kJ.

(ii) Once the jumper reaches the maximum stretched length of the bungee cord, the jumper will be at a point 2.5 m above the river. Thus, the jumper's gravitational potential energy above the river after the jumper reaches the maximum stretched length of the bungee cord. $U_{f,grav} = mgh_f = 70.0\times10\times2.5 = 1.750$ kJ

(iii) The maximum stretched length of the bungee cord is

$$y_{stretch} = (60.0 - 2.5) - 30.0 = 27.5 \text{ m}$$

(iv) The gravitational potential energy decrease of the jumper as he falls must equal the elastic potential energy increase of the bungee cord.

Thus, $\frac{1}{2}ky_{stretch}^2 = U_{i,grav} - U_{f,grav}$

$$\Rightarrow \quad k = \frac{2\left(U_{i,grav} - U_{f,grav}\right)}{y_{stretch}^2} = \frac{2(42\ 000 - 1750)}{27.5^2} = 106.45 \text{ N m}^{-1} \approx 106 \text{ N m}^{-1}$$

16. Take the bottom of the slides as the potential reference point. Therefore for each child, when they reach the bottom, their potential energy will be zero.

(i) Let m_A and m_M be the mass of Nathan and Mary, respectively. Applying the principle of conservation of mechanical energy for each child, we have

$$0 + m_A gh = \frac{1}{2}m_A v_A^2 + 0 \quad \Rightarrow \quad m_A gh = \frac{1}{2}m_A v_A^2 \quad \Rightarrow \quad v_A = \sqrt{2gh} \text{ and}$$

$$0 + m_M gh = \frac{1}{2}m_M v_M^2 + 0 \quad \Rightarrow \quad m_M gh = \frac{1}{2}m_M v_M^2 \quad \Rightarrow \quad v_M = \sqrt{2gh}$$

For each child, their initial potential energy is transformed into kinetic energy. Since their final speed at the bottom does not depend on their mass and they descend to the same vertical height, their speed will be the same.

(ii) During their sliding, Mary is consistently at a lower elevation than Nathan at any instant until the end. This means that Mary has converted her potential energy to kinetic energy earlier than Nathan. Therefore, she travels faster than Nathan for the whole sliding trip except towards the end, where Nathan catches up to the same speed when reaching the bottom. Since Mary is travelling faster than Nathan for the whole sliding trip and the distance travelled by Mary and Nathan are the same, Mary gets to the bottom first.

17. Take the bottom of the slides as the potential reference point.

(i) Applying the principle of conservation of mechanical energy, we have

$$0 + mgh = \frac{1}{2}mv_P^2 + 0 \quad \Rightarrow \quad v = \sqrt{2gh} = \sqrt{2\times10\times3.2} = 8.0\ \text{m s}^{-1}$$

(ii) As the object slides on the rough surface, we have, $f = \mu_k R_n$.

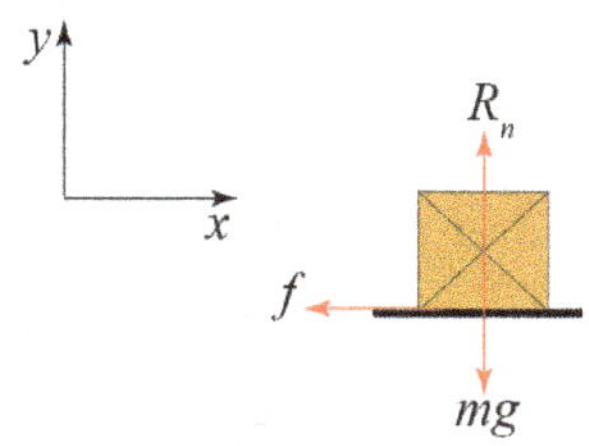

Applying Newton's 2nd Law the +ve y - direction, then $R_n = mg$.

Thus, $f = \mu_k mg = 0.100\times3.0\times10 = 3.0\ \text{N}$ in the –ve x direction.

As the force is constant, the work done by friction on the object is

$$W = -f\times d = -3.0\times30.0 = -90\ \text{J}$$

Therefore, energy dissipated by friction on the object is 360 J.

(iii) From the work – KE theorem, we have $W = \Delta K = \left(K_f - K_i\right)$

Therefore, $-90 = \frac{1}{2}\times3.0\times v_f^2 - \frac{1}{2}\times3.0\times8^2$

$$v_f = \sqrt{\frac{2}{3.0}\left(-90 + \frac{1}{2}\times3.0\times8^2\right)} = 2.0\ \text{m s}^{-1}$$

(iv) Work done by the spring on the buffer is $-150x^2$. From the work – KE theorem, we have $W = \Delta K = \left(K_f - K_i\right)$.

Thus, $-150x^2 = 0 - \frac{1}{2}\times3.0\times2.0^2 \quad \Rightarrow \quad x = 0.2\ \text{m}$

18. The mass of the blood $m = \left(7500 \times \frac{\quad}{1000}\right) \times 1050 = 7875$ kg.

 (i) The work performed in a single day is:

 $W = mgh = 7875 \times 10 \times 1.6 = 1.26 \times 10^5 \text{ J}$.

 (ii) The power output $P = \frac{W}{t} = \frac{1.26 \times 10^5}{24 \times 3600} = 1.46 \text{ W}$

 (iii) The heart actually puts out more power than the value calculated in part (ii) because two forms of energy were not considered; the heart also provides kinetic energy to the blood and generates thermal energy by working against the friction of the vein and vessel walls.

19. The free-body diagram

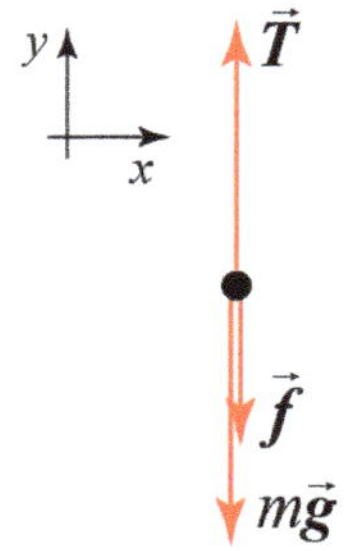

The total mass m of the system (elevator + passenger) $m = 1\ 920$ kg .

 (i) Applying Newton's Second Law in the upward direction we have

 $$T - mg - f = 0 \quad \Rightarrow \quad T = 1\ 920 \times 10 + 3\ 500 = 22\ 700 \text{ N}$$

 (ii) The power output of the motor required to lift the elevator and its passenger is $P = Tv = 22\ 700 \times 2.5 = 5.68 \times 10^4 \text{ W}$

 (iii) Applying Newton's Second Law in the upward direction we have

 $$T - mg - f = ma \quad \Rightarrow \quad T = 1\ 920 \times 1.00 + 1\ 920 \times 10 + 3\ 500 = 24\ 620 \text{ N}$$

 $$P = Tv = 24\ 620 \times 2.5 = 6.16 \times 10^4 \text{ W}$$

(iv) Larger than the power found in part (i). The motor must supply the force that pulls the elevator and passengers upward with an increasing speed and more power will be required to do that than in part (i) because the motor must now perform the additional task of accelerating the elevator and passengers.

20. The mass of a given volume of air $m = \rho V$. The kinetic energy of a mass m m moving with a velocity v is $\frac{1}{2}mv^2$. The amount of wind passing through the blades of the wind turbine in a given time is given by the volume of a cylinder that has a radius equal to the radius r of its blades and a height equal to the distance h the wind travels in the given amount of time. Thus, $V = \pi r^2 h$.

(i) The mass of 1.00 m^3 of air is $m = \rho V = 1.2 \times 1.00 = 1.2 \text{ kg}$.

For wind moving at 10.0 m s^{-1}, the kinetic energy associated with this volume is $K = \frac{1}{2}mv^2 = \frac{1}{2} \times 1.2 \times 10.0^2 = 60.0 \text{ J}$.

Thus, the wind energy per cubic metre is 60.0 J.

(ii) In one second, a cylindrical column of air with a radius of 40.0 m and a height of 10.0 mpasses through the wind turbine. This column has a volume

$$V = \pi r^2 h = \pi \times 40.0^2 \times 10.0 = 5.03 \times 10^4 \text{ m}^3$$

Thus, the amount of wind energy available to the turbines in 1 second is

$$E = \left(5.03 \times 10^4\right) \times 60.0 \text{ J} = 3.04 \times 10^6 \text{ J}$$

Since the wind turbine is 35% efficient, the energy output of the turbine per second would be $E_{ouput} = \left(3.04 \times 10^6\right) \times 0.35 = 1.06 \times 10^6 \text{ J}$.

Therefore, the power output of this turbine is $P_{ouput} = 1.06 \times 10^6 \text{ W}$.

Chapter 12

1. The kinetic energy of an object can be written as $\frac{p^2}{2m}$ in terms of momentum p and its mass m. Both objects have the same kinetic energy. Therefore, it is evident that the object with the greater mass has the larger magnitude of momentum.

2. **First Law:** The momentum of an object is constant unless acted upon by an external force.

   Second Law: The net force acting on an object is equal to the rate of change of the object's momentum.

   Third Law: When two objects interact, the changes in momentum that each imparts to the other are equal in magnitude and opposite in direction.

3. (i) The cricket ball itself does not carry force; force requires another object upon which it can exert its effect.

   (ii) It is not possible for the cricket ball to deliver more kinetic energy than it already possesses, as doing so would violate the principle of energy conservation.

   (iii) During a collision, the cricket ball can impart more momentum than it initially carries while in motion, achieved by bouncing off the object it strikes.

4. The total momentum before the collision is equal to the total momentum after the interaction. It is valid if the system is closed or there are no external forces acting.

5. The total momentum and the total kinetic energy.

6. Let v_i be the impact speed (downward direction).

   Using conservation of mechanical energy, we have $mgh_i = \frac{1}{2}mv_i^2$.

   Thus $v_i = \sqrt{2gh_i} = \sqrt{2\times10\times1.50} = 5.48 \text{ m s}^{-1}$

   Let v_f be the rebound speed (upward direction).

   Using conservation of mechanical energy, we have $mgh_f = \frac{1}{2}mv_f^2$.

   Thus $v_f = \sqrt{2gh_f} = \sqrt{2\times10\times0.85} = 4.12 \text{ m s}^{-1}$

   Take upward direction as + ve.

   The impulse the floor imparts to the ball is the change in the ball's momentum Δp.

   $\Delta p = m\{v_f(-v_i)\} = m(v_f + v_i) = 200\times10^{-3}(4.12+5.48) = 1.92 \text{ kg m s}^{-1}$

   upward.

7. Assume the ball is initially moving to the right, and let this be the $+x$ direction. The ball stops, so its final velocity is zero.

   (i) The impulse exerted on the wicketkeeper by the ball is
   $J_x = \Delta p_x = m(v_{f,x} - v_{i,x}) = 0.150\times(0-35.0) = -5.25 \text{ kg m s}^{-1}$.
   The minus signs of J_x shows that this is acting in the left, i.e. to the $-x$ direction.

   (ii) The average force imparted by the ball to the wicketkeeper is

   $$F_x = \frac{J_x}{\Delta t} = \frac{-5.25}{20\times10^{-3}} = -262.5 \text{ N} \approx -263 \text{ N}$$

   The minus signs of F_x shows that this is acting in the left, i.e. to the $-x$ direction.

8. (i) A gas molecule undergoes an elastic collision. Therefore, the magnitude of the molecule's velocity will be 350 m s^{-1}, and its direction will be opposite to the direction before the collision.

   (ii) Change in momentum

   $$\Delta p = m\left(v_f - v_i\right) = 8.0\times10^{-26}\left(-350-350\right) = -5.60\times10^{-23} \text{ kg m s}^{-1}$$

   (iii) The gas molecule undergoes a change in momentum during the collision, indicating that the wall exerts a force on the molecule to bring about this change. Therefore, according to Newton's Second/Third Law, the gas molecule must exert an equal but opposite force on the wall.

9. (i) The magnitudes of the pushing forces they experience are the same because according to Newton's 3rd Law, "for every action, there is an equal and opposite reaction". As they push against each other, their "pushing" forces are an "action-reaction pair".

   (ii) Treating Nathan and Mary as a "system", the pushing forces between them are internal forces. Therefore, there are no net external forces on them, so momentum is conserved.

   i.e. Initial total momentum = zero = final momentum.

   Therefore their momenta must be equal in magnitude and opposite in direction.

   (iii) Because the momentum is conserved, their final momentum must be equal but opposite. However, Nathan has a larger mass compared to Mary. Hence, Nathan's final velocity must be smaller than Mary's. Therefore, Mary will travel further in a given amount of time because she has a larger velocity.

10. The ball's velocity components before and after the kick

    $v_{i,x} = -25 \text{ m s}^{-1}$ and $v_{i,y} = 0$

    $v_{f,x} = 35\cos 45^\circ = \dfrac{35}{\sqrt{2}} \text{ m s}^{-1}$ and $v_{f,y} = 35\sin 45^\circ = \dfrac{35}{\sqrt{2}} \text{ m s}^{-1}$

(i) Thus, the components of impulse will be

$$J_x = \Delta p_x = m\left(v_{f,x} - v_{i,x}\right) = 400\times10^{-3}\left\{\frac{35}{\sqrt{2}} - (-25)\right\} = 19.90 \text{ kg m s}^{-1}$$

$$J_y = \Delta p_y = m\left(v_{f,y} - v_{i,y}\right) = 400\times10^{-3}\left\{\frac{35}{\sqrt{2}} - 0\right\} = 9.90 \text{ kg m s}^{-1}$$

$$J = \sqrt{J_x^2 + J_y^2} = \sqrt{19.90^2 + 9.90^2} = 22.2 \text{ kg m s}^{-1}$$

(ii) The average net force components will be

$$F_{avg,x} = \frac{J_x}{\Delta t} = \frac{19.90}{0.01} = 1\,990 \text{ N} \quad \text{and} \quad F_{avg,y} = \frac{J_y}{\Delta t} = \frac{9.90}{0.01} = 990 \text{ N}$$

Thus, $F_{avg} = \sqrt{F_{avg,x}^2 + F_{avg,y}^2} = \sqrt{1990^2 + 990^2} = 2.22\times10^3 \text{ N}$

$$\theta = \tan^{-1}\left(\frac{F_{avg,y}}{F_{avg,x}}\right) = \tan^{-1}\left(\frac{990}{1990}\right) = 26.5^\circ$$

(iii) The weight of the ball is only 4.0 N which is very small compared to the average force exerted by the player's foot ($2.22\times10^3 \text{ N}$) during the kick. Thus the influence due to gravity could be neglected.

11. Take the horizontal surface of the trampoline as a gravitational potential reference point and take the upward direction as the +ve y-axis.

(i) Using the conservation of mechanical energy, we have

$$0 + mgh = \frac{1}{2}mv^2 + 0 \quad \Rightarrow \quad v = \sqrt{2gh} = \sqrt{2\times10\times2.5} = 7.1 \text{ m s}^{-1}$$

(ii) Change in the momentum of the child artist is $\Delta p = m\left(v_f - v_i\right)$.

Thus, $\Delta p = 40.0\{6.0 - (-7.1)\} = 524 \text{ kg m s}^{-1}$.

(iii) The average force exerted by the trampoline on the girl is $F_{avg} = \frac{\Delta p}{\Delta t}$. Thus,

$$F_{avg} = \frac{524}{0.50} = 1048 \text{ N}$$

(iv) Let y be the maximum compression length. When the child artist lands on the trampoline, she has only kinetic energy (gravitational potential energy will be zero). Once she reaches the maximum compression length, she will have only gravitational potential energy and elastic energy (kinetic energy will be zero). Applying the principle of conservation of energy, we have

$$\frac{1}{2}mv^2 + 0 = 0 + mgy + \frac{1}{2}ky^2$$

$$\Rightarrow \; ky^2 + 2mgy - mv^2 = 0$$

i.e. $y = \dfrac{-2mg \pm \sqrt{4m^2g^2 + 4kmv^2}}{2k} = \dfrac{-mg \pm \sqrt{m^2g^2 + kmv^2}}{k}$

$$y = \frac{-40.0\times10 \pm \sqrt{40.0^2 10^2 + 6.2\times10^4 \times 40\times7.1^2}}{6.2\times10^4} = \frac{-400 \pm 111.89\times10^2}{6.2\times10^4}$$

$$y = 0.17 \text{ m} \;\text{ or }\; y = -0.19 \text{ m}$$

The gravitational potential energy at maximum compression position is negative, therefore $y < 0$. Thus, $y = -0.19$ m, i.e. the maximum compression length is 0.19 m below the surface of the trampoline.

12. Take the direction toward the east to be +ve. Let us assume the net force exerted on the car by the wall and friction from the ground is large compared with other forces on the car, such as air resistance. The gravitational force and the normal force exerted by the road on the car are perpendicular to the motion and therefore do not affect the horizontal momentum.

(i) (a) The impulse $\vec{J} = \Delta\vec{p}$, if we apply it in the direction of the east. We have,

$$J = \Delta p = mv_f - mv_i = 1200\times3.00 - 1200\times(-17.0) = 2.40\times10^4 \text{ kg m s}^{-1}$$

(b) The average net force exerted on the car and the dummy is ,

$\left(\vec{F}_{net}\right)_{avg} = \dfrac{\Delta\vec{p}}{\Delta t}$ if we apply it in the direction of the east. We have,

$$\left(F_{net}\right)_{avg} = \frac{\Delta p}{\Delta t} = \frac{24000}{0.150} = 1.60\times10^5 \text{ N}$$

(ii) (a) It requires a smaller force than that is obtained in part (i) (b).
i.e. smaller than 1.60×10^{5} N.

In the previous situation, the car rebounds, and the net force on the car does two things during the collision: (1) it stops the car, and (2) it causes the car to move away from the wall at 3.00 m s^{-1} after the collision. If the car does not rebound and comes to a complete stop after the collision, then the net force has to do only the first of these steps, i.e. to stop the car, which requires a smaller force.

(b) The change in momentum of the car and the dummy in the direction of the east. We have,

$\Delta p = mv_f - mv_i = 1200\times0 - 1200\times(-17.0) = 2.04\times10^{4}$ kg m s^{-1}

(c) The average net force exerted on the car is $\left(\vec{F}_{net}\right)_{avg} = \frac{\Delta\vec{p}}{\Delta t}$,

$$\left(F_{net}\right)_{avg} = \frac{\Delta p}{\Delta t} = \frac{20400}{0.150} = 1.36\times10^{5}\ \text{N},$$

which is less than the average force obtained in part (i) (b).

(d) Let F be the average force exerted on the dummy by the seat belt. Then the work done by the force on the dummy will be $W = F\times(-0.5)$. The change in the dummy's kinetic energy will be $\Delta K = 0 - \frac{1}{2}\times75.0\times17.0^{2}$.

Using the work–kinetic energy theorem, we have

$$W = F\times(-0.5) = \Delta K = 0 - \frac{1}{2}\times75.0\times17.0^{2}$$

i.e. $F = 75.0\times17.0^{2} = 21\ 675\ \text{N} \approx 2.17\times10^{4}$ N

13. Conservation of momentum in two dimensions requires conservation of both the x- and y-components of momentum. Thus,

In $x\rightarrow$ direction, $(m_F + m_P)v_{x,f} = m_F\left(v_{x,i}\right)_F + m_P\left(v_{x,i}\right)_P$

i.e. $(1.00 + 0.350)v_{x\,f} = 1.00\times20.0\cos45^{\circ} + 0.35\times10.0$

i.e. $$v_{x,f} = \frac{1.00 \times 20.0 \cos 45^\circ + 0.35 \times 10.0}{(1.00 + 0.350)} = 13.1 \text{ m s}^{-1}$$

In $y \downarrow$ direction, $(m_F + m_F)v_{y,f} = m_F\left(v_{y,i}\right)_F + m_P\left(v_{y,i}\right)_P$

i.e. $(1.00 + 0.350)v_{y,f} = 1.00 \times 20.0 \sin 45^\circ + 0$

i.e. $$v_{y,f} = \frac{1.00 \times 20.0 \sin 45^\circ}{(1.00 + 0.350)} = 10.5 \text{ m s}^{-1}$$

Therefore, the magnitude of the velocity is

$$v = \sqrt{v_{x,f}^2 + v_{y,f}^2} = \sqrt{13.1^2 + 10.5^2} = 16.8 \text{ m s}^{-1}$$

$$\theta = \tan^{-1}\left(\frac{v_{y,f}}{v_{x,f}}\right) = \tan^{-1}\left(\frac{10.5}{13.1}\right) = 38.7^\circ$$

Immediately after impact, the falcon, with its pigeon, is travelling at 16.8 m s^{-1} at an angle 38.7° below the horizontal.

14. (i) Conservation of mass implies that

$$10.0 = 3.0 + 3.0 + m$$

i.e. $m = 4.0$ kg

(ii) Conservation of momentum in two dimensions requires conservation of both the *x*- and *y*-components of momentum. Thus

In $x \rightarrow$ direction $3.0 \times 12 \cos 40^\circ + 3.0 \times (-10.0) + 4.0 \times v \cos\theta = 10.0 \times 2.0$

In $y \uparrow$ direction $3.0 \times 12 \sin 40^\circ + 0 + 4.0 \times (-v \sin\theta) = 0$

i.e. $v\cos\theta = 5.61$ and $v\sin\theta = 5.79$

i.e. $(v\cos\theta)^2 + (v\sin\theta)^2 = v^2 = 5.61^2 + 5.79^2$ and $\tan\theta = \dfrac{5.79}{5.61}$

Therefore, $v = 8.1 \text{ m s}^{-1}$ and $\theta = 45.9^\circ$.

The third fragment, with a mass of 4.0 g, will travel 45.9° south of east at a speed of 8.1 m s^{-1}.

15. Let the mass of the wooden block $M = 3.00 \text{ kg}$, the mass of the arrow $m = 0.055 \text{ kg}$ and its initial velocity $v_i = 40.0 \text{ m s}^{-1}$.

 (i) The kinetic energy possessed by the arrow just before the collision

$$K_{arrow,i} = \frac{1}{2}mv_i^2 = \frac{1}{2}\times 0.055\times 40.0^2 = 44.0 \text{ J}$$

 (ii) The momentum of the arrow just before the collision

$$p_{arrow,i} = mv_i = 0.055\times 40.0 = 2.20 \text{ kg m s}^{-1}$$

 (iii) Since momentum is conserved, the momentum of the combined system consisting of the block and the embedded arrow immediately following the collision will be the same as the initial momentum of the arrow (the initial momentum of the wooden block is zero).

 (iv) $p_{combined,f} = (M+m)v_f = 2.20$.Therefore, $v_f = \dfrac{2.20}{3.00+0.055} = 0.65 \text{ m s}^{-1}$

 (v) The kinetic energy just after the collision is

$$K_{block+arrow,f} = \frac{1}{2}(M+m)v_f^2 = \frac{1}{2}(3.00+0.055)\times 0.65^2 = 0.65 \text{ J}$$

Take the horizontal surface of the table as the gravitational potential reference point. Therefore, the gravitational potential before and after the collision will be zero. Also, the initial kinetic energy of the wooden block is zero. Thus, the change in energy is

$$\Delta E = K_{block+arrow,f} - K_{arrow,i} = 0.65 - 44.0 = -43.35 \text{ J}$$

So the energy lost is 43.4 J

16. The collision is elastic, so the initial and final kinetic energies of the system are equal. Let the masses of marbles A and B be m.

$$\frac{1}{2}mv_{A,i}^2 + 0 = \frac{1}{2}mv_{A,f}^2 + \frac{1}{2}mv^2$$

$$\Rightarrow |v| = \sqrt{\left(v_{A,f}^2 - v_{A,i}^2\right)} = \sqrt{\left(4.00^2 - 2.00^2\right)} = \sqrt{12} = 3.46 \text{ m s}^{-1}$$

Total momentum is conserved. Thus,

In $+x$ direction

$mv_{A,i} + 0 = mv_{A,f} \cos 60 \ + mv\cos\theta$

$$\Rightarrow \quad 4.00 = 2.00\cos 60^\circ + \sqrt{12}\cos\theta$$

i.e. $\theta = 30^\circ$

17. The mass of the tennis ball m, the mass of the basketball M and both balls are released from rest simultaneously to fall through a distance of h. Take the ground as the gravitational potential reference point.

(i) Using the conservation of mechanical energy for the basketball after it has fallen a distance h from rest. We have

$$0 + Mgh = \frac{1}{2}Mv^2 + 0$$

$$\Rightarrow \quad v = \sqrt{2gh}$$

Therefore, the magnitude of the downward velocity of the basketball is $\sqrt{\ gh}$.

(ii) From part (i), we know that the velocity does not depend on the mass of the object. Both balls were released from rest simultaneously to fall through a distance of h. Therefore, the magnitude of the velocity of the tennis ball will be $\sqrt{2gh}$, and its direction will be, downward.

(iii) The collision of the basketball of mass M with the ground floor is an elastic collision. It is like an elastic collision of a light object (basketball) with a stationary massive object (ground floor). Thus, the velocity of the basketball reverses direction without change in magnitude. After the collision, the basketball will travel upward with the magnitude of the velocity $\sqrt{2gh}$.

(iv) Now consider the collision between the tennis ball and the basketball.

Let $v_{m,i}$ is the initial velocity of the tennis ball before the collision $\downarrow \sqrt{2gh}$

$v_{m,f}$ is the final velocity of the tennis ball after the collision.

$v_{M,i}$ is the initial velocity of the tennis ball before the collision. $\uparrow \sqrt{2gh}$

$v_{M,f}$ is the final velocity of the basketball after the collision.

From equations 12.29 and 12.30 (see section 12.6), we have

$$v_{M,f} = \left(\frac{M-m}{M+m}\right)v_{M,i} + \left(\frac{2m}{M+m}\right)v_{m,i} \quad \text{and}$$

$$v_{m,f} = \left(\frac{2M}{M+m}\right)v_{M,i} + \left(\frac{m-M}{M+m}\right)v_{m,i}$$

To obtain the above equations, use the conservation of momentum and since the collision is elastic total kinetic energy is conserved.

Therefore,

$$v_{M,f} = \left(\frac{M-m}{M+m}\right)\sqrt{2gh} - \left(\frac{2m}{M+m}\right)\sqrt{2gh} = \left(\frac{M-3m}{M+m}\right)\sqrt{2gh}$$

$$\text{and } v_{m,f} = \left(\frac{2M}{M+m}\right)\sqrt{2gh} - \left(\frac{m-M}{M+m}\right)\sqrt{2gh} = \left(\frac{3M-m}{M+m}\right)\sqrt{2gh}$$

$$\text{i.e. } \uparrow v_{M,f} = \left(\frac{M-3m}{M+m}\right)\sqrt{2gh} \quad \text{and} \quad \uparrow v_{m,f} = \left(\frac{3M-m}{M+m}\right)\sqrt{2gh}$$

(v) Stopping the basketball's motion after the collision means that $v_{M,f}$ becomes zero. This will happen only when $m = \frac{1}{3}M$. Therefore, when $m = \frac{1}{3}M$ the basketball's motion will be stopped after the collision

(vi) When $m = \frac{1}{3}M$, $\uparrow v_{m\ f} = 2\sqrt{2gh}$.

Let h' be the height that the tennis ball reaches.

Use conservation of mechanical energy.

i.e. $\frac{1}{2}mv_{m,f}^2 = mgh'$

i.e. $h' = \frac{v_{m,f}^2}{2g} = \frac{\left(2\sqrt{2gh}\right)^2}{2g} = 4h$

18. Let m_1 and $v_{1x,i}$ be the mass and initial velocity of the car in +ve x direction and m_2 and $v_{2y,i}$ be the mass and velocity of the car in +ve y direction. Conservation of momentum in two dimensions requires conservation of both the x- and y-components of momentum.

In $x \rightarrow$ direction, $m_1 v_{1x,i} + 0 = (m_1 + m_2) v\cos\theta$

In $y \uparrow$ direction, $0 + m_2 v_{2y,i} = (m_1 + m_2) v\sin\theta$

i.e. $(v\cos\theta)^2 + (v\sin\theta)^2 = v^2 = \frac{(m_1 v_{1x,i})^2 + (m_2 v_{2y,i})^2}{(m_1 + m_2)^2}$.

i.e. $v = \sqrt{\frac{(m_1 v_{1x,i})^2 + (m_2 v_{2y,i})^2}{(m_1 + m_2)^2}} = \sqrt{\frac{1200^2 \times 20^2 + 2000^2 \times 18^2}{(1200 + 2000)^2}} = 13.5\ \text{m s}^{-1}$

and $\tan\theta = \frac{m_2 v_{2y,i}}{m_1 v_{1x,i}} = \frac{2000 \times 18}{1200 \times 20}$, i.e. $\theta = 56.3$

19. Average force applied to water $F_{avg} = \frac{\Delta P}{\Delta t} = \frac{m(v_f - v_i)}{\Delta t} = \frac{m}{\Delta t}(v_f - v_i)$

i.e. $F_{avg} = 10(0 - 5) = -50$ N, Negative sign indicates force applied to water is against the direction of water flow. Therefore, force of water on wall is 50 N. If the water bounces, a greater force is applied because of a greater change in momentum.

20. Conservation of momentum in two dimensions requires conservation of both the x- and y-components of momentum. Let us take the direction towards east as $+x$ and towards north as $+y$. Let $v_{P,f}$, m_P and $v_{Q,f}$, m_Q be the final velocities and masses of the balls P and Q and respectively. Futhermore, $m_P = m_Q$.

    (i) In $+x$ direction momentum is conserved. Thus,

    $$m_P v_{P,f} \cos 30^\circ + m_Q v_{Q,f} \cos 45^\circ = m_P \times 7.5 + 0$$

    i.e. $$\frac{v_{P,f}\sqrt{3}}{2} + \frac{v_{Q,f}}{\sqrt{2}} = 7.5$$

    In $+y$ direction momentum is conserved. Thus,

    $$m_P v_{P,f} \sin 30^\circ - m_Q v_{Q,f} \sin 45^\circ = 0 + 0$$

    i.e. $$\frac{v_{P,f}}{2} = \frac{v_{Q,f}}{\sqrt{2}} \quad \Rightarrow \quad v_{P,f} = \sqrt{2} v_{Q,f}$$

    Therefore, $v_{Q,f} = \dfrac{7.5 \times \sqrt{2}}{1+\sqrt{3}} = 3.9 \text{ m s}^{-1}$ and $\Rightarrow \quad v_{P,f} = \dfrac{7.5 \times 2}{1+\sqrt{3}} = 5.5 \text{ m s}^{-1}$

    (ii) Initial kinetic energy $K_i = \dfrac{1}{2} m_P v_{P,i}^2 + 0 = \dfrac{1}{2} m_P v_{P,i}^2$

    Final kinetic energy $K_f = \dfrac{1}{2} m_P v_{P,f}^2 + \dfrac{1}{2} m_Q v_{Q,f}^2$

    Therefore $$\frac{\Delta K}{K_i} = \frac{K_f - K_i}{K_i} = \frac{K_f}{K_i} - 1 = \frac{v_{P,f}^2 + v_{Q,f}^2}{v_{P,i}^2} - 1 = \frac{5.5^2 + 3.9^2}{7.5^2} - 1 = -0.192$$

    i.e. 19.2% of the original kinetic energy is dissipated during the collision.

Chapter 13

1. The angle in radians = angle in degrees $\times \frac{2\pi}{360}$ radians.

2. The angular speed is defined as the angle that an object rotates through per second.

3. The period is the time taken for a complete revolution. The frequency is the number of complete revoltions per second.

4. Second hand turns through 360° in 1 minute, i.e. 2π rad in 1 minute, i.e. $\frac{2\pi}{60}$ rad in 1 second. Therefore the angular speed $\omega = \frac{2\pi}{60} = 0.105 \text{ rad s}^{-1}$.

5. If an object is moving in a circle, centripetal acceleration is the acceleration of the object directed towards the centre of the circle.

   $a_r = r\omega^2$, where a_r is the centripetal acceleration, ω is the angular speed in rad s^{-1} and r is the radius of circular motion in metres.

6. An object moves round a circular track of radius 0.3 m at 2 revolutions per second

   (i) For 1 revolution, angle θ turned is 2π rad.

   Therefore, the anguar speed $\omega = 2\times 2\pi = 4\pi \text{ rad s}^{-1}$.

   (ii) Period $T =$ time for 1 rev $= \frac{2\pi}{\omega} = \frac{2\pi}{4\pi} = 0.5$ s

   (iii) Linear speed $v = r\omega = 0.3\times 4\pi = 3.8 \text{ m s}^{-1}$.

7. The Moon completes a full orbit around the Earth in approximately 28 days. Thus, the period $T = 28\times 24\times 3600 \text{ s} = 2.42\times 10^6 \text{ s}$.

   Angular speed $\omega = 2\pi f = \frac{2\pi}{T} = \frac{2\pi}{2.42\times 10^6} = 2.5972\times 10^{-6} \approx 2.60\times 10^{-6} \text{rad s}^{-1}$

   Linear speed $v = r\omega = 3.84\times 10^8 \times 2.60\times 10^{-6} = 998.4 \text{ m s}^{-1}$.

8. We have $\omega^2 = \omega_o^2 + 2\alpha\Delta\theta$, i.e. $50^2 = 100^2 + 2\alpha(2\pi \times 2n)$.

If the angular velocity after n rotation is ω_n then we have $\omega_n^2 = 100^2 + 2\alpha(2\pi \times n)$

Thus, $\dfrac{50^2 - 100^2}{\omega^2 - 100^2} = \dfrac{2\alpha(2\pi \times 2\ \ )}{2\alpha(2\pi \times\ \ )} = 2$, $\Rightarrow \quad \omega_n = 25\sqrt{10}$ rad s^{-1}.

9. A pot maker rotates the potter's wheel from rest to 240 rpm in a time of 0.90 s

(i) Initially the wheel is at rest, so the initial angular velocity is zero, i.e. $\omega_i = 0$

Final angular velocity $\omega_f = \dfrac{240}{60} \times 2\pi = 8\pi$ rad s^{-1}.

Since angular acceleration α is constant, we can calculate it by finding the average angular acceleration for the time interval. Thus,

$$\alpha = \frac{\omega_f - \omega_i}{t_f - t_i} = \frac{8\pi - 0}{0.90 - 0} = 27.9 = 28 \text{ rad s}^{-2}$$

(ii) The angular displacement is $\Delta\theta = \omega_i t + \frac{1}{2}\alpha t^2 = 0 + \frac{1}{2} \times 27.9 \times (0.90)^2 = 11.3$ rad

Since $2\pi =$ one revolution, the number of revolutions is $\dfrac{11.3}{2\pi} = 1.8$ rev.

(iii) At 200 rpm, the angular velocity is $\omega = \dfrac{210}{60} \times 2\pi = 7\pi$ rad s^{-1}

The radial acceleration component is $a_r = r\omega^2 = 0.10 \times (7\pi)^2 = 48$ m s^{-2}

The tangential acceleration component is $a_t = r\alpha = 0.10 \times 27.9 = 2.8$ m s^{-2}

10. No. There will be almost no frictional force between the road and the wheels. Friction between the tyres and the road provides the centripetal force, and the friction force will not be enough to push the car around the desired curve. If the driver turns the steering wheel, the car will skid off the road.

11. (i) At the position shown, the weight of the truck.

(ii) Incorporating the weight and the centripetal acceleration into Newton's Second Law we have, $mg = m\dfrac{v^2}{r}$, where m is the mass of the truck. Thus,

$$v = \sqrt{gr} = \sqrt{9.8 \times 5.0} = 7.0 \text{ m s}^{-1}$$

12. Let T' be the tension in the string and θ be the angle between the string and the vertical.

(i) The free body diagram for the ball

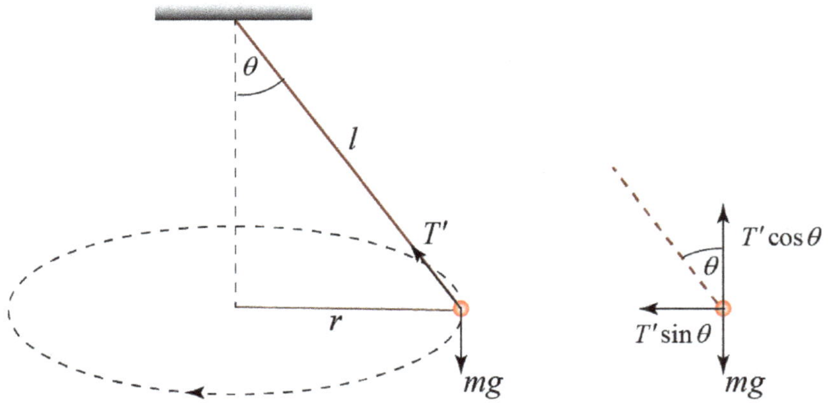

(ii) The θ is the angle between the string and the vertical.

$$\sin\theta = \frac{r}{l} = \frac{0.15}{1.00} = 0.15 \qquad \Rightarrow \quad \cos\theta = \sqrt{1-\sin^2\theta} = 0.99$$

The vertical component of the force of tension in the string is $T'\cos\theta$ and must equal the downward pull of gravity mg. Thus,

$$T'\cos\theta = mg \quad \Rightarrow \quad T' = \frac{mg}{\cos\theta} = \frac{0.050\times 10}{0.99} = 0.51 \text{ N}$$

(iii) The horizontal component of that tension will give the centripetal force. Therefore, we have

$$T'\sin\theta = m\frac{v^2}{r} \Rightarrow \quad \tan\theta = \frac{v^2}{rg}$$

$$\Rightarrow v = \sqrt{rg\tan\theta} = \sqrt{0.15\times 10\times \frac{0.15}{0.99}} = 0.45 \text{ m s}^{-1}$$

(iv) Let T be the period. Then $v = r\omega = \dfrac{2\pi r}{T}$. Therefore,

$$T = \frac{2\pi r}{v} = \frac{2\pi r}{0.45} = 2.1 \text{ s}$$

13. Communications satellites are launched in such a way that they remain above the same point on the Earth's surface and move in a circular orbit above the equator.

    (i) The time it takes for the communication satellite to complete one orbit around Earth is same as the rotational period of the Earth, which is 1 day. Thus,

    1 day is 24 hrs and it is 8.64×10^4 s

    Therefore the time it takes for the communication satellite to complete one orbit around Earth is 8.64×10^4 s

    (ii) Free-body diagram of the satellite in orbit

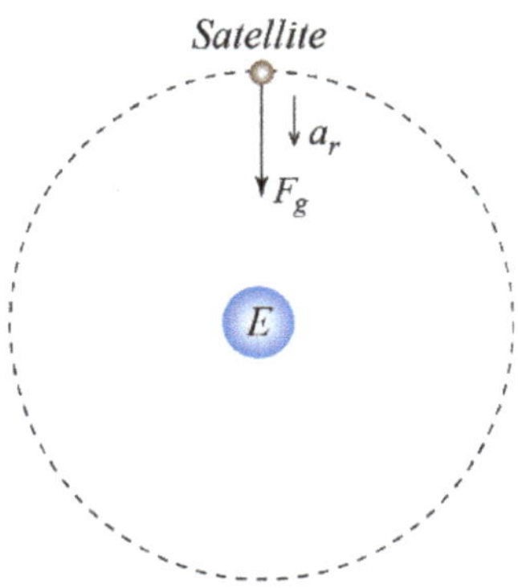

    (iii) The radius of the satellite's path is $r = h + R_E$, where h is its height above the surface of the Earth and R_E is the Earth's radius.

$$F_g = ma_r \quad \Rightarrow \quad \frac{Gmm_E}{r^2} = m\frac{v^2}{r} \quad \Rightarrow \quad v^2 = \frac{Gm_E}{r}$$, but we have $= \frac{2\pi}{\ }$

Therefore, $$\frac{4\pi^2 r^2}{T^2} = \frac{Gm_E}{r} \qquad \Rightarrow \qquad r = \left(\frac{Gm_E T^2}{4\pi^2}\right)^{\frac{1}{3}}$$

$$r = \left(\frac{6.673\times10^{-11}\times5.97\times10^{24}\times\left(8.64\times10^4\right)^2}{4\pi^2}\right)^{\frac{1}{3}} = 4.22\times10^7 \text{ m}$$

$$h = 4.22\times10^7 - 6.38\times10^6 = 3.58\times10^7 \text{ m}$$

14. An airplane with its wings tilted at an angle is in a circular motion and the centripetal force is involved. The free-body diagram for the airplane of mass m is shown below. Let F_L is the force of aerodynamic lift and r is the radius of the circle.

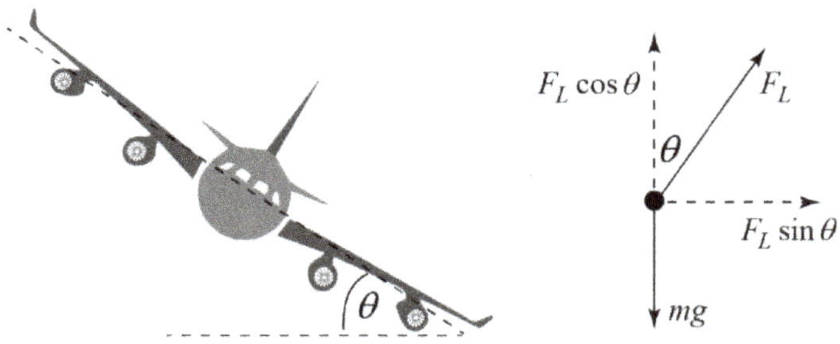

Applying Newton's Law to the axes of the problem (+x rightward and +y upward) we obtain $F_L \sin\theta = m\frac{v^2}{r}$ and $F_L \cos\theta = mg$

$$\Rightarrow \quad r = \frac{v^2}{g\tan\theta} = \frac{120^2}{10\tan 40^\circ} = 1.7\times10^3 \text{ m}$$

15. (i) The free-body diagram of the vehicle

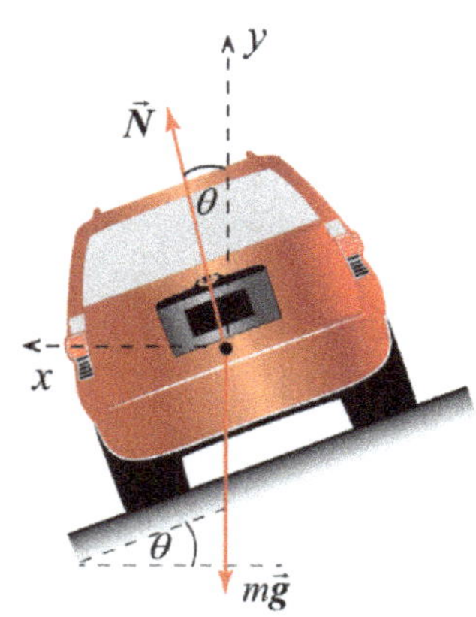

(ii) The force that causes the centripetal acceleration is the force of static friction between tyres and the road. If the road is banked at an angle θ, the magnitude of the normal force N has a horizontal component toward the centre of the curve. Since the road is to be designed so that the force of static friction is zero, the component $N\sin\theta$ is the only force that causes the centripetal acceleration. Thus, $N\sin\theta = m\frac{v^2}{r}$ and $N\cos\theta = mg$

$$\text{i.e. } \tan\theta = \frac{v^2}{rg}, \quad \Rightarrow \quad \theta = \tan^{-1}\left(\frac{v^2}{rg}\right) = \tan^{-1}\left(\frac{\left(\frac{35\times10^3}{3600}\right)^2}{40\times10}\right) = 13.3^\circ$$

Chapter 14

1. Using the right-hand rule, point the fingers in the direction of the Earth's rotation, from west to east. Then, the thumb points north. Thus, the Earth's angular velocity points towards the North Star along its axis of rotation.

2. The angular momentum of the turntable and the person system will be conserved since no external torques are applied as the person walks to the centre. As the person walks to the centre, the overall moment of inertia of the system becomes smaller since the person is closer to the axis of rotation. Since the angular momentum is constant, the angular velocity must increase. Therefore, the turntable will begin to rotate faster as you walk to the centre. This is similar to the spinning ice skater who pulls her arms to increase her angular speed.

3. If the object is free to rotate about any axis, the object will start to rotate if the two forces act along different lines of action. Then the torques of the forces will not be equal in magnitude and opposite in direction.

4. Momentum and angular momentum are conserved for closed systems, i.e. the systems in which no external forces or torques are applied to them. In real-world situations, no macroscopic systems on Earth are truly closed, and so external forces and torques, such as those applied by air friction, affect the systems over time and result in the system stopping.

5. When the angular momentum of the main propeller of the helicopter changes, the body of the helicopter experiences an opposite change in angular momentum.

Therefore, an external torque about a vertical axis must be created in the helicopter to counteract this effect. The small propeller attached to the tail of the helicopter is used to produce this external torque to keep the helicopter body from rotating. Attaching the propeller at the tail produces the longest lever arm, and therefore the maximum torque can be applied about the vertical axis through its centre of mass.

6. The moment of inertia of an object depends on how its mass is distributed with respect to the axis of rotation. When the mass is distributed far away from the axis, the object will have a larger moment of inertia. In the case of animals, their leg muscles are concentrated close to the hip joint and their legs have relatively small moments of inertia. This makes it easier for them to rotate their legs, allowing them to run faster.

7. The best place is as far from the hinge as possible so that it has the highest possible moment arm for the torque exerted by the stopper on the door. In this way, the force required by the stopper to keep the door open will be as small as possible, making the stopper less likely to slip on the floor.

8. Let the direction of rotation of the merry-go-round be positive.

(i) Apply $\sum \tau = I\alpha$, $\Rightarrow \ \alpha = \frac{\sum \tau}{I} = \frac{25.0 \times 2.00}{250} = 0.200 \text{ rad s}^{-2}$

Using the constant acceleration equation $\omega = \omega_0 + \alpha t$, we get

$\omega = 0 + 0.200 \times 20.0 = 4.00 \text{ rad s}^{-1}$

(ii) Using the equation $\Delta\theta = \omega\ t + -\alpha t$, we get the angular displacement after 20.0 s as, $\Delta\theta = 0 + \frac{1}{2}0.200 \times 20.0^2 = 40.0 \text{ rad}$. Thus, the work done by the child on the merry-go-round will be $W = \tau\Delta\theta$. Therefore,

$$W = \tau\Delta\theta = (25.0 \times 2.00) \times 40.0 = 2.00 \times 10^3 \text{ J}$$

(iii) The average power supplied by the child is $P_{avg} = \frac{W}{\Delta t} = \frac{2000}{20.0} = 100 \text{ W}$

9.(i) (a) Consider a tooth on the front sprocket. If we consider interaction of chain with front sprocket, we could find the speed, relative to the frame. Thus,

$$v = r\omega = \left(\frac{15.0\times10^{-2}}{2}\right)\times\left(75.0\times\frac{2\pi}{60}\right) = 0.589 \text{ m s}^{-1}$$

(b) If we consider the interaction of the chain with rare sprocket, we can find the angular speed of the bicycle wheels. Thus,

$$\omega = \frac{v}{r} = \frac{0.589}{\left(\frac{7.50\times10^{-2}}{2}\right)} = 15.7 \text{ rad s}^{-1}$$

(c) If we consider a point on the wheel's tyre, we can find the bicycle wheels' speed relative to the road. Thus,

$$v' = r\omega = \left(\frac{62.0\times10^{2}}{2}\right)\times15.7 = 4.87 \text{ m s}^{-1}$$

(ii) The length of the pedal cranks is 17.5 cm long. Hence, the linear speed of the pedal is $v'' = r\omega = \left(17.5\times10^{-2}\right)\times\left(75.0\times\frac{2\pi}{60}\right) = 1.37 \text{ m s}^{-1}$

10. The magnitude of the torque acting on the wheel about its axis of rotation

$$\sum\tau = TR = I\alpha$$

i.e. $\alpha = \frac{TR}{I}$ .... (A)

Applying Newton's Second Law to the motion of the object along the downward direction, we have

$$mg - T = ma$$

i.e. $a = \frac{mg - T}{m}$ .... (B)

When the cord unwinds from the wheel, it does not slip. Therefore, the translational acceleration of the suspended object is equal to the tangential acceleration of a point on the wheel's rim. Thus,

$$a = R\alpha \quad \text{................} \quad (C)$$

(i) The gravitational force exerted by the Earth on the wheel and the normal force exerted by the axle on the wheel both pass through the axis of rotation and therefore produce no torque.

(ii) From equation B and C we have

$$a = R\alpha = \frac{mg - T}{m} \quad \Rightarrow \quad T = mg - mR\alpha$$

Substituting this in equation A we get

$$\alpha = \frac{TR}{I} = \frac{R}{I}(mg - mR\alpha) \quad \Rightarrow \quad \alpha = \frac{mgR}{mR^2 + I}$$

(iii) Substituting the value of α in equation C we get

$$a = \left(\frac{mgR}{mR^2 + I}\right)R = \frac{mgR^2}{mR^2 + I}$$

(iv) Substituting the value of α in equation A we get

$$\alpha = \frac{mgR}{mR^2 + I} = \frac{TR}{I} \quad \Rightarrow \quad T = \frac{mgI}{mR^2 + I}$$

11. The flywheel is a solid disk of mass M and radius R with axis through its center.

(i) The moment of inertia of the flywheel

$$I = \frac{1}{2}MR^2 = \frac{1}{2} \times 75.0 \times \left(60.0 \times 10^{-2}\right)^2 = 13.5 \text{ kg m}^2$$

(ii) The net torque on the flywheel $\tau_{net} = I\alpha = 13.5 \times 1.50 = 20.3 \text{ N m}$

(iii) $\tau_{net} = T_1 r - T_2 r = I\alpha \quad \Rightarrow \quad T_2 = \dfrac{T - I\alpha}{r}$

$$T_2 = 125 - \frac{13.5 \times 1.50}{20.0 \times 10^{-2}} = 23.8 \text{ N}$$

(iv) $\alpha = \dfrac{\Delta\omega}{\Delta t} \quad \Rightarrow \quad \Delta t = \dfrac{\Delta\omega}{\alpha} = \dfrac{100 \times \left(\dfrac{2\pi}{60}\right)}{1.50} = 6.98 \text{ s}$

(v) Use the equation $\Delta\theta = \omega_0 t + \dfrac{1}{2}\alpha t^2$.

The number of revolutions $\Delta\theta = 0 + \dfrac{1}{2}\alpha t^2 = \dfrac{\dfrac{1}{2} \times 1.50 \times 6.98^2}{2\pi} = 5.82 \text{ rev}$

12. The moment of inertia of the rigid bodies through their respective centre of mass; for a solid sphere is $I_{cm} = \dfrac{2}{5}MR^2$, a solid cylinder is $I_{cm} \; \dfrac{1}{2}MR^2$, thin-walled hollow sphere is $I_{cm} = \dfrac{2}{3}MR^2$ and thin-walled hollow cylinder is $I_{cm} = MR^2$ where, M and R are the mass and radius of the respective rigid bodies.

(i) If you look at the moment of inertia of the rigid bodies through their respective centre of mass, it can be written as $I_{cm} = \beta MR^2$, where β is a number less than or equal to 1 that depends on the shape of the rigid body. The solid sphere will have $\beta = \dfrac{2}{5}$, the solid cylinder will have $\beta = \dfrac{1}{2}$, the thin-walled hollow sphere will have $\beta = \dfrac{2}{3}$, and the thin-walled hollow cylinder will have $\beta = 1$.

Since the rigid bodies roll without slipping, we have $\omega = \dfrac{v_{cm}}{R}$, where ω, v_{cm} and R are the angular velocity, the speed of the centre of mass and the radius of the respective rigid bodies. Each rigid body starts from rest at the top of an incline with height h. From the conservation of energy, we have

$$K_i + PE_i = K_f + PE_f$$

i.e. $0 + Mgh = \dfrac{1}{2}Mv_{cm}^2 + \dfrac{1}{2}I\omega^2 = \dfrac{1}{2}Mv_{cm}^2 + \dfrac{1}{2}\beta MR^2\left(\dfrac{v_{cm}}{R}\right)^2$

i.e. $Mgh = \frac{1}{2}(1+\beta) Mv_{cm}^2$

Therefore, $v_{cm} = \sqrt{\frac{2gh}{1+\beta}}$

This shows that for a given value of β, the speed of the centre of mass after descending a distance h is independent of the respective rigid body's mass M and radius R. For example, for a solid sphere $\beta = \frac{2}{5}$, thus $v_{cm} = \sqrt{\frac{10gh}{7}}$. Furthermore, the smaller the β higher the v_{cm}.

(ii) From part (i), we see that the values of β tell us that the order of reaching the bottom of the inclined will be as follows: the 1st solid sphere having $\beta = \frac{2}{5}$, the 2nd solid cylinder having $\beta = \frac{1}{2}$, the 3rd thin-walled hollow sphere having $\beta = \frac{2}{3}$ and the 4th thin-walled hollow cylinder having $\beta = 1$. The rigid bodies having small β values always beat the rigid bodies having large β values because less of their kinetic energy is tied up in rotation and so more is available for translation.

13. Yo-yo is made using two uniform disks, each with mass M and radius R. The moment of inertia $I_{cm} = 2\times\left(\frac{1}{2}MR^2\right) = MR^2$. The moment arm for the tension T on the string is r.

(i) Apply $\sum \vec{F}_{ext} = m\vec{a}_{cm}$ to the motion of the centre of mass, we have

$$2mg - T = 2ma_{cm} \text{ ....[A]}$$

Apply $\sum \tau = I_{cm}\alpha$ to the rotation about the centre of mass, we have

$$Tr = I_{cm}\alpha = I_{cm}\left(\frac{a_{cm}}{r}\right) \text{ ....[B]}$$

Divide the equation [B] by r and add to the equation [A] we get

$$2mg = 2ma + I_{cm}\left(\frac{a}{r^2}\right)$$

$$\Rightarrow \quad 2mg = 2ma + mR^2\left(\frac{a}{r^2}\right) \quad \Rightarrow \; 2g = a\left(2 + \frac{R^2}{r^2}\right)$$

$$\Rightarrow \quad a = \frac{2gr^2}{2r^2 + R^2}$$

The angular acceleration α is $\alpha = \dfrac{a}{r}$. Thus,

$$\alpha = \frac{2gr}{2r^2 + R^2}$$

(ii) The tension in the string is

$$T = 2mg - 2ma_{cm} = 2mg\left(1 - \frac{2r^2}{2r^2 + R^2}\right)$$

$$\Rightarrow \quad T = 2mg\left(\frac{R^2}{2r^2 + R^2}\right)$$

(iii) Let the mass of the yo-yo will be M

(a) $I_{cm} = \dfrac{1}{2}MR^2 . \Rightarrow \; R = \sqrt{\dfrac{2 \times I_{cm}}{M}} = \sqrt{\dfrac{2 \times 1000 \times 10^{-3} \times 10^{-4}}{120 \times 10^{-3}}} = 4.0 \text{ cm}$

(b) $|a_{cm}| = \dfrac{2gr^2}{2r^2 + R^2} = \dfrac{2g}{2 + \left(\dfrac{R}{r}\right)^2} = \dfrac{2 \times 10}{2 + \left(\dfrac{4.0}{0.35}\right)^2} = 15.1 \text{ cm s}^{-2}$

Direction of the acceleration in downward.

(c) Take the coordinate origin at the initial position and the positive y axis as upward. Initial vlocity is zero. Displacement of the centre of mass during the travel $y_{cm} = -120 \times 10^{-2}$ m.

Thus, using $s = ut + \dfrac{1}{2}at^2$, we have $y_{cm} = 0 + \dfrac{1}{2}a_{cm}t^2$

$$\Rightarrow \quad t = \sqrt{\frac{2y_{cm}}{a_{cm}}} = \sqrt{\frac{2\times\left(-120\times10^{-2}\right)}{-15.1\times10^{-2}}} = 3.99 \text{ s}$$

(d) (α) As the yo-yo reaches the end of the string, its centre of mass velocity is given $v_{cm} = 0 + a_{cm}t$

$$\Rightarrow \quad v_{cm} = -15.1\times10^{-2}\times3.99 \text{ m s}^{-1} = -60.3 \text{ cm s}^{-1}$$

i.e. $|v_{cm}| = 60.3 \text{ cm s}^{-1}$

(β) The translational kinetic energy of the yo-yo is

$$K_{trans} = \frac{1}{2}mv_{cm}^2 = \frac{1}{2}\times125\times10^{-3}\times\left(60.3\times10^{-2}\right)^2 = 2.27\times10^{-2} \text{ J}$$

(γ) The angular speed is $\omega = \frac{|v_{cm}|}{r}$. Thus, the rotational kinetic energy is

$$K_{rot} = \frac{1}{2}I_{cm}\omega^2 = \frac{1}{2}I_{cm}\left(\frac{v_{cm}}{r}\right)^2$$

$$\Rightarrow K_{rot} = \frac{1}{2}\times1000\times10^{-3}\times10^{-4}\times\left(\frac{60.3\times10^{-2}}{3.5\times10^{-3}}\right)^2 = 1.48 \text{ J}.$$

14. Since the torque is constant, it is equal to the change in angular momentum divided by the time interval. Thus, $\tau = \frac{\Delta L}{\Delta t}$. $\Rightarrow \quad \tau = \frac{100-223}{3.0} = -41 \text{ N m}$.

The torque applied is 41 N m, opposite the rotation of the wheel.

15. The earth has mass $m_E = 5.97\times10^{24}$ kg 24, radius $R_E = 6.38\times10^6$ m and orbit radius $r = 1.50\times10^{11}$ m. The earth completes one rotation on its axis in 24 hours $= 86\ 400$ s and one orbit in 1 year $= 3.156\times10^7$ s.

(i) $L = I\omega = m_E r^2\omega = \left(5.97\times10^{24}\right)\times\left(1.50\times10^{11}\right)^2\times\left(\frac{2\pi}{3.156\times10^7}\right)$

$L = 2.67\times10^{40} \text{ kg m}^2 \text{ s}^{-1}$

(ii) $L = I\omega = \left(\frac{2}{5}m_E R_E^2\right)\omega = \frac{2}{5}\left[\left(5.97\times10^{24}\right)\times\left(6.38\times10^{6}\right)^2\right]\times\left(\frac{2\pi}{86\ 400}\right)$

$$L = 7.07\times10^{33}\ \text{kg m}^2\ \text{s}^{-1}$$

16. (i) From conservation of angular momentum for the isolated system of two disks:

$$\left(I_0 + 3I_0\right)\omega = I_0\omega_0 \quad \Rightarrow \quad \omega = \frac{1}{4}\omega_0$$

(ii) $K_f = \frac{1}{2}\left(I_0 + 3I_0\right)\omega^2$ and $K_i = \frac{1}{2}I_0\omega_0^2$

Therefore, $\frac{K_f}{K_i} = \frac{\frac{1}{2}\left(I_0 \quad 3I_0\right) \quad ^2}{\frac{1}{2}I_0 \ {}^2_0} = \frac{4 \quad ^2}{{}^2_0} = \frac{1}{4}$

Thus, 3/4 of the initial rotational kinetic energy is lost to the coupling of the two flywheels.

17. (i) Total moment of inertia of the turntable and the bird after the bird lands on the outermost edge of the turntable is

$$I_f = I_{turntable} + mR^2 = \frac{1}{2}MR^2 + mR^2 = \left(\frac{1}{2}\times 86.0 + 7.0\right)\times 2.0^2 = 200\ \text{kg m}^2$$

(ii) Initial moment of inertia $I_i = I_{turntable} = \frac{1}{2}MR^2 = \frac{1}{2}\times 86.0\times 2.00^2 = 172\ \text{kg m}^2$

From conservation of angular momentum $L_i = L_f$

Therefore, $I_i\omega_i = I_f\omega_f \quad \Rightarrow \quad \omega_f = \frac{I_i}{I_f}\omega_i$

$$\Rightarrow \quad \omega_f = \frac{172}{200}\times 5.00 = 4.3\ \text{rad s}^{-1}$$

18. (i) Yes, the bullet has angular momentum about an axis through the hinges of the door before it strikes the door.

The bullet strikes the door at $l = 0.45$ m from the hinge. Its initial angular momentum is therefore $L_i = m_B v_B l = 18.0\times10^{-3}\times 500\times 0.45 = 4.05\ \text{kg m}^2\ \text{s}^{-1}$.

(ii) No. Although the torque created by the weight is an external torque, it does not cause the door to rotate.

Torque is defined as the force multiplied by the perpendicular distance from the line of action of the force to the axis of rotation. In the case of a door swinging about a vertical axis, the force of gravity acts along the axis of rotation and, therefore, the perpendicular distance from the line of action of the force to the axis of rotation is zero. As a result, the torque created by the weight of the door is zero. Therefore, it does not cause the door to rotate. Hence, the net external torque about the vertical axis of rotation will be zero, and we can apply the conservation of angular momentum.

(iii) Final angular momentum (door and bullet together)

$$L_f = \left(\frac{1}{3}Md^2 + ml^2\right)\omega = \left(\frac{1}{3}\times 12.0\times 0.9^2 + 18.0\times 10^{-3}\times 0.45^2\right)\omega$$

Apply conservation of angular momentum, $L_i = L_f$

$$\Rightarrow \quad 4.05 = \left(\frac{1}{3}\times 12.0\times 0.9^2 + 18.0\times 10^{-3}\times 0.45^2\right)\omega = 3.24\omega$$

Therefore, $\omega = 1.25 \text{ rad s}^{-1}$

(iv) The kinetic energy of the door-bullet system immediately after impact is

$$K_f = \frac{1}{2}I_f\omega^2 = \frac{1}{2}\left(\frac{1}{3}\times 12.0\times 0.9^2 + 18.0\times 10^{-3}\times 0.45^2\right)1.25^2 = 2.53 \text{ J}$$

The kinetic energy (of the bullet) just before impact was

$$K_i = \frac{1}{2}m_B v_B^2 = \frac{1}{2}\times 18.0\times 10^{-3}\times 500^2 = 2.25\times 10^3 \text{ J}$$

Kinetic energy is not conserved – collision is completely inelastic.

19. (i) The free-body diagram

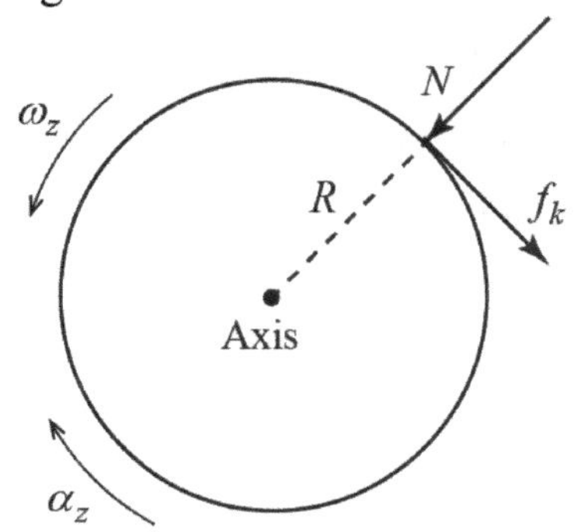

(ii) The grindstone comes to rest in $10.0\ \mathrm{s}$. Use the $\omega_z = \omega_{z_i} + \alpha_z \Delta t$

$$\Rightarrow \alpha_z = \frac{0-85.0}{10} = -8.5\ \mathrm{rad\ s^{-2}}.$$

(iii) The grindstone is rotating counter - clockwise, and let that be the positive sense of rotation.

The normal force N has a zero moment arm for rotation about an axis at the centre of the grindstone. Therefore, zero torque. The only torque on the grindstone is due to the friction force f_k exerted by the knife. For this force, the moment arm is R. Thus, $\tau = -f_k R = -\mu_k NR$. But we have $\tau = I\alpha_z$.

Hence $-\mu_k NR = I\alpha_z$,

$$\mu_k = -\frac{I\alpha_z}{NR} = -\frac{\frac{1}{2}MR^2\alpha_z}{NR} = -\frac{MR\alpha_Z}{2N} = -\frac{20.0\times\left(15.0\times10^{-2}\right)\times(-8.5)}{2\times50.0} = 0.255$$

20. (i) The expression for the precession angular speed of the gyroscope is $\Omega = \frac{mgd}{I\omega}$.

(ii) (a) When the angular speed ω of the spinning flywheel is doubled the precession angular speed of the gyroscope will be halved.

(b) When the weight mg the spinning flywheel is doubled the precession angular speed of the gyroscope will be doubled (assuming that the added weight is distributed in such a way that the distance d from the pivot to the centre of gravity and moment of inertia I are not changed)

(c) When the moment of inertia I about the axis of the spinning flywheel is doubled the precession angular speed of the gyroscope will be doubled (assuming that mg and the distance d from the pivot to the centre of gravity are not changed)

(d) When the distance d from the pivot to the centre of gravity is doubled the precession angular speed of the gyroscope will be doubled.

(iii) If all four variables ω, mg, I, and d are doubled the precession angular speed of the gyroscope will not change.

(iv) The precession angular speed of the gyroscope

$$\Omega = \frac{mgd}{I\omega} \quad \Rightarrow \quad \frac{\Omega}{g} = \frac{md}{I\omega} = \text{constant}$$

Therefore, $\dfrac{\Omega_{Earth}}{g_{Earth}} = \dfrac{\Omega_{Moon}}{g_{Moon}}$.

$$\Rightarrow \quad \Omega_{Moon} = \Omega_{Earth}\left(\frac{g_{Moon}}{g_{Earth}}\right) = 0.6 \times \frac{\;}{6} = 0.1 \text{ rad s}^{-} \cdot$$

Index

D

E

F

G

H

I

J

K

L

M

N

P

Q

R

X

Y

Z

www.ingramcontent.com/pod-product-compliance
Lightning Source LLC
Chambersburg PA
CBHW080535250926
51846CB00016B/984
9786246709198